HOLISM IN PHILOSOPHY OF MIND AND PHILOSOPHY OF PHYSICS

SYNTHESE LIBRARY

STUDIES IN EPISTEMOLOGY,

LOGIC, METHODOLOGY, AND PHILOSOPHY OF SCIENCE

VOLUME 298

HOLISM IN PHILOSOPHY OF MIND AND PHILOSOPHY OF PHYSICS

by

MICHAEL ESFELD
University of Konstanz, Germany, and
University of Hertfordshire, England

SPRINGER-SCIENCE+BUSINESS MEDIA, B.V.

A C.I.P. Catalogue record for this book is available from the Library of Congress.

DOI 10.1007/978-94-017-1787-8

Printed on acid-free paper

TABLE OF CONTENTS

ACKNOWLEDGEMENTS

This book was accepted as *Habilitationsschrift* by the Faculty of Humanities of the University of Konstanz in February 2000. The subject of this book goes back to a summer school of the *Schweizerische Studienstiftung* by Gereon Wolters, Jürgen Audretsch and Eric Kubli in 1993. I have pursued the topic of holism ever since – first with Hans Primas and his group at the Swiss Federal Institute of Technology in Zurich from 1994 to 1996, then with Michael Redhead and Jeremy Butterfield at the University of Cambridge in 1996/97, and finally, owing to the generosity of Gereon Wolters, at the University of Konstanz since 1997. A leave from Konstanz in winter 1998/99 enabled me to discuss holism in philosophy of mind with Philip Pettit in Canberra and Bob Brandom in Pittsburgh. Over and above the criticism and encouragement of these persons, I have benefitted from comments on parts of various versions of this thesis by Christian Beyer, Thomas Breuer, Martin Carrier, Rob Clifton, Pascal Engel, Steven French, Ken Gemes, Jane Heal, Wolfram Hinzen, Ludger Jansen, Markus Knecht, Cathy Legg, John McDowell, Hugh Mellor, Peter Mittelstaedt, Peter Rohs, Alois Rust, Wolfgang Spohn, Niko Strobach, Paul Teller, Marcus Willaschek and Christian Zemlin.

My work on holism was supported first by a postdoctoral fellowship of the *Studienstiftung des deutschen Volkes* in cooperation with the *Dr. Meyer-Struckmann-Stiftung* and then by a fellowship for Habilitation of the *Deutsche Forschungsgemeinschaft*.

This book draws in part on material that is published in the following papers: "Holism and Analytic Philosophy", *Mind* 107 (1998), pp. 365–380 (Chapter 1); "Ein Argument für sozialen Holismus und Überzeugungs-Holismus", *Zeitschrift für philosophische Forschung* 54 (2000), pp. 387–407 (Chapter 3); "How a Social Theory of Meaning can be connected with Realism", forthcoming in *Facta Philosophica* **3** (2001), first issue (Chapter 5); "Holism in Cartesianism and in Today's Philosophy of Physics", *Journal for General Philosophy of Science* 30 (1999), pp. 17–36 (Chapter 6); "Der Holismus der Quantenphysik: seine Bedeutung und seine Grenzen", *Philosophia Naturalis* 36 (1999), pp. 157–185 (Chapter 8); "Quantum Holism and the Philosophy of Mind", *Journal of Consciousness Studies* 6

(1999), pp. 23–38 (Chapter 9); “Holismus und Verbindungen zwischen der Philosophie des Geistes und der Philosophie der Physik”, *Conceptus* 31 (1998), issue no. 79, pp. 153–173 (Chapter 10).

Konstanz, November 2000

Michael Esfeld

INTRODUCTION

The Scope of the Project

The concept of holism is at the centre of far-reaching changes in various areas of philosophy in the second half of the twentieth century. Holism in epistemology and the philosophy of mind is widespread among analytic philosophers subsequent to the work of the later Wittgenstein and to Quine's "Two Dogmas of Empiricism". Roughly speaking, the claim is that (a) for a person to have beliefs, a social, linguistic community is required and that (b) single beliefs have a meaning only within a whole system of beliefs. Furthermore, holism is discussed in science, in particular in the interpretation of quantum physics. In fact, the term "holism" goes back to Smuts (1926), who introduces this term in a biological context.[1] Holism in any of these areas has considerable consequences for our philosophical view of the world and ourselves. Holism in quantum physics is said to refute atomism, which has been predominant in modern philosophy of nature. Holism in epistemology and the philosophy of mind is seen as an alternative to what is known as the Cartesian tradition, which dominated modern thought down to logical empiricism.

A good deal of research has been done on holism in each of the mentioned areas. But there is no research work available that provides a critical overview of holism in general. That is to say: we lack research that focuses on a general conception of holism, enquires into possible connections between forms of holism in different areas, and evaluates the significance of such connections for contemporary philosophy. The aim of this book is to contribute to filling this gap. Given this background, the scope of this book is threefold:

(1) At first I develop a *general conception of holism* that is intended to be applicable to candidates for holism in any area. The purpose is to show that holism is not opposed to rational analysis. What holism is can be captured by the conceptual tools of analytic philosophy.

---

1 For an overview of positions that are known as holism and their historical development see Mayer (1997).

(2) This general conception is then employed in order to examine whether some of the forms of holism in different areas that are currently discussed have a *common conceptual content.* I do not mean that the forms of holism that I shall consider are in any way logically dependent on each other. It is possible that the truth value attributed to one of them does not determine the truth value of any of the other ones. Nonetheless, what "holism" means in each of these areas may exhibit significant common features. I shall concentrate on holism about beliefs and social holism on the one hand and holism in space–time theories and in quantum physics on the other hand. This book thus is an attempt to bring together two discussions across the divide between the sciences and the humanities. It shows the following: what "holism" means in philosophy of mind and what "holism" means in philosophy of physics has a common conceptual content. This result refutes a widespread impression according to which these cases have nothing more than the mere name "holism" in common.

(3) Although it is an important result to reveal a common conceptual content of holism in philosophy of mind and holism in philosophy of physics, we can and should go further and ask: *What is the significance of this result?* I argue that we cannot melt these two sorts of holism together in order to achieve one comprehensive holism. Starting from holism in philosophy of mind, it is possible to reach a holism that includes the physical realm – but, paradoxically enough, only at the price of not being able to include that form of holism which originates in the physical realm. And, starting from holism in quantum physics, it is possible to reach a holism that extends to the mental realm – but, paradoxically enough, only at the price of not being able to include that form of holism which originates in the philosophy of mind. The significance of the common conceptual content of holism in philosophy of mind and philosophy of physics is not that we end up with one comprehensive holism. Instead, this common conceptual content is significant, because (a) it shows that holism in philosophy of mind fits into a view of the world that bases itself on scientific realism, and because (b) it leads to common consequences of both these sorts of holism for topics that are at the centre of today's philosophy such as supervenience and realism.

The Structure of the Book

This book provides an overview of holism in philosophy of mind and philosophy of physics. However, it is not a neutral description of the different forms of holism in these areas. Instead, it focuses on what I take to be the best arguments for holism in each of the cases considered. I try to make these arguments as strong as possible and develop their consequences as clearly as possible. As far as the philosophy of mind is concerned, any theory of meaning faces the problem of rule-following. Starting from the rule-following considerations, an argument is set out that covers both social holism and holism about beliefs. The result is an inferential role semantics that conceives the inferential role of beliefs as determined by social practices. I argue that social holism and holism about beliefs relate only to our epistemic access to a world that is ontologically and causally independent of our beliefs. Thus conceived, holism in philosophy of mind can go with realism. It is a social theory of meaning, but not a social theory of truth. Furthermore, although holism in philosophy of mind has an anti-reductionist implication as far as epistemology is concerned, it is parsimonious in its ontological commitments: it is not committed to any sort of a dualistic ontology.

As regards holism in quantum physics, the pros and cons of the proposals for a philosophical characterization of this holism are reviewed. I bring these proposals together in a suggestion that shows how quantum holism fits into a general conception of holism. I thereby argue that holism in quantum physics and holism in philosophy of mind have a significant common conceptual content. Contrary to a widespread opinion that regards quantum holism as amounting to a total break with the ontology of classical physics, it is shown that the ontology of classical physics can admit of a holism. This is a holism about space or space–time. This holism is employed as a background to evaluate what exactly is the impact of quantum holism.

Chapter 1 introduces conceptual tools for examining cases of holism in different areas. I set out a general conception of holism and two types of holism within that conception. Chapters 2 to 5 treat holism in philosophy of mind. I begin with an overview of arguments for holism about beliefs (Chapter 2). Then an argument is developed that covers both social holism and holism about beliefs (Chapter 3). Chapter 4 starts evaluating the impact of holism in philosophy of mind: I explain in what respect holism amounts to a revision of what is known as the Cartesian tradition in epistemology (and in what respect it does not amount to such a revision). Chapter 5 shows that the proposed holism can go with realism. I conclude with elaborating on the claim that the proposed holism is parsimonious in its ontological commitments.

Chapters 6 to 9 are about holism in philosophy of physics. Again I use the Cartesian tradition as a background. Chapter 6 sets the stage for examining holism in quantum physics by arguing that Cartesian physics admits of a holism. Consequently, from that perspective, the shift that we face today is a shift from one sort of physical holism to another sort of physical holism. Chapter 7 introduces the conceptual basis for holism in quantum physics for those readers who are not familiar with quantum physics. In Chapter 8, I set out my suggestion for a philosophical characterization of quantum holism. Chapter 9 then relates quantum holism to the philosophy of mind: it is argued that it is possible to build on quantum holism a holism that includes the mental, but only at the price of not being able to accommodate that sort of holism which originates in the philosophy of mind. In the concluding Chapter 10, I argue in favour of a mutual enrichment of quantum holism and holism in philosophy of mind: the holism in philosophy of physics shows that the holism in philosophy of mind fits into a world view that bases itself on scientific realism. Furthermore, both these sorts of holism have a significant common consequence for topics such as supervenience and realism.

Nonetheless, this book can only touch upon such broader consequences. There are at least three areas which require further elaboration that goes beyond a comparison of two sorts of holism: (a) the possible implications of the proposed social holism and holism about beliefs for the metaphysics of mind (Chapter 5.4); (b) the possible implications for the metaphysics of nature that result from applying quantum holism to the microphysical realm as a whole (Chapter 8.4); and (c) the possible connection between holism in both these areas and a pragmatic realism as well as a pragmatic conception of rationality (Chapters 5.2.3, 5.3.2 and 10.4).

Each chapter begins with a summary of the main points. The footnotes are limited to references. They do not add any content to the main text. The addressees of this book are philosophers in the broad sense of people who care about our view of the world and ourselves in the spirit of an argumentative examination of different positions. I try to circumvent technicalities as far as possible without compromising argumentative precision. In particular, no familiarity with quantum physics is presupposed. Especially in my examination of holism in philosophy of physics, I shall lay stress on avoiding to pass disputable interpretative steps off as established facts and mention alternative options.

CHAPTER 1
WHAT IS HOLISM?
PROPOSAL FOR A GENERAL CONCEPTION

Summary

After setting out three requirements for a general conception of holism [1.1], a proposal for such a conception is developed in two steps:

1) What sort of dependence subsists among the parts of a holistic system? The suggestion set out in this chapter is that generic ontological dependence is the way in which the parts of a holistic system depend on each other. Having a general conception of holism in view, I formulate generic ontological dependence in such a way that it does not refer to the existence of the parts as such, but to the parts insofar as they have certain properties.
2) Which properties? I introduce the notion of parts by which a system of the kind *S* is constituted, i.e., constituent parts, or constituents for short. For every constituent of an *S*, there is a family of properties that make something a constituent of an *S* provided that there is a suitable arrangement with other things. Generic ontological dependence among the constituents of a holistic system relates to things insofar as they have some of the properties that belong to such a family of properties.

These two steps yield the following proposal for a general conception of holism: Consider a system of the kind *S* and its constituent parts. For every constituent of an *S*, there is a family of qualitative properties that make something a constituent of an *S* provided that there is a suitable arrangement. An *S* is holistic if and only if the following condition is satisfied by all the things which are its constituents: with respect to the instantiation of some of the properties that belong to such a family of properties, a thing is ontologically dependent in a generic way on there actually being other things together with which it is arranged in such a way that there is an *S* [1.2]. Discussing the application of this proposal to candidates for holistic systems, it is shown that it contains a general and substantial criterion for holism which excludes trivial cases. In particular, it is argued that functional definitions of parts are not sufficient to make a case for holism [1.3]. Finally, two types of holism within this general conception are introduced: a bottom-up conceptualization of holism that begins with the constituents and their properties, and a top-down conceptualizations that begins with the properties of the whole. It is shown that the proposed general view of holism covers both these conceptualizations [1.4].

1.1 Requirements for a General Conception of Holism

At first glance holism seems to be opposed to analytic philosophy. It is often associated with a treatment of wholes that is not accessible to the conceptual tools of rational argument. Some philosophers therefore have strong reservations about holism. Thus, Mario Bunge, for instance, characterizes holism in general by theses such as "The whole precedes its parts", "The whole acts on its parts", "Totalities cannot be explained by analysis: they are irrational", "The whole is better than any of its parts". He concludes: "Holism, in a nutshell, is antianalytic and therefore antiscientific".[1] Given such a characterization of holism, it is easy to refute that position. However, as the discussions on holism in philosophy of mind and philosophy of physics show, we cannot dispense with holism that easily: supporting holism in any of these areas is not at all opposed to rational argument. Before the merits and demerits of holism – in specific cases – can be evaluated, we should have a precise and substantial characterization of holism at our disposal.

A general conception of holism should meet the following three requirements:

1) *As far as possible, it should show that it is an issue of substantial philosophical argument whether holism obtains in specific areas.* A general conception should capture the cases that are recognized as candidates for holism without either trivializing holism or turning it into an absurd position.
2) *It should construe holism in the first place as a thesis about the way in which certain systems are organized.* A general conception has to make a suggestion on what holism means for the organization of certain systems. Positions which count as epistemological holism, such as the Duhem–Quine thesis or coherentism, can be construed as claims about the way in which systems of beliefs are organized with respect to confirmation or justification. Furthermore, a methodological holism, which is about the conditions of our epistemic access to certain systems, can be conceived as making claims about the way in which we have to represent the organization of certain systems in order to gain access to them.

---

1 Bunge (1979), pp. 39–41. Nonetheless, Bunge grants what is known as holism in quantum physics, i.e., nonseparability of the states of two or more systems; however, based on his definition of holism, he refuses to speak of holism in that case. See Bunge (1985), Chapter 2, section 6.2, in particular p. 215.

3) *It should characterize a holistic system by a certain sort of mutual dependence among its parts*. Anything that can count as a serious candidate for a holistic system must have an internal structure. Specifying that structure means taking parts in some sense into account. The things that are parts may not be capable of existing independently of the whole. But there is no reason to recognize a thing that is considered to have no internal structure as a holistic system (such as the One of Parmenides under some readings of his poem). Such a thing would rather be an atom in the literal sense. Holism is therefore not opposed to the analysis of a complex system in terms of its parts.

My strategy will be to characterize a holistic system by means of a necessary and sufficient condition that the constituent parts of such a system have to satisfy. This condition will be about certain properties of these parts. Consequently, my proposal will not only tell us what it is for something to be a holistic system, but also what it is for something to be a part of such a system, and it will enable us to conceive holistic properties.

None of the candidates for holism can be treated in this chapter. I mention examples only to illustrate what the conceptual tools that are introduced mean and how they contribute to a substantial and general conception of holism. My foremost example is *holism about beliefs*: the system of beliefs of each person is a holistic system; for a belief has properties such as meaning, confirmation, justification only if there are other beliefs together with which it constitutes the system of beliefs of a person. Hence, a person cannot have just one belief. If she has one belief, she has a whole system of beliefs. To avoid complicating the example, I assume that it is beliefs which have meaning in the first place. Furthermore, for the sake of simplicity and convenience, I work with the notion of individual tokens of beliefs that are parts of a system of beliefs, and I use the notion of properties of beliefs, although beliefs and belief systems are themselves attributed to persons. When the focus is on the property of meaning, I speak of semantic holism. The second example which I employ to illustrate my claims is *social holism*: a social community of thinking beings is a holistic system; for having beliefs with a determinate meaning is tied to social relations. A person thus has the property of having beliefs only if there are other persons with whom she constitutes a social community and shares a language.

Given the three mentioned requirements, let us settle a few terminological matters. The notion of holism is about a whole that has to satisfy certain conditions in order to count as a holistic system. According to the terminology used here, anything of which properties are predicated is a *system*. No specific way of being organized is thus presupposed. Hence, even an atom in the literal sense is a system, as long as properties can be predicated of it. Furthermore, a system – e.g., a system of beliefs – may itself be predicated as a property of another system, e.g., a person. The concept of a whole is related to the concept of parts.[2] I use the term "*whole*" in such a way that anything that is a whole has parts. By a *part*, I always mean a proper part. Nothing is a proper part of itself. If a thing *x* is a proper part of another thing *y*, then *y* is not a proper part of *x*. I further specify the notion of a part which I use in [1.2.2][3].

Holism is seen as being opposed to atomism. Holism on the one hand and atomism on the other hand are associated with the following idea: roughly speaking, if one starts with parts, maintains that these things have the properties that are characteristic of them independently of each other, and conceives a whole as nothing but an aggregate of such things, then one advocates atomism. Such things can be considered as atoms. In the context of holism versus atomism, it is irrevelevant whether they are atoms in the literal sense, i.e., ultimate parts, or whether they can be divided up into smaller units. One supports holism in contrast to atomism if, roughly speaking, one claims that the things which are parts of the whole have the properties that are characteristic of them only within the whole.

Even a heap of sand, for example, is a whole. It is appropriate to regard a heap of sand as an aggregate of grains of sand and hence as an atomistic system. Each of the things that contribute to a heap of sand has the properties that make it a grain of sand independently of whether or not there are other things together with which it constitutes a heap. If something is a grain of sand, it has properties such as a certain molecular structure as well as a shape, size, and mass within a certain margin. These properties do not depend on the grain of sand being a part of a heap of sand. There is a possible world in which there is only one grain of sand. This grain of sand has a certain molecular structure and it is spherical. Hence, an individual which is the only physical object in a possible world can have the properties that are characteristic of a part of a heap of sand. If one were to propose a holistic view of a heap of sand, one would have to maintain that the parts

---

2 Concerning the different meanings of "whole" and "part", compare Nagel (1961), pp. 381–383.

3 All the references to other parts of this work are made by indicating the chapter in question in square brackets.

which constitute the heap have the properties that make them grains of sand only within a heap of sand.

If something is a part of an atomistic system, the properties which are characteristic of it are thus non-relational. In contrast, if something is a part of a holistic system, the properties which are characteristic of it are relational. I shall employ only the notions of non-relational and relational properties, since there is no general consensus on how terms such as "intrinsic properties" and "extrinsic properties" should be used.[4] As regards the demarcation between non-relational and relational properties, recall the idea that something can have non-relational properties irrespective of whether or not there are other contingent things in the same world. (The restriction to contingent things is necessary, since, if there are abstract objects that exist in all possible worlds, this does not make all properties relational). In a recent paper, Rae Langton and David Lewis (1998) make this idea more precise by starting from the notion that *having or lacking a property which is intrinsic is independent of accompaniment or loneliness*.[5] Consequently, by contrast, *having or lacking a relational property depends on accompaniment or loneliness*.[6] In the common cases, one thing can have a relational property only if there is at least one other contingent thing with which the first thing has no common parts. Thus, in the common cases, if something has a relational property, there is at least one other contingent thing relative to which the first thing has the property in question. (The restriction "in the common cases" is necessary, because properties that consist in the lacking of relations can be regarded as relational properties as well. Thus, Langton and Lewis (1998) consider the property of being lonely as extrinsic (p. 341). However, in such cases there do not have to be other contingent things in the same world).

I prefer speaking of relational properties to speaking of relations in order to bring out what sort of properties are characteristic of the parts of a holistic system. For instance, at the start of a discussion on semantics, a holist can grant to her opponents that meaning is a characteristic property of beliefs. The point of the holist then is that something has the property of meaning only relative to there being other things together with which it constitutes a system of beliefs.

---

4 For an overview and helpful distinctions see Humberstone (1996).

5 For another recent suggestion of defining intrinsic properties see Vallentyne (1997).

6 Langton and Lewis (1998), pp. 343–344, however, apply the distinction between intrinsic and extrinsic to relations as well.

To point out the opposition to atomism, I thus regard a system as holistic if and only if its parts have some of the properties that are characteristic of these things only within the whole. With respect to the instantiation of these properties each of these things is dependent on there being other things together with which it constitutes a whole of the kind in question. In order to shape this starting point into a substantial and general conception of holism, we have to replace the notion of properties that are characteristic of the parts with a more precise notion, and we have to explain what sort of dependence holds among the parts of a holistic system. I will first develop a proposal for this sort of dependence. In a second step, I shall elaborate on the properties of the parts and thereby also sharpen up the notion of a part of a holistic system.

1.2 Proposal for a General Conception of Holism

1.2.1 Step 1: Generic Ontological Dependence

Let us start by making clear that the dependence that holds among the parts of a holistic system is distinct from causal dependence. Everything that is a candidate for a holistic system is not a mere mereological sum of parts. However, an atomistic system does not have to be a mere mereological sum of parts either. For instance, not any sum of grains of sand is a heap of sand. There have to be certain spatio–temporal and causal relations among the grains of sand if they are to constitute a heap. Causal relations are not sufficient to characterize the sort of dependence that is at issue among the parts of a holistic system. Causal relations can provide for an arrangement of things so that these things constitute a holistic system. But the adjective specifying the kind of dependence among the parts of a holistic system should have a different meaning than the adjective "causal".

Consider social holism: no one denies that the development of thought and rationality in a human being causally depends on her social environment. But that commonplace does not amount to social holism on pain of trivializing holism and turning social atomism into an absurd position.[7] Pointing out relations of circular or mutual causal dependence, such as those ones studied by systems theory, does not change the matter. Such relations are not sufficient to establish that the system in question is holistic.

---

7 Compare Pettit (1996), pp. 169–172.

For any particular grain of sand in our world, there may be a chain of causes which lead to that particular grain of sand and which comprise an unknown number of other things. But this does not hinder that there is a possible world in which there is only one grain of sand. Propositions which describe relations of causal dependence are commonly taken to entail counterfactual conditionals such as "x would not have been F if y had not been G". But such propositions alone do not imply that an individual which is the only thing in a possible world cannot be F. Relations to other things which cause a thing to have a certain property do not exclude that this property can be non-relational. Therefore, the dependence among the parts of a holistic system cannot be simply causal dependence. By way of consequence, nothing like downward causation is necessary for something to be a holistic system.

In their criticism of semantic holism, Jerry Fodor and Ernest Lepore (1992) characterize holism in terms of holistic properties: "Holistic properties are properties such that, if anything has them, then *lots* of other things must have them too." (p. 2) They speak of metaphysical dependence: one thing is, insofar as it has a holistic property, metaphysically dependent on there being a lot of other things which have that property too (pp. 1–2). Metaphysical dependence is a candidate for the sort of dependence that holds among the parts of a holistic system. It ensures that the properties in question are relational. However, I prefer to speak of *ontological dependence.* Fodor and Lepore employ that term too, and they use it interchangeably with "metaphysical dependence".[8] Sometimes the term "logical dependence" is used as well. But this can lead to confusion: logical relations obtain only among propositions and not among the things to which propositions refer. Only a proposition can be logically dependent on other propositions, e.g., by being entailed by other propositions.

Let us go into the notion of ontological dependence. In his book on parts, Peter Simons elaborates on this notion.[9] He introduces a distinction between rigid and generic ontological dependence. Ontological dependence is rigid if and only if a particular individual a cannot exist unless another particular individual b exists too. Rigid ontological dependence is not the way in which

---

8 Fodor and Lepore (1992), pp. 6–7; see also p. 221 note 26.

9 Simons (1987), Chapter 8.3. On ontological dependence, see also Lowe (1994) and Fine (1995). On the background of that notion in Husserl, compare Husserl's *Third Logical Investigation* (Husserl (1984), pp. 227–300; English translation: Husserl (1970)) and Smith and Mulligan (1982).

the parts of a holistic system depend on each other. It is not to be supposed that an individual that is a part of a holistic system is ontologically dependent on those particular other individuals together with which it constitutes a holistic system at a given time. If some parts of a holistic system are removed or replaced, it is nevertheless possible that the remaining parts are the same individuals and that even the system as a whole remains the same system. Members of a social community die, and new members enter into the community, but all the other members remain the same persons, and the community remains the same community. The social holist claims that having beliefs is ontologically dependent on social relations; but these do not have to be relations to particular individuals such as, for instance, relations to one's family, or one's tribe. We should therefore regard generic ontological dependence as the way in which the parts of a holistic system depend on each other. My formulation of generic ontological dependence for the task of achieving a general conception of holism differs from the one that Simons proposes in his book on parts.

Generic ontological dependence is a relation among individuals too. The idea is that there can be no individual of a certain kind unless there is some other individual of a certain kind. Thus, no individual is ontologically dependent on there being any particular other individual. Instead, any individual that is *F* is, with respect to its being *F*, ontologically dependent on there being some other individual that is *G*:

(1.1) *F*s are ontologically dependent on *G*s in a generic way if and only if

$$\Box \forall x \{Fx \supset \exists y (Gy \wedge y \neq x)\} \wedge \Diamond \exists x Fx \wedge \sim \Box \exists x Gx$$

In this formula, □ is the modal operator for necessity, and ◊ is the modal operator for possibility. The main clause says: it is necessary that if there is something which is *F*, there is some other thing which is *G*. The first addition states: it is possible that there is something which is *F*. This clause is to exclude the vacuous case in which there is nothing that is *F* in any possible world. The second addition means: it is not necessary that there is something which is *G*. If it were allowed that there is something which is *G* in every possible world, the notion of generic ontological dependence would be trivialized: every individual in every possible world which is not *G* would be ontologically dependent on there being something which is *G* just because there exists something which is *G* in every possible world anyway.

Formula (1.1) states what ontological dependence among individuals means insofar as they have certain properties. Let me add some remarks to explain further this conception, having its application to the parts of a holistic system in view:

- The proposed formulation of generic in contrast to rigid ontological dependence is not to be confused with rigid designators in the sense of Saul Kripke (1980): *F* and *G* can be proxies of rigid designators. Terms that designate natural kinds can be put in for *F* and *G*.
- It is allowed that *F* and *G* stand for the same property. In this case, there cannot be only one thing which is *F*. If there is one thing which is *F* in a possible world, there is at least one other thing which is also *F* in that world.
- Generic ontological dependence is transitive. Assume that it is necessary that if there is an *x* which is *F*, there is a *y* which is *G*. Assume furthermore that it is necessary that if there is a *y* which is *G*, there is a *z* which is *H*. In this case, it is necessary that if there is an *x* which is *F*, there is a *z* which is *H*.
- If we apply (1.1) to the parts of a holistic system, we have to conceive generic ontological dependence as a symmetric relation: it is not only necessary that if there is an *x* which is *F*, there is some other individual *y* which is *G*; but it is also necessary that if there is a *y* which is *G*, there is some other individual *x* which is *F*. What we are looking for is the sort of dependence that captures the way in which the parts of a holistic system are dependent on *each other*.
- I wish to leave open whether the necessity operator in (1.1) stands in all cases for metaphysical necessity and thus concerns all metaphysically possible worlds or whether it stands in some cases only for physical necessity and thus concerns merely all physically possible worlds.[10] I will come back to this point at the end of [1.3].

The most important feature that makes (1.1) different from the way in which Simons in his book on parts among others conceives generic ontological dependence is that (1.1) does not impose the condition that something which is *F* is necessarily *F*. If one imposes this condition, one regards not only rigid, but also generic ontological dependence as concerning the existence of individuals: in this case, any individual which is *F* cannot exist unless there is some other individual which is *G*.[11] Considering (1.1) as expressing a sort

---

10 Simons (1987), however, construes the term "ontological dependence" in such a way that it automatically concerns all metaphysically possible worlds (especially pp. 300–301).

11 See Simons (1987), p. 297.

of ontological dependence without imposing this condition is not odd. For instance, Fodor and Lepore employ the notions of ontological dependence and metaphysical dependence as in (1.1). They introduce the latter notion by considering the property of being a sibling. They write: "My being a sibling is thus, as one says, *metaphysically dependent* upon someone else's being a sibling (and so too, of course, is my sibling's being a sibling)."[12] By speaking of metaphysical dependence, they do not intend to imply that someone who is a sibling necessarily is a sibling and thus cannot exist without being a sibling. Nonetheless, it is sensible to use the concepts of metaphysical dependence or ontological dependence in such cases: what is at issue is not a linguistic matter of description, but an ontological matter of the dependence of an individual on the existence of other individuals insofar as this individual has certain properties.

Generic ontological dependence as set out in (1.1) is not tailor-made for holism. It is not sufficient to characterize what it is for something to be a part of a holistic system, because it applies to individuals insofar as they have any relational property. I will now first give two reasons for conceiving generic ontological dependence in this broad way. In the next section, a further notion will be introduced that picks out the properties on which we have to focus in the case of the parts of a holistic system.

There is at least one important form of holism that cannot be construed as a thesis about ontological dependence among individuals with respect to properties that these individuals necessarily have. If one espouses social holism, one does not claim that, say, Carol, a human, thinking individual, necessarily has the property of having beliefs and thus cannot exist unless there are other thinking beings with whom she constitutes a social community. The claim is only that she is ontologically dependent on there being other thinking beings insofar as she has beliefs with a determinate meaning. But nothing precludes that she can exist – and thus be the same individual – without having beliefs. Think of Carol as a baby. Carol as a baby and Carol as an adult are the same individual; but as a baby, she does not yet have beliefs with a determinate meaning. Hence, if we are to capture social holism, we cannot conceive holism in such a way that the *existence* of a thing which is a part of a holistic system depends on the *existence* of other things together with which that thing constitutes a system of the kind in question.

---

12 Fodor and Lepore (1992), p. 1. As regards ontological dependence, compare pp. 6–7.

Over and above paying tribute to social holism, conceiving ontological dependence in the proposed way avoids committing holism to essentialism. Essentialism is commonly construed as the claim that some individual has some of its properties necessarily.[13] These properties are its essence. It cannot exist without having these properties. Many philosophers who are regarded as prominent holists do not endorse essentialism. To take the most extreme example, Willard Van Orman Quine goes as far as rejecting modal notions altogether, because he thinks that quantified modal logic is committed to essentialism.[14] However, when it comes to elaborating a precise, substantial, and general conception of holism, we cannot dispense with modal notions. But if we apply the notion of ontological dependence to individuals only insofar as they have certain properties, we avoid essentialism as long as we do not require that an individual has any properties necessarily.

By refraining from employing the notion of essential properties, I propose not to characterize what it is for something to be a part of a holistic system in terms of identity conditions for that thing. Questions of the identity of individuals across different possible worlds do not arise in the context of my formulation of generic ontological dependence: the reference to possible worlds that is contained in (1.1) does not involve propositions about the identity of individuals across different worlds. The propositions concerned are only propositions about relations among individuals *in* one world.

Avoiding a commitment to essentialism is not to be confused with a denial of essentialism. Above all, avoiding such a commitment does not imply the position that an individual is a bare particular that could have all its properties changed and nevertheless be the same individual. My suggestion is not committed to any specific theory of individuals.

1.2.2 Step 2: Properties that make Something a Constituent Part

According to my proposal, generic ontological dependence is the way in which the parts of a holistic system depend on each other. But, given the manner in which I construed generic ontological dependence in the last

---

13 Compare, for instance, the essays in Linsky (1971) and in French, Uehling and Wettstein (1986). For a view that this claim is only a necessary, but not a sufficient condition for essentialism see Fine (1994).

14 "Reference and Modality" in Quine (1980), pp. 155–156; Quine (1960), § 41. For a rebuttal of Quine's objections to quantified modal logic see Barcan Marcus (1993), essays 1, 3, 4, 14.

section, the decisive question is: If a thing that is a part of a holistic system is ontologically dependent on there being certain other things insofar as it has certain properties, what are these properties? In the rough sketch of the opposition between holism and atomism at the end of [1.1], I spoke of properties that are characteristic of the parts of a whole of a certain kind. If the parts have these properties only within the whole, then the whole is a holistic system. Taking the argument in the preceding section into account, we cannot explain "characteristic properties" in terms of essential properties. What shall we say instead?

Consider an individual which is a system of the kind *S*. From now on, I employ "*S*" as abbreviation for "individual which is a system of the kind *S*". I use the capital *S*, because the emphasis is on the qualitative kind. *For every S, there is a family of properties that make something an S*. These properties have the following features:

- They are qualitative or pure properties, because they make something a thing of a certain kind. Qualitative or pure properties are such that their instantiation does not depend on the existence of any particular individual. Properties such as being that individual or having a certain proper name hence do not count as properties of this family. Moreover, when I speak of a system of the kind *S*, I mean a qualitative kind. This kind does not have to be a natural kind; but, for instance, a system of the kind "mereological sum of atoms" is excluded by the requirement that the system is a system of a qualitative kind.
- They are not disjunctive. That is to say, properties such as "being round or square" are excluded.
- They can include both non-relational and relational properties.
- The property of *being an S* does not belong to the family of properties that *make* something an *S*.

The boundaries of such a family of properties may be vague. Nonetheless, if and only if something has more or less all these properties, it is a thing of the kind in question.

Let us transfer this notion of a family of properties that make something an *S* to the parts of an *S*: *There is a family of qualitative, non-disjunctive properties that make something a part of an S in case this thing is arranged with other things in a suitable way*. Having all or nearly all the properties that belong to such a family of properties is a necessary condition for something to be a part of an *S*; it is a sufficient condition in conjunction with the condition of a suitable arrangement with other things. The suitable arrangement can consist in causal relations. For example, let *S* be a social community of thinking beings. It is not sufficient for there to be a social community in a possible world that there are a number of humans in that world. They have to interact with each other. Furthermore, to take a clearly non-holistic example, let *S* be a heap of sand: there have to be certain spatio–temporal and causal relations among grains of sand if there is to be a heap of sand.

The condition of a family of properties that make something a part of an *S* tightens up the notion of a part of an *S*: it excludes inappropriate candidates for parts of an *S*. Consider again the case of a social community of thinking beings. Carol, a human, is a part of such a community. The property which makes her a part of such a community, provided that there is a suitable arrangement with other humans, is her having beliefs in the first place. Bones, in turn, are parts of Carol. But bones are not parts of a community of thinking beings. They do not have properties that make something a part of such a community. Furthermore, something can have the properties that make it a part of a community of thinking beings without being a human and having bones. Imagine Martians who have beliefs with a determinate meaning, but who do not have bones. Hence, *what I mean by a part of an S is such that being a part of an S is not transitive*: if *a* is a part of an *S*, and *b* is a part of *a*, it does not follow that *b* is a part of that *S* as well. Since I thus use a notion of parts which is more restricted than the one that is common in mereology, I shall employ from now on the expression "*constituent parts of an S*", or simply "*constituents of an S*".

The notion of a family of properties that *make* something a constituent of an *S* excludes properties such as *being* a constituent of an *S*. However, this notion on its own is not sufficient to select the properties which we are looking for. That is why the further condition of a suitable arrangement with other things is imposed. For instance, a human being can exercise a social role – such as being a judge, or being a salesman – only by being part of a social system. Social roles are arrangement properties. Pointing out social roles is not sufficient to make a case for social holism, again on pain of trivializing social holism and turning social atomism into an absurd position. The question is whether properties that are the prerequisite for exercising a social role – such as the property of thought and rationality in the sense of having beliefs with a determinate meaning – require social relations. By excluding from the mentioned family of properties those relational properties in which the arrangement with other things consists, we pick out the properties that underlie the arrangement. We are up to a substantial case of holism if and only if these underlying properties are relational as well – in the sense that one thing can have these properties only if there are other things together with which this thing is arranged in such a way that there is an *S*.

For any constituent of any system of a qualitative kind, there is a family of qualitative properties that make something a constituent of such a system provided that there is a suitable arrangement. If *S* is a heap of sand, there is a family of properties that make something a constituent of a heap of sand in case there is a suitable arrangement with other things that also have these properties. These properties are the properties that make something a grain of sand. A certain molecular structure as well as a shape, size, and mass within a certain margin count among these properties.

The family of properties that make something a constituent of an *S* in case there is a suitable arrangement does not have to be the same for each constituent of an *S*. For instance, if *S* is an organism, there are various families of such properties. The family of properties that make something a heart differs from the family of properties that make something a kidney, etc. If an organism is a holistic system, then this is an example of a holistic system whose constituents have properties of different families. This example shows that the characterization of holism in terms of holistic properties which Fodor and Lepore (1992) set out in their book on semantic holism (p. 2) and which I quoted at the beginning of [1.2.1] needs further elaboration if it is to be applicable to all candidates for holism. As it stands, their proposal excludes by definition that an organism can be a holistic system.

Turning to a clear candidate for a holistic system, let *S* be a system of beliefs. Take a constituent of such a system. It has properties like meaning, having a truth value, a degree of confirmation, a degree of justification, etc. These are the properties that make something a belief and a constituent of a system of beliefs in case a suitable arrangement with other beliefs is realized. The holist and her opponent can agree on this family of properties. The point of their disagreement is whether these properties are relational in the following sense: can anything have these properties independently of whether or not it is arranged with other beliefs so that there is a system of beliefs?

According to some versions of semantic holism, a belief is not only ontologically dependent on there being other beliefs with respect to its having meaning; what is more, the specific, determinate meaning of each belief depends on the specific, determinate meaning of all the other beliefs with which it is arranged in a system of beliefs: if the meaning of one belief in a system of beliefs changes, the meaning of all the other beliefs in the system changes as well. However, the foremost point of semantic holism is not that the meaning of one belief may change if the meaning of other beliefs changes. Whatever the specific meaning of a given belief is and in whatever way its meaning can change, the point is that something cannot have any meaning at all unless there are a number of other beliefs together with which it is arranged in a system of beliefs.

Hence, by a family of properties that make something a constituent of an *S* given a suitable arrangement, I mean *a family of generic, determinable properties* in distinction to specific, determinate properties. Suffice it to say that a property is generic or determinable if and only if nothing can have this property *simpliciter*, but only in a way that admits of further specification. Accordingly, a property is specific or determinate if and only if it admits of no further specification. For instance, like meaning, mass is a generic property. Nothing can have mass *simpliciter*. If something has mass, it has a specific quantity of mass. The same goes for all other physical magnitudes. Thus, each constituent of a concrete system has the properties that belong to the mentioned family of properties in a specific, determinate way. However, for an individual to be a constituent of an *S*, the point is not the specific, determinate way in which it has these properties, but simply its having the properties that belong to such a family of properties in the generic, determinable sense.[15]

1.2.3 Result: Proposal for a General Conception of Holism

Let us apply the notion of generic ontological dependence to the notion of things insofar as they have properties of a family that make something a constituent of an *S* provided there is a suitable arrangement. The idea is: in case something can have the properties that belong to such a family only if there actually are other things together with which this thing is arranged in such a way that there is an *S*, the systems of the kind *S* are holistic systems. The adverb "actually" simply emphasizes the crucial point. We can now set out the following characterization of a holistic system:

General characterization of a holistic system

> Consider a system of the kind *S* and its constituent parts. For every constituent of an *S*, there is a family of qualitative, non-disjunctive

15 Fodor and Lepore (1992), pp. 2, 5, also mean generic properties when they speak of holistic properties.

> properties that make something a constituent of an *S* provided that there is a suitable arrangement. An *S* is holistic if and only if the following condition is satisfied by all the things which are its constituents: with respect to the instantiation of some of the properties that belong to such a family of properties, a thing is ontologically dependent in a generic way on there actually being other things together with which it is arranged in such a way that there is an *S*.

This proposal characterizes holism by explaining what it is for something to be a holistic system. The distinctive feature of a holistic system is a certain sort of dependence among its constituent parts. Therefore, this proposal also characterizes what it is for something to be a constituent part of a holistic system. Since the sort of dependence in question concerns individuals insofar as they have certain properties, we can consider these properties themselves as holistic and thus introduce the notion of a *holistic property*. Holistic properties are relational properties. A relational property is holistic if and only if it satisfies the following two conditions: (1) Instead of being an arrangement property, it belongs to the family of properties that make something a constituent of an *S* provided that there is a suitable arrangement. (2) Nothing can have this property unless there actually are other things together with which this thing is arranged in such a way that there is an *S*. For example, meaning is a holistic property if semantic holism is right and a system of beliefs is a holistic system.

According to this proposal, it is an ontological matter whether or not something is a holistic system. This is not relative to the way in which someone chooses to describe a system. But whether or not an *S* is a holistic system can be relative to some of the properties that belong to the family of properties that make something a constituent of an *S*. The proposed characterization demands only that ontological dependence relates to individuals insofar as they have *some* properties that belong to such a family. It is thus not demanded that all the properties that belong to such a family of properties are holistic properties. We thereby gain a fine-grained conception of holism: this qualification enables us to admit different forms and different degrees of holism with respect to the same kind of systems. One and the same system can be holistic with respect to some of the properties of the mentioned family, but atomistic with respect to others of them.

For example, one can maintain that a system of beliefs is a holistic system only with respect to confirmation, but not with respect to meaning. That is to say: a belief is ontologically dependent on there being other beliefs with which it is arranged in a system of beliefs insofar as it has a degree of confirmation. But confirmation holism does not imply semantic holism, i.e., the thesis that a belief is ontologically dependent on there being other beliefs insofar as it has meaning. If and only if one maintains that, conversely, semantic holism implies confirmation holism, one can say that these different forms of holism in the same field amount to different degrees of holism in that field: in this case, semantic holism is a stronger form of holism than confirmation holism.

A widespread expression that may seem appropriate for the relations of ontological dependence among the constituents of a holistic system is the term "internal relations". This term has been common since the British Idealists of the beginning of the twentieth century.[16] But internal relations are considered as essential properties, whereas the conception of holism suggested here is not committed to essentialism. Note, however, that the proposed view of a holistic system does not require that something which is a constituent of a holistic system has non-relational properties too. It is not excluded that in some candidates for holism, the constituents of a holistic system do not have any qualitative non-relational properties. The system of the natural numbers may be an example; for one may maintain that all the qualitative properties of any natural number consist in relations to other natural numbers. Consequently, I do not presuppose that speaking of constituents of a holistic system requires that the constituents have non-relational properties.

1.3 Is the Proposed Conception Substantial?

The proposed conception of a holistic system not only precludes trivial cases like the heap of sand; it also contains a substantial criterion for holism which is able to turn a social community of thinking beings and a system of beliefs into significant candidates for holistic systems. This conception excludes that every functional definition of constituents amounts to a case of holism. To bring this point out, consider the following example: assume for the sake of argument that a heart is functionally defined as the blood pump in an organism. Consequently, something can be a heart only if there is blood and other organs such as kidneys together with which that thing is arranged in

---

16 See, for instance, the examinations of Ewing (1934), Chapter 4, and Horstmann (1984), especially the introduction and Chapter 2. On internal versus external relations, see the overview by Rorty (1967). Langton and Lewis (1998), pp. 343–344, use the term "internal relations" in another sense.

such a way that there is an organism. It can be maintained that the heart that a butcher sells is a heart in name only like a corpse is a man in name only.

Does a functional definition of organs suffice to establish that an organism is a holistic system? Reservations are appropriate: a thing that is a piece of flesh with certain non-relational properties can merely not exercise the function of a blood pump if there is no blood. But, independently of its being a constituent of an organism, such a thing has a number of properties that make it function as a heart if it is arranged with other things in a suitable way. We can produce an artificial heart in isolation. Such a thing has to be endowed with very specific properties in the production process. As soon as it is implanted into an organism, it functions as a heart by virtue of the properties that it already has. Being a heart is an arrangement property. Therefore, a functional characterization of organs is not sufficient to make a case for holism. I do not intend to exclude that organisms may be holistic systems. But the dependence on other things in order that the properties that a thing already has are exercised in such a way that this thing fulfills a certain function is not sufficient for a substantial case of holism.

Another example from a different area may strengthen that claim. It can be maintained that a traffic light is functionally defined as a thing that controls traffic. Thus, something can be a traffic light only if it exercises a certain function. However, such a thing has a number of properties that make it fulfill that function independently of its being a part of the system of traffic. An individual that is the only physical object in a possible world can have those properties. We can produce such a thing in isolation from traffic. In virtue of the properties that it has acquired by that production, it functions as a traffic light as soon as it is installed in a suitable position on a road with traffic. Being a traffic light is an arrangement property. A substantial conception of holism clearly has to preclude that cases of depending on other things for fulfilling a certain function automatically count as cases of holistic systems.

The proposed conception of holism excludes these cases. An organism would be a holistic system only if the following requirement were satisfied: with respect to some of those properties that make something function as a blood pump in case that thing is implanted into an organism, a heart would have to be ontologically dependent on there actually being other things together with which it is arranged in such a way that there is an organism. The same goes for the traffic light. This requirement furthermore excludes that a family in the biological sense automatically is a holistic system: an individual which is the only physical object in a possible world can in principle have all the biological properties that make an individual a human being. Provided that there is a suitable arrangement, a human being becomes a member of a family in the biological sense in virtue of those biological properties.[17]

By excluding that cases of a mere functional definition of constituents count as cases of holism, the proposed conception turns the case of a social community of thinking beings and the case of a system of beliefs into significant candidates for holistic systems. The property that makes something a constituent of a social community of thinking beings is the property of having beliefs in the first place. According to social holism, no finite being can have this property unless there are other beings with whom this being is arranged in such a way that there is a social community of thinking beings. Thus, the relations to other humans make a thinking being of a human being. Social holism is a substantial form of holism if and only if it is claimed that this formation is not paralleled by the way blood, kidneys and other organs turn a thing into a heart according to the view of organisms sketched above.

How could such a claim be established? The epistemic criterion of irreducibility can serve as a guide for excluding a relational property from being a mere arrangement property: for something to be a holistic property, it is necessary that the description of the property cannot be reduced to a description of non-relational properties and the description of a suitable arrangement. Hence, to establish social holism it is necessary to show that the description of thinking in the sense of having beliefs with a determinate meaning cannot be reduced to a description of human beings and their interrelations in terms of the natural sciences.

In the case of holism about beliefs, the proposed conception imposes the following two conditions: (1) Properties like meaning, confirmation, justification *irreducibly* count among the family of properties that make something a constituent of a system of beliefs, given suitable connections with other entities that also have these properties.[18] (2) An entity can have some of these properties only if there actually are other entities which also

17 See also the examples of a cluster of gnats and a picnicking family that Popper uses as counter-examples to holism ("Of Clouds and Clocks" in Popper (1989), pp. 209–210).

18 According to Davidson, irreducibility is an immediate consequence if a holistic theory applies to beliefs. See "Mental Events" (1970) in Davidson (1980), pp. 217, 222–223, and "Belief and the Basis of Meaning" (1974) in Davidson (1984), p. 154.

have these properties and with which this entity is arranged in such a way that there is a system of beliefs of a person.

If it were possible to reduce the description of types of beliefs to the description of types of brain states (or other physical states), the case of beliefs would be parallel to the case of the heart. Being a belief would be an arrangement property of the neural material that makes up the brain. This material has properties that its constituents can have in isolation, and these properties make that material exercise certain causal functions in case a suitable arrangement is realized. There is no holism on the level of that material. Even if there were a holism on the level of that material, it would not be a holism about beliefs. However, granting that holism about beliefs is incompatible with a reduction of the description of belief states to a description of physical states, one can nonetheless subscribe to the proposition "If there are intentional properties such as meaning, then holism about beliefs is true" and reject the antecedent, i.e., endorse eliminativism with respect to intentional properties. Quine, among others, can be seen as adopting that attitude.[19]

The epistemic condition of irreducibility says nothing against claims to the effect that having beliefs (1) supervenes on physical properties and (2) has a physical realization.[20] (To give a rough idea of supervenience, if and only if having beliefs supervenes on physical properties, then any variation of intentional properties requires a variation of physical properties; consequently, if the physical properties are the same, then the intentional properties are the same too. I shall consider supervenience in [5.4.1]). A substantial conception of holism does not commit us to a dualistic ontology in the sense of having to admit anything that exists beyond the physical when it comes to applying that conception to the philosophy of mind. Although a social community of human thinking beings certainly is realized as an arrangement of bones and other material that make up humans in a suitable physical environment and although one can maintain that a system of beliefs of a person is realized in brain states and other physical states, this does not make bones and neural material constituent parts of a social community or a system of beliefs. For this does not imply that intentional

---

19 Compare Quine (1960), §§ 45, 54; "Epistemology Naturalized" in Quine (1969), essay 3.

20 As to suggestions for a precise formulation of this notion of realization, see, for instance, Papineau (1993), pp. 24–25, as well as Heil (1999).

properties are identical with and thus the same properties as certain physical properties.

Hence, the proposed conception of holism includes an account of the relation between holism and reductionism. If there are holistic properties in the sense explained in [1.2.3] at some level of complex systems in nature, the description of these properties cannot be reduced to a description of lower level properties plus a description of a suitable arrangement of objects at the lower level in question. However, this anti-reductionism does not speak against the possibility that each instantiation of holistic properties at some level (1) supervenes upon a lower level of properties and (2) has a lower level realization. Throughout this work, I use the term "reductionism" only in an epistemic sense, i.e., a sense that concerns the relation between descriptions or theories. Holism is incompatible with reductionism in an epistemic sense. But it is not opposed to what is sometimes called "ontological reductionism", namely the rejection of dualist ontology and in particular the claim that there are no forces over and above those ones that are acknowledged in physics.[21] As mentioned at the beginning of [1.2.1], holism is not committed to anything like downward causation.

On the other hand, if there are cases of holism in the domain of the natural sciences, such cases do not exclude that the description of higher level properties can be reduced to a description of the lower level holistic systems in question. For instance, if there is holism in quantum physics, nothing is decided against a reduction of theories of, say, chemistry or biology or even psychology to quantum physics by quantum holism alone [9.3.2].

When candidates for holism in the domain of the natural sciences are examined, it is reasonable to limit ontological dependence to physical necessity, since, on most views, natural laws are not metaphysically necessary. For example, a world in which microphysical systems such as photons, electrons, protons, and neutrons are atoms in the sense of classical, Democritean or Newtonian atomism is metaphysically possible. However, if the laws of quantum physics apply to the microphysical level, such a world is not physically possible.

If we qualify ontological dependence in such a way that it applies only to physically possible worlds when it comes to candidates for holism in the domain of the natural sciences, we have to make sure that we do not give up the distinction between ontological and causal dependence. For instance, the argument for holism in quantum physics is based on what is known as

21 See, for instance, Stöckler (1992).

entanglement of quantum systems. It can be maintained that entanglement is not a causal relation. Quantum systems can be entangled independently of whether or not there is interaction between them [8.1.3]. If one speaks of ontological dependence among the constituents of a holistic system insofar as they have certain properties, the point is not that ontological dependence is often construed as concerning all metaphysically possible worlds in all cases, whereas causal dependence is usually seen as being related to natural laws that are not metaphysically necessary. The point is rather: we have to name in each candidate for a holistic system properties of its constituents to which ontological dependence applies and to show the distinction from causal dependence for each case.

Of course, holism about beliefs and social holism can also be qualified: in defending these claims, one can limit their application to physically possible worlds or other suitably qualified worlds – provided that one can still maintain the conceptual distinction between relations of ontological dependence and mere causal relations. But in cases such as holism about beliefs and social holism, it makes sense to consider an application of these claims to all metaphysically possible worlds.

To conclude this sub-chapter, let us briefly consider whether the proposed view of holism is not only substantial, but also able to pay tribute to our intuitions. Think of the saying "The whole is more than the sum of its parts". In this saying, it is claimed that a whole or its parts have significant properties whose description cannot be reduced to a description of the parts as they are independently of being organized to constitute the whole. The proposed conception is able to suggest a precise holistic meaning for this saying. If we interpret this saying in line with the proposed view, we have to maintain: each of the constituent parts is, insofar as it has some of the properties that make it a constituent of an *S*, given a suitable arrangement, ontologically dependent on there actually being other things with which it is arranged in such a way that there is an *S*. These properties cannot be taken into account by a description of the properties which the things that are the constituents of an *S* can have in isolation. What is more, these properties cannot be taken into account by just adding a description of a suitable causal and spatial arrangement to that description either. Therefore, the holistic sense which the proposed conception can provide for the saying "The whole is more than the sum of its parts" is not just that the "more than the sum" amounts to the parts being causally arranged in a certain way. This conception can suggest a significant holistic sense for that saying: the "more than the sum" of the organization of the system consists in the parts having themselves holistic properties.

1.4 Two Types of Holism

The view of holism set out in the last sub-chapters is a general conception that is intended to be applicable to all candidates for holistic systems. The proposed characterization is a guideline that brings to the point conceptual tools that we need in dealing with candidates for holistic systems. We can regard this conception as describing the genus "holistic system". Let me now introduce two species or types of holism within this general conception; these types will be helpful when it comes to examining candidates for holistic systems.

Developing the proposed conception of holism, I assumed that each constituent of a holistic system has some holistic properties. Consider again holism about beliefs and focus on meaning as a property of beliefs. A holist does not have to deny that each belief has a meaning. She can make her point by saying that each belief is, insofar as it has a meaning, ontologically dependent on there being other beliefs with which it is arranged in such a way that there is a system of beliefs of a thinking being. If a whole system of beliefs has the property of meaning too, the meaning of the whole derives from the meanings of its constituents.

However, one can also conceptualize holism about beliefs in a different manner. One can maintain that it is a whole system of beliefs which has properties such as meaning, confirmation, justification in the first place. To the extent that the constituents of a system of beliefs have these properties too, they have them derivatively from the whole. This conceptualization of holism about beliefs owes much to Quine's work. In his "Two Dogmas of Empiricism", Quine coins the slogan "The unit of empirical significance is the whole of science".[22] In later papers, he qualifies this slogan by maintaining that it is a cluster of statements (or beliefs) which has meaning and that this cluster encompasses only in the last resort our whole system of statements (or beliefs).[23] In the spirit of Quine's holism, one can therefore say that beliefs have the property of meaning only taken together; strictly speaking, only a whole system of beliefs has meaning. Nonetheless, meaning as a property of a whole system of beliefs indicates a differentiation within the whole: the meaning of the whole indicates the way in which the parts are

22 In Quine (1980), p. 42.

23 For instance, Quine (1991), pp. 268–269.

related with respect to meaning, although the parts do not have a meaning each.

These two types or species of holism differ in the way in which the phrase "with respect to the instantiation of some of the properties that belong to such a family of properties" in the proposed general characterization of a holistic system is elaborated on. On the reading of this phrase that corresponds to the way in which I have introduced the notion of ontological dependence in [1.2.1], this phrase is about relational properties that one thing has and that can be instantiated by this thing only if there are other things which fulfill a certain condition. This is a *bottom-up conceptualization of holism*: we begin with constituents and properties that make something a constituent. We come to the conclusion that something can have some of these properties only if there are other things with which it is arranged in such a way that there is a whole of the kind in question.

On the type of holism now under consideration, the phrase in quotation marks is about properties which can be instantiated in the last resort only by a whole of the kind in question, but which indicate a differentiation within the whole with respect to the properties in question. Although these properties are instantiated in the last resort only by the whole, they belong to the family of properties that make something a constituent of an *S*. This is a *top-down conceptualization of holism*: we begin with the whole and properties of the whole. Some of the properties of the whole specify a differentiation within the whole by introducing constituents of the whole and thus including relations among the constituents.

These two conceptualizations of holism are not disjunctive. The proposed general conception of holism covers both these types, since it focuses on the internal structure of a whole that is a holistic system. The only difference between these two types of holism is the way in which they specify the phrase quoted at the beginning of the last but one paragraph. Thus, there are two different types of holism; but they both come under one general and substantial conception of holism.

Let us sum up these two types:

1) Each of the constituents has the properties of the family that make something a constituent of an *S*. These properties are primarily instantiated by the constituents. If the whole has these properties too, it has them because its constituents have them.
2) Some of the properties that make something a constituent of an *S* are instantiated only by several constituents taken together. Strictly speaking, these properties are instantiated only by the whole. But these

properties of the whole introduce constituents by indicating the way in which constituents are related with respect to the properties in question.

As my sketchy remarks on semantic holism intend to show, semantic holism admits of a conceptualization along the lines of both these types. There are thus two types of semantic holism as far as the structure of a holistic system of beliefs is concerned. Social holism, by contrast, is tied to the first type: each individual thinking being has the property of having beliefs, and individual thinking beings are at least the primary, if not the only systems to which this property can be attributed. Quantum holism, by contrast, is an example of the second type of holism [8.3].

The distinction between these two types of holism roughly corresponds to the two characterizations of holism which Fodor and Lepore (1992) propose in their criticism of semantic holism. As quoted in [1.2.1], they characterize holism in terms of holistic properties in the first chapter of their book: "Holistic properties are properties such that, if anything has them, then *lots* of other things must have them too." (p. 2) The first type of the proposed general conception of holism can to a certain extent be considered as an elaboration on this idea. In the preface to their book, however, Fodor and Lepore take up the vein of Quine's formulation of holism and characterize semantic holism as "the doctrine that only whole languages or whole theories or whole belief systems *really* have meanings, so that the meanings of smaller units ... are merely derivative" (p. X).[24] This is an account of semantic holism in the spirit of the second type of the proposed general conception.

To conclude this chapter, let us review the proposed steps. Intending to achieve a general conception of holism, I maintained that two points have to be taken into account: (1) the sort of dependence among the constituents of a holistic system, and (2) the group of properties of the constituents to which holism extends. In order that the intended conception of holism is substantial, I argued that (1) the sort of dependence among the constituents of a holistic system is generic ontological dependence and that (2) the properties which holism concerns are those ones that make something a constituent of a system *S* provided that there is a suitable arrangement. Fixing the matter in this way nevertheless leaves open a number of factors in

---

24 On the difference between the two accounts of semantic holism, see Heal (1994), in particular pp. 326–328. Compare already Heal (1989), pp. 86–87, as well as the discussion between Cohen (1999a) and (1999b) and Heal (1999). See furthermore Dummett (1974), pp. 358–359; Brandl (1993), pp. 3–4.

which holistic systems can vary. The following three factors are of particular importance:

a) the properties that holism concerns. These properties can vary at least in the following three ways:

- their number: There is a family of properties that make something a constituent of a holistic system *S* provided that there is a suitable arrangement. Holism touches some of the properties of this family; but it does not have to touch all of them. For instance, one can regard a system of beliefs as a holistic system only insofar as confirmation is concerned. But one can also consider a system of beliefs as a holistic system insofar as both meaning and confirmation are concerned.
- their importance: all the properties which holism concerns are important in that they belong to the family of properties that make something a constituent of a holistic system *S* provided that there is a suitable arrangement. Nonetheless, I left the question of essential properties open in [1.2.1]. Consequently, the properties that are touched by holism can be more or less important as far as the identity of the system and its constituents is concerned. For instance, in the case of a thinking human being who is a constituent of a holistic social system, I argued that the property of having beliefs with a determinate meaning is not necessary for the identity of the human being concerned. If Carol develops from a baby into a normal adult and then becomes insane, she nevertheless remains the same human being throughout this development [1.2.1]. Consequently, a change in the constituents in the sense that some individuals enter a social system of thinking beings and others drop out of the system does not affect the identity of the system. By contrast, consider the case of holism about beliefs with respect to meaning: if a constituent of a system of beliefs changes its meaning, it is no longer the same belief. Furthermore, one may claim that a change in the meaning of one belief induces a change in the meaning of all the other beliefs in the system and thus touches upon the identity of the system as a whole.
- the number of times a family of properties that make something a constituent of an *S* is instantiated in each *S*: For instance, if an organism is a holistic system, more or less each constituent has its own family of properties that make it a constituent of the system under consideration. The family of properties that make something a heart differs from the family of properties that make something a kidney, etc. Each of these families of properties is instantiated only once or twice in each organism. By contrast, in the cases of holism about beliefs and social

holism, each constituent of a system of beliefs or a social system has the same family of properties. This family of properties consequently is instantiated several times in each system.

b) the way in which the constituents of a holistic system have the properties to which holism extends. The variation concerns the two types of holism that have been introduced in this sub-chapter:

- Each of the constituents has the properties of the family that make something a constituent of an *S* given a suitable arrangement.
- Some of the properties that make something a constituent of an *S* are instantiated only by several constituents taken together (or only by the whole).

c) the scope of the necessity which ontological dependence implies: the necessity can be metaphysical in that it concerns all possible worlds. But one can also maintain that in some cases of holism the scope of ontological dependence is restricted so that it applies only to, for instance, all physically possible worlds.

In conclusion, I submit that the proposed view shows how a satisfactory conception of holism can be reached by employing conceptual tools of analytic philosophy. This view pays tribute to our intuitions on holism. It is general, since it is applicable to candidates for holism in any area. It is substantial, because it not only excludes trivial cases, but also indicates a strong criterion for paradigmatic cases like holism about beliefs and social holism as well as holism in quantum physics. It makes it a matter of real argument whether there is holism in any of these areas. It thereby provides a basis not only for examining whether the candidates for holism in different areas have a common conceptual content, but also for considering their implications for our overall philosophical view of the world and ourselves.

CHAPTER 2
ARGUMENTS FOR HOLISM ABOUT BELIEFS

Summary

Beginning with Quine's "Two Dogmas of Empiricism", the main arguments for holism about beliefs are considered. Three forms of this holism are distinguished: holism about (a) confirmation, (b) justification, and (c) meaning, i.e., semantic holism. It is suggested that (a) is best construed along the lines of the second type of the general conception of holism set out in the first chapter; (b) and (c), by contrast, can be construed along the lines of both types of this conception. I shall not be concerned with (b). As regards (a), I endorse Quine's confirmation holism in "Two Dogmas". However, I do not approve of Quine's step to (c), because it presupposes verificationism [2.1]. I then go into the argument for semantic holism that bases itself on the conditions for the interpretation of persons. This argument on its own is neither necessary nor sufficient to establish semantic holism [2.2]. Therefore, the focus of this chapter is on the direct argument for semantic holism, namely inferential role semantics. The tasks for such a semantics are set out. Furthermore, the major objection against such a semantics is considered, i.e., that it does not allow for shared beliefs [2.3]. The strategy to counter this objection pursued here is to conceive inferential role as role in a social, linguistic community. In elaborating on this strategy, I base myself on Brandom's "Making It Explicit" and follow Brandom in grounding an inferential role semantics on a normative pragmatics. According to this pragmatics, "meaning" is a normative notion; nonetheless, a descriptive theory of meaning is possible: In ascribing beliefs to a person, one describes the commitments that the person in question undertakes [2.4].

2.1 The Argument from Confirmation

2.1.1 The Duhem–Quine Thesis

Holism in philosophy of mind is a thesis about the intentional states of persons. Its focus are belief states or beliefs. Therefore, I shall limit myself to holism about beliefs. This is the claim that a person can have single beliefs only within a whole system of beliefs. The reason is that a belief has some of the properties which make something a belief – such as meaning, confirmation, justification – only if there are other beliefs with which it constitutes the system of beliefs of a person. The claim thus is that the system of beliefs of each person is a holistic system. In this chapter, different types of arguments for this claim are evaluated.

A popular starting-point for considering holism about beliefs is Willard Van Orman Quine's seminal paper "Two Dogmas of Empiricism" (1951); quoted according to the edition in Quine (1980)). This paper is intended to be a criticism of logical empiricism and, in particular, the work of Rudolf Carnap. Quine attacks two theses that he attributes to logical empiricism:

- Scientific statements divide up in analytic statements of logic and mathematics as well as definitions, which are true or false independently of the way the world is, and synthetic, empirical statements about the way the world is.
- All empirical statements can be reduced to logical constructs upon statements which describe sense experience. Each of these latter statements can be directly confirmed by sense experience.

At first, Quine argues that there is no separation between analytic and synthetic or empirical statements (pp. 20–37; see also Quine (1960), § 14). Then he maintains that the programme to reduce all statements of empirical science to logical constructs upon statements of sense experience is doomed to failure. He considers the claim that experience confirms single statements to be a residue of this programme (pp. 37–41). These two theses are interconnected: if there is no separation between analytic and synthetic, empirical statements, then there are no statements which are true come what may. Experience touches upon all statements, including those that are regarded as logical laws. Quine suggests that these two theses have the same root: the supposition that the truth of each statement divides up in a linguistic and an empirical component (pp. 41–42).

Quine's alternative to these two theses – or dogmas, as he calls them – starts from the following claim: "Taken collectively, science has its double dependence upon language and experience; but this duality is not significantly traceable into the statements of science taken one by one." (p. 42) Consequently, experience cannot confirm or refute any statement taken in isolation. What is confirmed or refuted by experience – be it scientific experience, be it common sense experience – is a whole system of statements (p. 41). In "Two Dogmas", Quine says: "Any statement can be held true come what may, if we make drastic enough adjustments elsewhere in the system." (p. 43) Consequently, no statement is immune to revision as a result of experience. Even statements that are regarded as logical laws can be abrogated in order to accommodate new experience. The demarcation between what counts as logic and what counts as empirical science can hence be subject to change.[1]

Quine suggests regarding our system of knowledge as a – seamless – web. This web touches experience at its edges. Empirical statements such as "There are brick houses on Elm Street" are on the periphery of this web. Statements of logic are located in its centre. This web is not determined by experience: if a conflict with experience occurs, we have several options for adjusting the web to experience (pp. 42–45). Already in "Two Dogmas", Quine proposes a pragmatic attitude: it is rational to opt for those changes which imply the slightest perturbation within our system of knowledge as a whole in order to accommodate this system to new experience (pp. 43–44, 46). Hence, Quine's position implies not only that experience cannot confirm or refute any statement taken in isolation, but also that there is no separation between science and philosophy in the sense of metaphysics. Quine says: "Ontological questions, under this view, are on a par with questions of natural science." (p. 45)

This view of knowledge is related to a position that goes back to the French scientist and philosopher of science Pierre Duhem (1861–1916). In his work on the object and the structure of the physical theory (part 2, Chapter 6, quoted according to the edition Duhem (1981), English translation Duhem (1954)), Duhem maintains that it is not possible to put a hypothesis of physics to the test in isolation. Every experiment involves assumptions about the way in which the measuring instruments function, and these assumptions, in turn, imply physical laws. Therefore, Duhem claims that an ensemble of hypotheses or theories is put to the test in any experiment (§§ 1–2). If the experimental results do not agree with the predictions, we only know that at least one of the hypotheses in question is false. But we do not know which hypothesis or which hypotheses are false. We have a number of options for changing our theory in such a way that it accords with the experimental results (§ 2). Duhem proposes *bon sens* as a criterion for deciding which option should be endorsed (§ 10).

Consequently, even the first principles of physics are subject to empirical control in the same way as all the other statements that are contained in a physical theory. If a conflict with experimental evidence occurs, it is not logically determined which statements are to be rejected. It may be reasonable to change even fundamental hypotheses of physics subsequent to the results of experiments (§ 9). Duhem concludes that the entire physics is

---

1 See also Wittgenstein, *On Certainty*, § 98.

one theory that is confronted as a whole with the whole of the experimental facts (§§ 2, 5, 7–8).

Quine mentions Duhem in his "Two Dogmas" (p. 41). However, when this paper was first published in 1951 (Quine (1951)), the reference to Duhem had not yet been included. After that paper had been finished, colleagues drew Quine's attention to Duhem.[2] Duhem's philosophy of physics thus is not Quine's background. Nonetheless, the criticism of logical empiricism which Quine proposes is widely seen as radicalizing Duhem's position.[3] Duhem merely refers to physics. He excludes common sense knowledge as well as sciences such as physiology on the one hand (§ 1) and logic and mathematics on the other hand from his thesis about confirmation. Quine, by contrast, generalizes this thesis in such a way that it applies to all knowledge, including common sense knowledge as well as logic and mathematics. Nonetheless, paying tribute to Duhem, this latter thesis is known as the *Duhem–Quine thesis*.

2.1.2 From Confirmation Holism to Semantic Holism

Instead of speaking of statements, as Quine does in "Two Dogmas", I prefer speaking of thoughts or beliefs, because I intend to take the Duhem–Quine thesis as a starting-point for considering holism about beliefs. Beliefs are attributed to persons. To be more precise, what is attributed to persons is instances of beliefs, i.e., tokens of beliefs. When I talk about beliefs, I always mean tokens of beliefs. A person has beliefs by being in a state of believing that, for example, snow is white. Beliefs exist only insofar as persons are in states of believing something. I use the term "belief" in such a way that only what a person is prepared to acknowledge counts as a belief of hers. When I speak about the consequences of certain acknowledged beliefs, I talk in terms of what a person is committed to believe. Having beliefs in this sense is tied to being able to have beliefs about beliefs; for having beliefs involves the capacity to assess whether the beliefs that one has cohere with or are incompatible with one another. For instance, if one has a belief of the type "This is *F*" and a belief of the type "This is *G*", one also has the capacity to form the meta-belief "The belief that this is *F* and the belief that

2 See Quine (1991), p. 269.

3 See Quine (1986b), p. 619, and, for instance, Gochet (1977), p. 121; Vuillemin (1986), pp. 599–601, 609, as well as the distinctions which Moulines (1986), pp. 313–319, makes. For views that qualify the received opinion see Krips (1982) and Flügel (1996). As to the recent state of the discussion, see Needham (2000).

this is *G* cohere with one another" (or are incompatible with one another), etc.

A state such as the state of believing that snow is white is an intentional state. Its characteristic feature is intentionality, that is, being about something. Belief states are a sub-class of intentional states. States of the type of desiring that something is the case are also intentional states. According to most accounts, all intentional states include belief states. For instance, one can be in the intentional state of desiring an ice-cream only by having beliefs about ice-creams. Therefore, when I speak of intentional states, I mean states that involve belief states. Belief states are epistemic states. I use the term "epistemic state" in a broad sense according to which all and only those states that have a conceptual content are either themselves epistemic states or include epistemic states. Hence, being in an epistemic state does not imply having knowledge in the sense of true and justified beliefs. I shall limit myself to holism about belief states, although one can argue that holism in the philosophy of mind encompasses all types of intentional states.[4]

I assume that it is beliefs that have meaning and conceptual content in the first place. To the extent that holism about beliefs is set out as a plausible semantics, this assumption is vindicated. By "content", I always intend "conceptual content". I do not distinguish between meaning and conceptual content. When I employ the term "meaning" in the following, I take this term to cover conceptual content, and *vice versa*. Since beliefs exist only insofar as persons are in states of believing that something is the case, I shall switch between talking in terms of beliefs having meaning and talking in terms of belief states having content. What a person believes can be expressed as a sentence in a language. Perhaps a person believes something only insofar as she is able to express what she believes in a language. But this point should be left open at the beginning of an examination of holism about beliefs.

As soon as a person has several beliefs, she has a system of beliefs, since all these beliefs are interconnected by logical relations. Because of the logical relations among all the beliefs of a person, each mentally sane person has only one system of beliefs, at least at a time. Thus, speaking of a system of beliefs is innocuous. It does not beg the question of holism. The question is whether the system of beliefs of each person satisfies the criteria for a holistic system which were set out in the preceding chapter.

---

4 See, for instance, Davidson (1995), pp. 210, 217–218.

The properties that Quine attributes to statements in "Two Dogmas" – such as having meaning, being confirmed or refuted, being justified or unjustified – can be attributed to beliefs too. Meaning is a generic, determinable property [1.2.2]. Nothing can have meaning *simpliciter*, but only in a way that admits of further specification. If something has meaning, it has a specific meaning. One can maintain that confirmation and justification are generic properties, too. I use these terms in such a way that they cover the whole spectrum of confirmation and justification. Thus, if something has the property of confirmation or the property of justification and if these properties are generic properties, everything to which confirmation or justification is attributable has a specific value or degree of confirmation or justification. This degree can be zero; in that case the belief in question is completely refuted or completely unjustified. Meaning, confirmation, justification are hence properties of beliefs. But beliefs are themselves attributed to persons. Beliefs occur only insofar as persons are in states of believing something. Therefore, speaking in terms of properties of beliefs is innocuous; it is no reason to reify beliefs. I use the terminology of properties of beliefs for the sake of simplicity and convenience in setting out holism about beliefs.

Let us now translate what Quine says about the confirmation of statements or sentences into the terminology of beliefs that are attributed to persons. The claim then is that single beliefs cannot be confirmed by experience. What is confirmed by experience is the whole system of beliefs of a person. This position is known as epistemological holism[5] or *confirmation holism*[6]. I prefer the term "confirmation holism"; this term makes clear that the target of this view is the property of confirmation.

Confirmation holism is a form of holism about beliefs. However, holism about meaning, i.e., *semantic holism*, is regarded as the core of holism about beliefs. Quine not only proposes confirmation holism in "Two Dogmas", but also semantic holism. He takes confirmation holism to lead to semantic holism. At least this is the received reading of "Two Dogmas".[7] The reason is that Quine adopts a verification theory of meaning from logical

---

5 Gochet (1977), p. 120. Fodor and Lepore (1992), p. 182, use the term "epistemic holism".

6 Fodor and Lepore (1992), Chapter 2.

7 See, for instance, Gochet (1977), p. 120. Fodor and Lepore (1992), Chapter 2, also build their criticism of Quine's argument on this reading. See Brandl (1993), pp. 12–13, against this reading and the criticism of Fodor and Lepore (1992).

empiricism: the meaning of a sentence (or a belief) consists in the method of its confirmation (p. 37). Consequently, if a sentence cannot be confirmed in isolation, it does not have a meaning in isolation either. If one takes a verification theory of meaning for granted and if one maintains that the property of confirmation applies only to a whole system of sentences (or beliefs), then one is committed to the conclusion that only a whole system of sentences (or beliefs) has meaning. Quine thus says in "Two Dogmas": "The unit of empirical significance is the whole of science." (p. 42)

In later publications, Quine qualifies the holism of "Two Dogmas": it is not the whole of science at once that is confronted with experience, but only a cluster of sentences. Consequently, a cluster of sentences is sufficient for meaning.[8] Saying that any sentence can be held true come what may is right from a legalistic point of view.[9] But, in practice, change usually occurs only within a salient cluster of sentences. Nevertheless, Quine maintains that such a cluster finally encompasses the whole of science. It may be plausible to change even logical laws in order to accommodate our system of knowledge to experience. For instance, Quine claims that it can turn out to be reasonable to abrogate the law of the excluded middle consequent upon the results of experiments in quantum physics.[10] In the last resort, it is hence only a whole system of sentences (or beliefs) to which confirmation and meaning apply. Quine's qualifications in later papers do not amount to a change of position, but to making his position more precise.

Even in "Two Dogmas", Quine does not deny that there is a difference with respect to meaning and confirmation between sentences such as "There are brick houses on Elm Street" and sentences such as the law of the excluded middle. There are cases in which it is reasonable to abrogate only sentences such as "There are brick houses on Elm Street" consequent upon new experience. Although the web that Quine suggests as a model for a system of knowledge is seamless, it has an internal structure: sentences such as "There are brick houses on Elm Street" are located at its periphery, and

---

8 Quine (1975), pp. 313–315; Quine (1991), pp. 268–269. See already Quine (1960), pp. 12–13, and also the foreword to the 1980 edition of *From a Logical Point of View*, Quine (1980), p. VIII.

9 Quine (1975), pp. 313–315; Quine (1986a), p. 427; Quine (1991), pp. 268–269. See also the discussion between Vuillemin (1986), in particular pp. 608–609, and Quine (1986b).

10 "Two Dogmas of Empiricism" in Quine (1980), p. 43; Quine (1991), pp. 268–269. See also "What Price Bivalence?" in Quine (1981), essay 3.

sentences such as the law of the excluded middle are located in its centre. This difference in location is possible only because of a difference with respect to meaning and confirmation.

Consequently, even if it is strictly speaking only a system of beliefs as a whole which has properties such as meaning and confirmation, these properties of the whole indicate a differentiation within the whole: they indicate the way in which its constituent parts, i.e., the single beliefs, are related with one another as regards meaning and confirmation. That is to say: The logical interconnections among sentences or beliefs are not only syntactic relations, but also semantic ones. They are the way in which beliefs are related with one another with respect to meaning. Otherwise any change within a system of beliefs would amount to replacing this system with a new one. It would not be possible to speak of the truth of single sentences or beliefs and thus to speak of continuing to hold certain single sentences true; for saying of some sentences or beliefs that they are true requires a distinction with respect to meaning between them and other sentences or beliefs in the system.

Quine's holism hence fits into the second type of the proposed general conception of holism [1.4].[11] Properties such as meaning and confirmation belong to the family of properties that make something a belief and a constituent of a system of beliefs, given other beliefs of the same person. According to Quine, these properties are only instantiated by the constituents of a system of beliefs taken together – and in the last resort only by all the constituents taken together, i.e., the system as a whole; but insofar as these properties are properties of the whole, they indicate an internal semantic structure: they indicate the way in which the constituents are related with each other with respect to meaning and confirmation.

2.1.3 Is Quine's Argument for Holism Convincing?

The main argument for confirmation holism can be construed as follows: If a conflict between a system of beliefs and experience occurs, the logical relations among the constituents of the system always open up a number of options for accommodating the system to new experience. For any one belief of which it is claimed that it is refuted by experience, there is the logical possibility to retain this belief and to change other beliefs in such a way that

---

11 But see Wagner (1986), who argues that Quine's confirmation holism should be conceived along the lines of what amounts to the first type of the conception of holism which I proposed in [1.4].

the conflict with experience is removed. Therefore, the logical relations among the constituents of a system of beliefs propagate confirmation (in the sense of the generic property explained above) from one belief to other beliefs. If there is no separation between synthetic and analytic beliefs, the logical relations among the constituents of a system of beliefs propagate confirmation to the system as a whole. Consequently, (a) strictly speaking, it is the whole system of beliefs that has the property of confirmation, but (b) the property of confirmation of the whole shows how its constituents are related with one another as regards confirmation.

However, why should one maintain that there is no separation between analytic beliefs and synthetic, empirical beliefs? The case of quantum physics, which Quine mentions, is a weighty argument: One can question whether it is reasonable to abrogate sentences which are regarded as logical laws – such as the law of the excluded middle – consequent upon experimental evidence in quantum physics. But the point is: claiming that the law of the excluded middle is an analytic truth is not an argument that counters a suggestion to that effect. As soon as there is a concrete suggestion for a change to logical laws which contributes to adjusting a system of beliefs to new experience, one has to evaluate this suggestion in the light of its consequences for the system as a whole in comparison with other suggestions. However, as soon as one agrees to such an evaluation, one concedes that it is in principle possible to change sentences that are regarded as logical laws subsequent to new experience. That is to say: as soon as one agrees to such an evaluation, one concedes that it is in principal possible to draw the line between logical laws and empirical sentences differently consequent upon new experience. Therefore, the mere availability of such a concrete suggestion can be regarded as a sufficient reason for the claim that there is no separation between logical and empirical beliefs.[12]

Hence, the point of confirmation holism is not the trivial claim that it is always logically possible to invent an *ad hoc* hypothesis by means of which one can retain any one sentence or change any one sentence so that the system as a whole agrees with experience. Rather, the claim is that for any one sentence including any one logical law, circumstances may turn up in which it is *reasonable* to abrogate the sentence in question in order to adjust a system of beliefs to new experience. Nonetheless, any change to a system of beliefs is only intelligible against the background of a vast amount of

---

12 Compare also Putnam, "'Two Dogmas' Revisited" in Putnam (1983), pp. 92–97, and furthermore Hookway (1988), pp. 41–42.

beliefs that remain unchanged. For any one logical law, abrogating this one law will not deprive a system of beliefs of meaning and intelligibility, because a number of other beliefs will remain unchanged and continue to count as logical laws. Thus, if one endorses confirmation holism, one is, in the case of new experience, only committed to admitting local changes to a system of beliefs against a background of a vast amount of beliefs that remain unchanged. The point of confirmation holism is that there are no beliefs for which it is rational to claim that they are immune to revision in the light of new experience.

The outlined argument for confirmation holism implies that the empirical evidence does not determine a system of beliefs. But note that this indeterminacy does not exclude (a) that there are rational criteria for a choice among different options and (b) that these criteria may distinguish one option. Hence, this argument for confirmation holism is not in itself an argument for any sort of relativism. I will come back to this point in [10.2.1], using the interpretation of quantum physics as an example.

According to the outlined argument, confirmation holism fits into the second type of the proposed general conception of holism: confirmation belongs to the family of properties that make something a belief and a constituent of a system of beliefs, given other beliefs of the person in question. A belief has this property only taken together with other beliefs and in the last resort only taken together with all the other beliefs of the person in question. But the resulting property of the whole system of beliefs indicates the way in which its constituents are related with one another as regards confirmation.

A similar argument can be developed with respect to the justification of beliefs. One can claim (a) that only other beliefs can justify a belief and (b) that a belief is justified if and only if it coheres with other beliefs. Claim (b) implies that there are no beliefs which are a foundation for other beliefs in the sense that they justify other beliefs, but are themselves either self-evident or justified by something which is not a belief. As a result of these two claims, the logical relations among the constituents of a system of beliefs propagate justification (again in the sense of the generic property explained above) from one belief to other beliefs. Since the logical relations that support coherence encompass all beliefs in a system, these relations propagate justification to the system as a whole. If the justification of a belief consists in showing that this belief is integrated into a coherent system of beliefs, then one can maintain that justification is, like coherence, a property which is, strictly speaking, a property only of a whole system of beliefs. Nonetheless, the property of justification of the whole indicates the way in which the constituents are related with each other as regards justification; this property of the whole indicates in how far the constituents cohere with each other. This argument makes a case for a holism about justification along the lines of the second type of the proposed general conception of holism. Thus, Laurence Bonjour maintains in his book *The Structure of Empirical Knowledge* (1985) that "beliefs are justified by being inferentially related to other beliefs in the overall context of a coherent system" (p. 90). He continues to set out an account of justification which moves from the coherence of the overall system to the justification of that system and from there to the justification of particular beliefs by virtue of their membership in the system (p. 92 and the whole Chapter 5).

However, one can also develop a holism about justification which claims that coherence is not a property of a whole system in the first place, but the way in which single beliefs fit together with other beliefs in a system of beliefs. Thus, justification is a property of single beliefs in the first place. Keith Lehrer sets out such a position in his book *Theory of Knowledge* (1990, Chapter 5). His argument makes a case for holism about justification along the lines of the first type of the proposed conception. Hence, holism about justification can be conceived along the lines of either type of this conception. We thus have to distinguish between two types of holism about justification.

Although there are close connections among all forms of holism about beliefs, confirmation holism and holism about justification do not entail semantic holism. One can endorse holism about confirmation (or holism about justification) and reject semantic holism. Such a position is adopted by Fodor and Lepore in their criticism of semantic holism (1992, Chapter 2) as well as, for instance, by Paul Horwich in his book *Meaning* (1998, pp. 150–153). Semantic holism, in turn, does not simply imply confirmation holism or holism about justification either. Given a formulation of semantic holism which grants that each belief has a meaning, it is logically possible to maintain the following: whereas a belief is ontologically dependent on there being other beliefs with respect to its meaning, its confirmation or its justification is not relative to there being other beliefs with which it constitutes a system of beliefs.[13]

In the following, I accept the sketched argument for confirmation holism. Furthermore, I assume holism about justification, but leave open whether

---

13 Compare Heal (1989), pp. 107–108.

this holism is best construed along the lines of the first or the second type of the proposed general conception of holism. I shall focus on semantic holism. The sort of semantic holism that I shall propose includes an argument for holism about justification. This will become evident in [5.3.3].

According to received opinion, Quine's argument in "Two Dogmas" from confirmation holism to semantic holism is based on a verification theory of meaning. However, such a theory of meaning cannot be taken for granted. We need a better argument for semantic holism than the one which Quine offers in "Two Dogmas". Consequently, even if we accept that confirmation holism – and possibly justification holism – fit into the second type of the proposed general conception of holism, it is questionable whether semantic holism is best conceived as belonging to that type too; for the Quinean route from confirmation holism to semantic holism via a verification theory of meaning is the only argument which is committed to regarding holism about meaning as belonging to that type.

Moreover, we should bear in mind two problems that concern Quine's confirmation holism. These problems will be pertinent to the discussion of semantic holism too. When I translated what Quine says in "Two Dogmas" about statements into the talk of properties of beliefs, I characterized confirmation holism in terms of confirmation being a property of the system of beliefs of a person. Confirmation holism has thus been characterized in terms of a system of beliefs of a person. Quine writes at the very end of "Two Dogmas":

> In repudiating such a boundary [a boundary between the analytic and the synthetic] I espouse a more thorough pragmatism. Each man is given a scientific heritage plus a continuing barrage of sensory stimulation; and the considerations which guide him in warping his scientific heritage to fit his continuing sensory promptings are, where rational, pragmatic. (p. 46)

Hence, in the closing sentence of "Two Dogmas", Quine formulates his thesis in terms of what is described in this book as the system of beliefs of each person and each person accommodating her system of beliefs to her experience.

Fodor and Lepore criticize Quine's argument from confirmation holism to semantic holism by maintaining that "whereas the natural objects of semantic interpretation are linguistic entities like formulas, the natural bearers of confirmation relations are trans-linguistic entities like propositions. ... confirmation holism and verificationism are true *of different things*" (1992, p. 53; see the whole section pp. 44–54). However, if we start with beliefs, we can conceive both confirmation and meaning as applying to the system of beliefs of a person. Both confirmation and meaning belong to the family of properties that make something a belief. Note that the problem for Quine's holism which is under discussion here is different from the objection that Fodor and Lepore develop.[14]

Whereas Quine uses in the quoted closing passage of "Two Dogmas" the vocabulary of sentences to which single persons assent, the main target of his claims is science. For instance, Quine proposes that "The unit of empirical significance is the whole of science" (p. 42). Science is a collective enterprise; it concerns knowledge that is shared among a whole culture. I therefore talked in terms of *our* system of knowledge when I exposed the Duhem–Quine thesis in [2.1.1]. For example, the question how to cope with the results of the experiments of quantum physics is not a question of each man "warping his scientific heritage to fit his continuing sensory promptings"; it is a question of what is the most plausible option for a change to our shared knowledge in order to adapt this knowledge to this new shared experience. There is thus a tension in Quine between on the one hand confirmation concerning knowledge shared among a whole culture and on the other hand confirmation concerning the system of beliefs of each person.

This tension is particularly acute in Quine's argument from confirmation holism to semantic holism. In this argument, semantic holism too is construed along the lines of the second type of the proposed general conception of holism. It is each person's system of beliefs as a whole that has meaning in the first place. No two persons have the same system of beliefs, because no two persons share all their beliefs: no two persons have tokens of all the same types of beliefs. If it is a whole system of beliefs that has meaning in the first place, one can raise the problem how two persons can share any belief at all, given that they do not share all their beliefs. Saying that insofar as the whole system has the property of meaning, this property indicates the way in which its constituents are related with respect to meaning does not counter this objection: these relations hold between any two constituents of such a system; no two systems are identical as far as these relations are concerned, because no two systems are identical as far as their constituents are concerned. Hence, conceiving semantic holism in terms of the second type of the proposed general conception of holism threatens the very possibility of shared knowledge.

---

14 See Okasha (2000) against Fodor and Lepore's criticism of Quine.

There is another problem for Quine's conception of holism that concerns confirmation holism in particular: How can experience confirm sentences or beliefs? In "Two Dogmas", Quine talks vaguely in terms of the tribunal of sense experience (p. 41). How shall we understand metaphors such as this one? We face the following dilemma: If, on the one hand, experience is considered to be conceptual, then experience itself consists in beliefs or sentences. In this case, experience is not outside a system of beliefs. It is nothing with which a system of beliefs can be confronted. Instead, experience is included in the holism about beliefs. If, on the other hand, experience is not considered to be conceptual, it is unclear how it can exert a *rational* constraint on a system of beliefs. In his later essay "On the Very Idea of a Third Dogma", Quine says: "The tribunal, to worry another of my old metaphors, is just the firing of the receptors." (in Quine (1981), p. 40) Experience in this sense causes the acquisition of beliefs. But if experience is a mere sensual happening, it is not intelligible how it can be a *reason* for changing one's beliefs and contemplating different options for changes to one's system of beliefs. John McDowell elaborates on this problem in *Mind and World* (1996, pp. 129–135).

The position that experience itself is conceptual is endorsed by philosophers such as Hanson and Kuhn.[15] They come to the conclusion that observation is laden with theory. They go as far as claiming that a system of beliefs cannot face any rational check from outside. Different systems of beliefs are incommensurable. From *Word and Object* (1960) on, Quine also maintains that the experience which confirms scientific sentences is conceptual. It consists in observation sentences. But Quine vehemently opposes the relativistic conclusions of Hanson and Kuhn.[16] Instead, he claims that observation sentences are excluded from semantic holism. They have a meaning each independently of one another.[17] Thus, according to Quine, in the case of observation sentences or beliefs, meaning is not a

---

15 Hanson (1958), Chapters 1 and 6; Kuhn (1970), Chapters 10 and 12. See also Rorty (1980), parts 1 and 2. On the link between semantic holism and this position, compare Fodor and Lepore (1992), pp. 11–13.

16 See, in particular, "Epistemology Naturalized" in Quine (1969), pp. 87–88; Quine (1993), p. 107.

17 Quine (1960), § 10; "Epistemology Naturalized" in Quine (1969), p. 89; Quine (1975), pp. 313–318; Quine (1986a), pp. 427–428; Quine (1993). For an interpretation of Quine's distinction between observation sentences and theoretical sentences in terms of attributing instrumentalism to Quine, see Heal (1989), Chapter 4.

property whose instantiation requires relations to other sentences or beliefs. However, it is unclear how there can be a strict separation between sentences or concepts of observation and sentences or concepts of theory.

To sum this sub-chapter up, Quine's confirmation holism has been endorsed including the thesis that there is no separation between analytic and synthetic, empirical sentences or beliefs as far as confirmation is concerned. But we need a better argument for semantic holism than Quine's move that is based on a verification theory of meaning. Furthermore, when making a case for semantic holism, we should bear the following in mind: we have to pay tribute to the fact that different persons share some of their beliefs, and we have to explain the relation between experience and beliefs.

2.2 The Argument from Interpretation

A second prominent type of arguments for semantic holism focuses on the conditions for the interpretation of the intentional states of a person. The idea is that (a) these conditions are such that one cannot but attribute a whole system of beliefs to a person and that (b) the conditions of interpretation determine the meaning of the beliefs of the interpreted person (to the extent that meaning is determined at all).[18]

The paradigmatic context for setting out the conditions of interpretation is considered to be the situation in which a person seeks to interpret the utterances of another person whereby these persons do not share any language. Subsequent to the work of Donald Davidson, this situation is known as *radical interpretation*.[19] Radical interpretation is the project to find out the beliefs of a person solely on the basis of a shared physical environment and the behaviour of the person in question in this environment including her verbal behaviour, non-intentionally described. (Davidson moreover includes the sentences that the interpreted person holds true among the data that the radical interpreter has).[20] It is argued that it does not make sense to attribute only one belief to the interpreted person. Given the shared physical environment and the behaviour of the interpreted person, one has to attribute either a whole system of beliefs to her or no beliefs at all.

---

18 As to (b), compare what Child (1994b), Chapter 1, describes as interpretationism. As to the argument considered in this sub-chapter, compare Scholz (1999), part II, Chapter 8, in particular pp. 205–227.

19 See, in particular, "Radical Interpretation" (1973) in Davidson (1984), essay 9. See also David Lewis, "Radical Interpretation" (1974), in Lewis (1983a), essay 8.

20 "Radical Interpretation" (1973) in Davidson (1984), essay 9, pp. 134–135.

The sort of semantic holism that this argument seeks to establish can be of either type of the proposed conception of holism: one can say that meaning primarily is a property of the whole system of beliefs that is attributed to the interpreted person. But one can also maintain that the particular beliefs which are attributed to the interpreted person primarily have meaning – insofar as they constitute a system of beliefs of the interpreted person.

Thus far, however, the argument merely supports a methodological holism: one has access to the beliefs of another person only by attributing a whole system of beliefs to that person. Fodor and Lepore present a detailed critical analysis of the arguments for this claim (1192, Chapters 3 to 5). I shall not go into these arguments. Instead, I propose to focus on (b), because (b) is the nerve of the argument from interpretation when it comes to establish semantic holism in distinction to a methodological holism: if and only if the conditions for the attribution of beliefs to a person are at the same time the conditions which determine the meaning of the beliefs of that person, semantic holism ensues.

The reason for this link may be that we are able to determine the meaning of the beliefs of the interpreted person only via the conditions for the attribution of beliefs to this person. However, in this case the argument from interpretation faces an objection that is similar to the objection against Quine's argument from confirmation holism to semantic holism: verificationism is assumed as going without saying.[21] To avoid this objection, one can maintain that the argument for this link is a transcendental one.[22] One may propose a full-fledged transcendental argument in the original Kantian sense, that is, an argument in the vein of Kant's highest principle of all synthetical judgements (Critique of Pure Reason, A 158 / B 197): the necessary conditions for the attribution of beliefs to a speaker by an interpreter are at the same time the conditions for the speaker to have beliefs. This argument leads to the following thesis: being interpreted by another person is a necessary condition for a person to have beliefs.

This thesis cannot go without saying. We need an argument for it. To avoid the charge of verificationism, this has to be an argument that proceeds from the conditions for having beliefs to showing that these conditions include mutual interpretation. The most prominent candidate for an

---

21 Lance and O'Leary-Hawthrone (1997), pp. 275–280, 328, for instance, object to Davidson that he is committed to verificationism about interpretation.

22 But see Stroud (1968) who maintains that transcendental arguments imply a verificationist premise.

argument that ties having beliefs to mutual interpretation is the claim that meaning is public. The most prominent argument for this claim starts from considering the case which Wittgenstein makes according to the interpretation of Kripke (1982) against private rule-following and private language in his *Philosophical Investigations.* I will go into this case in the next chapter when the focus is on a link between social holism and holism about beliefs. What is important here is only that the case against private rule-following and private language is in itself sufficient to argue for social holism and holism about beliefs. That is to say: *If we follow the argument for the link between having beliefs and mutual interpretation that bases itself on Wittgenstein, this argument alone will be sufficient to establish semantic holism. We do not need in this case an argument from the conditions for attributing beliefs to a person when it comes to arguing for semantic holism.*

Donald Davidson can be regarded as the philosopher who offers the most sophisticated version of the argument for semantic holism from interpretation. However, he does not pursue the strategy that starts from Wittgenstein's case against private rule-following and private language.[23] The holism about beliefs which Davidson advocates falls within the first type of the proposed general conception of holism: each belief has a meaning; but for its having meaning, other beliefs are required with which the belief in question constitutes a whole system of beliefs. Davidson's argument for the claim that mutual interpretation is necessary for having beliefs is this one: having beliefs requires having the notion of objective truth, and having this notion, in turn, requires mutual interpretation. In "Thought and Talk" (1975), Davidson sums up his argument in these words:

> Someone cannot have a belief unless he understands the possibility of being mistaken, and this requires grasping the contrast between truth and error—true belief and false belief. But this contrast, I have argued, can emerge only in the context of interpretation, which alone forces us to the idea of an objective, public truth.[24]

---

23 Compare, for instance, Davidson (1990), pp. 310–311, where he voices misgivings about the argument that takes its lead from the interpretation of Kripke (1982) of Wittgenstein's *Philosophical Investigations*. But see also Davidson (1991b), p. 157 note 1, where he expresses basic agreement with Kripke's broad reading of "private language".

24 "Thought and Talk" in Davidson (1984), p. 170. See also Davidson (1983), p. 437; Davidson (1991a); Davidson (1995).

How do we get from here to semantic holism? When it comes to arguing from the claim that mutual interpretation is necessary for having beliefs to the claim that the conditions for interpretation imply that a whole system of beliefs has to be attributed to the interpreted person, Davidson takes for granted that a belief gets its meaning by its position in a whole system of beliefs.[25] This is a further premise in Davidson's epistemology in addition to the premise that concerns grasping the notion of objective truth. An examination of what Davidson builds on grasping the notion of objective truth is beyond the scope of this book. Instead, I propose to focus on the thesis that a belief has meaning owing to its position in a whole system of beliefs. In the following, I shall consider the direct argument for this thesis. Nonetheless, the position that will be defended here has much in common with Davidson's position.

2.3 The Argument from Inferential Role

2.3.1 Tasks for an Inferential Role Semantics

An argument that seeks to establish semantic holism directly is one that is formulated in terms of the conditions for having beliefs instead of being formulated in terms of the conditions for gaining access to the beliefs of a person. One can also speak of a metaphysical argument in distinction to an epistemic one. There is only one type of direct argument for semantic holism, characterized as the thesis that, as far as meaning is concerned, a belief is ontologically dependent on there being other beliefs with which it constitutes the system of beliefs of a person. The argument is that the meaning of a belief is its relations to other beliefs within a system of beliefs. The semantics which elaborates on this argument is known as conceptual, or, more precisely, *inferential role semantics*. The idea is that the meaning of a belief of the type p is the inferential context in which p stands. What inferential role semantics says about meaning is parallel to what holism about justification [2.1.3] says about justification: according to this latter holism, the justification of a belief consists in its relations of coherence with other beliefs within a whole system of beliefs. According to inferential role semantics, the meaning of a belief consists in its inferential relations to other beliefs within a whole system of beliefs.

---

25 See, for instance, "Thought and Talk" (1975) in Davidson (1984), p. 168; Davidson (1991a), pp. 193–194; Davidson (1995), pp. 210–218. As to Davidson's holism, see Malpas (1992), Chapter 3.

However, whereas holism about justification can be construed along the lines of either type of the proposed general conception of holism, inferential role semantics is tied to the first type: a whole system of beliefs can have the property of justification. But a whole system of beliefs does not have an inferential role. If meaning is inferential role, then it is single beliefs to which meaning has to be attributed.

Inferential role semantics is not automatically a holism about beliefs. It is a holism about beliefs if and only if one subscribes to the following claim, which can be based on Quine's criticism of a separation between analytic and synthetic statements:[26] the inferential context that is the meaning of a belief *p* is not closed by means of those beliefs which *p* entails analytically (or by another means).[27] If there were such a closure, then it would not be the case that, as far as meaning is concerned, a belief is ontologically dependent on there being other beliefs with which it constitutes *a whole system of beliefs of a person*. The system of beliefs of a person would not be a holistic system with respect to meaning, because there would be parts of such a system that are self-sufficient as far as meaning is concerned. Hence, inferential role semantics is a holism about beliefs if and only if the inferential context of a belief *p* which is its meaning is conceived as being open-ended, comprising in an indirect way in the last resort all the other beliefs in the system. Thus, by its very nature, an inferential role semantics that is a semantic holism has to leave somewhat open what exactly is the inferential context that constitutes the meaning of a belief.

However, this openness does not imply any objectionable indeterminacy of meaning. For any two concepts *F* and *G* – even concepts such as "rabbit" and "temporal stage of a rabbit", to take up Quine's famous example from *Word and Object* (1960, Chapter 2, in particular pp. 51–53) – the inferential context of a belief of the type "This is *F*" differs from the inferential context of a belief of the type "This is *G*". From "This is a temporal stage of a rabbit", but not from "This is a rabbit", one can infer "This has temporal parts", whereby one can explain what it is to have temporal parts without mentioning rabbits. But these considerations are not an argument against Quine's claim of the indeterminacy of translation. Nonetheless, these considerations show the following: *An inferential role semantics can take*

26 "Two Dogmas of Empiricism" in Quine (1980), pp. 20–37. See also Quine (1960), § 14.

27 For an argument against this premise, see Devitt (1995), Chapter 1, as well as Jacob (1997), pp. 195–204.

into account the distinction between meaning (intension) and reference (extension) without falling back into any mentalism about meaning. Beliefs that have the same reference, but not the same meaning differ in their inferential contexts.

Given the mentioned necessary and sufficient condition under which inferential role semantics is a semantic holism, it is clear that the inferential context which determines the meaning of a belief of the type *p* does not only consist in those beliefs which are entailed by *p* in the sense of being deductively entailed. The inferential context of a belief of the type *p* consists in three sorts of other beliefs:

a) There are beliefs that are *entailed* by *p* in the sense that they can be deduced from *p*. For instance, the belief that it is raining in London entails the belief that London exists.
b) There are beliefs that are *supported* by *p* in the sense that *p* supports an induction to them. The belief that it is raining in London supports the belief that the temperature will go down in London.
c) There are beliefs that are *excluded* by *p* in the sense that endorsing *p* precludes one from endorsing these beliefs. The belief that it is raining in London excludes the belief that the sun is shining in London.

If the inferential role and thus the meaning of a belief is spelled out along the lines of these three sorts of beliefs, a direct or an indirect link holds between any two beliefs in a system of beliefs: such a link is at least a direct or an indirect relation of providing or excluding support. Furthermore, the notion of beliefs that are entailed by *p* does not have to be construed in such a way that these are analytic beliefs in Quine's sense; for we do not have to maintain that relations of entailment are immune to revision.

To elaborate on such inferential role semantics, we need the notion of *material inferences*.[28] This notion is necessary in order to explain the notion of support, which is crucial for inferential role semantics. Material inferences are commonly conceived as those inferences that are based on the meaning of the premises and conclusions. However, inferential role semantics cannot start with such a conception of material inferences; for this semantics sets out to explain the meaning of a belief *p* in terms of the material inferences which *p* supports. Hence, *one of the tasks for inferential role semantics is to develop a conception of material inferences that does not presuppose meaning*.

---

28 See Sellars (1953) and "Some Reflections on Language Games", §§ 28–29, p. 331 in Sellars (1963) as well as Brandom (1994), pp. 97–104.

Inferential role semantics implies the thesis of a primacy of belief[29] in the sense that it is beliefs (or sentences for that matter) which have meaning in the first place. Accordingly, conceptual content is primarily propositional content, since the meaning or content of beliefs is propositional content. Only beliefs or sentences have an inferential role, because only they can figure as premises and conclusions. Consequently, if meaning is inferential role, then it is whole beliefs or whole sentences that have meaning in the first place. Nonetheless, one can say the following: meaning as a property of a whole sentence includes an internal semantic structure in the sense that there are sub-sentential components which contribute to the meaning of the whole. One can thus apply the second type of the proposed conception of holism to a sentence as a whole. Making such a move enables an inferential role semantics to take compositionality into account in the sense of the thesis that the meaning of a sentence depends on its components and their arrangement.[30] *A further task for inferential role semantics therefore is to explain the contribution of sub-sentential components to the meaning of a sentence in terms of their contribution to the inferential role of that sentence.*

2.3.2 Inferential Role and Shared Beliefs

Although inferential role semantics grants that single beliefs primarily have meaning, this semantics is confronted with a problem that came up already in the discussion of Quine's holism in [2.1.3]: How can two or more persons have some beliefs that have the same meaning, given that no two persons share all their beliefs? This problem can be called the *inter-personal problem of communication*, since communication presupposes some shared beliefs. If the meaning of a belief is its inferential role in the system of beliefs of a person and if there is a link in terms of an inferential context between any two beliefs in the system of beliefs of a person, then there are no beliefs of two persons which have exactly the same meaning; for there are no beliefs which stand in exactly the same inferential context in the systems of beliefs

29 I adopt the term "primacy of belief" from Fodor and Lepore (1992), p. 114.

30 Fodor and Lepore (1992), pp. 174–186, argue that an inferential role semantics which implies semantic holism cannot pay tribute to compositionality. See also the objection which Dummett (1991), pp. 224–225, develops. For an argument that semantic holism is compatible with compositionality see Pagin (1997). Brandom (1994), pp. 354–355, argues that compositionality is neutral between bottom-up and top-down explanatory strategies; as regards Brandom's account, see below [2.4.2].

of two persons. Fodor and Lepore regard this point as the major objection against semantic holism.[31]

Furthermore, there is an intra-personal version of this problem: If the meaning of a belief is its inferential role in a system of beliefs and if there is a direct or an indirect inferential connection between any two beliefs in a system of beliefs, then any change within a system of beliefs – say, simply adding the belief that it is raining – changes the inferential role of all the beliefs in the system, however slightly. Consequently, any change to a belief system changes the meanings of all the beliefs in the system. This can be called the *intra-personal problem of belief change.*

Note the consequences of this problem. When discussing confirmation holism the thesis that for any one belief circumstances may turn up in which it is reasonable to abrogate this belief was considered [2.1.3]. I stressed that any revision of single beliefs is intelligible only against the background of a vast amount of other beliefs that remain unchanged. The intra-personal problem of belief change, however, undermines this requirement. Quine's famous claim in "Two Dogmas" that any statement can be held true come what may, if we make drastic enough adjustments elsewhere in the system (p. 43), would be pointless if such an operation had the consequence that the meaning of the statement in question had been altered.

There are a number of strategies to cope with both these problems. Some of these strategies can be combined with one another. One famous strategy is to suggest some sort of a two factor semantics: Ned Block (1986) and (1995), for instance, argues in favour of an inferential role semantics that determines a narrow content. That content depends only on the physical and psychological make-up of the person in question. However, Block proposes as a second factor a relation of reference to something in the world that is not touched by holism. Narrow, inferential content plus reference result in wide content. Because of the relation of reference that is included in wide content, wide content is not subject to the inter-personal problem of communication.

---

31 Fodor and Lepore (1992), pp. 9–10. See also Fodor (1998), Chapter 2, as well as Dummett (1991), p. 237. For critical discussions of the objections to semantic holism which Fodor and Lepore (1992) develop, see the papers in Fodor and Lepore (1993) and Lepore et al. (1993).

That problem thus is avoided by shifting from meaning, inferentially construed, to reference.[32]

Another prominent strategy is to replace identity of meaning with similarity of meaning and thereby alleviate the mentioned problems. As regards the inter-personal problem, one may reply along the following lines: in order that two persons share a belief it is not necessary that they have tokens of a belief that have strictly the same meaning; far-reaching similarity of meaning is sufficient.[33] Thus, my belief that it is raining has not exactly the same meaning as your belief that it is raining, because the inferential role of my belief in my system of beliefs is not exactly the same as the inferential role of your belief in your system of beliefs. But these inferential roles exhibit a far reaching similarity, because my conclusions from my belief that it is raining by and large agree with your conclusions from your belief that it is raining. An analogous consideration applies to the intra-personal problem of belief change. Although any change within a system of beliefs has an effect on the inferential role of all the beliefs in the system, this effect is only very slight. The inferential role of nearly all beliefs after the change is very similar to their inferential role before the change. Fodor and Lepore, however, argue that similarity cannot stand on its own: similarity between two items presupposes qualitative identity between them in some respect; inferential role semantics, by contrast, cannot provide for any qualitative identity (1992, pp. 17–22 and Chapter 7).

A further strategy is to embrace the claim that change of belief is change of meaning, but to propose compensatory moves in a system of beliefs. Thus, Davidson (1995) suggests that in any case of a change within a system of beliefs, there are many possibilities for an adjustment that restores for most of its elements their old positions (pp. 215–216). Consequently, despite the change in the system, for most beliefs there is no change of meaning after all.

Having mentioned these strategies, I shall not enter into an examination of them. Instead, I intend to counter the objection from shared beliefs and belief change by advocating an inferential role semantics that broadens inferential

---

32 See also the two factor position of Bilgrami (1992), pp. 10–13, 141–150, who grants that persons hardly ever share concepts, but then introduces local contents that explain action. These contents are shared.

33 Compare Block (1986), pp. 624, 629, 633. Consider also the strategy of Lormand (1996), section 2, who employs the notion of multiple meanings in the sense that each inference from a given belief *p* constitutes a meaning of *p*.

role in such a way that the social environment of a person is included. Such semantics avoids the inter-personal problem of communication as well as the intra-personal problem of belief change at its roots by combining holism about beliefs with social holism.

My main reason for favouring a social solution to these problems, however, is that a semantics which conceives inferential role as a social role replies to another serious problem, namely the problem of rule-following. An inferential role semantics that is limited to the systems of beliefs of single persons including all the mentioned strategies face this problem. If the rule-following problem leads us to social holism, then it is an important reason for adopting an inferential role semantics that conceives the inferential role of beliefs as a social role. My aim in linking up inferential role semantics with the rule-following considerations is to develop an argument that covers both holism about beliefs and social holism. The next sub-chapter treats a social, inferential semantics. The next chapter then moves on to social holism.

2.4 Inferential Role as Social Role

2.4.1 Meaning through Use in a Community

Ludwig Wittgenstein's *Philosophical Investigations*[34] are one of the main sources for semantic holism. One of Wittgenstein's central concerns is to argue against the view that the meaning of a belief is some sort of an entity in the mind of a person (or an abstract object beyond spatio–temporal reality). The alternative to this view which he proposes is that the meaning of a belief is determined by the way in which the sentence that expresses the belief in question is used in a social, linguistic community. If we apply this position to inferential role semantics, we come to the conclusion that the inferential role of a sentence or belief that constitutes its meaning is a social role. It is crucial for a social use theory of meaning that the use of sentences is not conceived as expressing beliefs or other items that are considered to be meaningful independently of whether or not they are expressed in a language. The use of sentences in a language is a necessary and sufficient condition for there to be something which has meaning.

Nonetheless, a social use theory of meaning does not imply the thesis that beliefs can be reduced to or eliminated in favour of the expression of sentences in a public language. Such a thesis would entail the task of

34 Wittgenstein (1952) (English translation: Wittgenstein (1958)).

developing an account of meaningful sentences that does not have recourse to belief states of persons. The only thesis to which a social use theory of meaning is committed is that, as far as meaning is concerned, having beliefs and mastering a public language are on an equal footing: neither has conceptual priority over the other, and neither can be explained independently of the other.[35]

Davidson, who goes as far as considering mutual interpretation to be a necessary condition for having beliefs, argues for not more than this position in "Thought and Talk" (1975).[36] Brandom, who develops an inferential role semantics in terms of social practices in his *Making It Explicit* (1994),[37] also adopts this position. Brandom classifies this position as a relational version of a linguistic approach to intentionality; it is relational in the sense that beliefs can only be understood in relation to their expression in a language, and meaningful expressions in a language can only be understood in relation to beliefs or belief states of a person (pp. 150–153, 230). I will therefore switch between talking in terms of beliefs and talking in terms of sentences that express beliefs, since, according to the position under consideration, it is a necessary condition for a person to have beliefs that she masters a language. When I speak of beliefs or sentences, I always mean to restrict myself to assertible beliefs or sentences, that is, beliefs or sentences that have a content which can be asserted such as "This is a table". Furthermore, I use the term "claim" in such a way that claims are assertible beliefs as expressed in a language. For lack of space, I do not consider degrees of belief.

What is the use of a sentence? In his interpretation of Frege's philosophy of language, Michael Dummett offers an account in terms of two factors:

> Learning to use a statement of a given form involves, then, learning two things: the conditions under which one is justified in making the statement; and what constitutes the acceptance of it, i.e. the consequences of accepting it. Here 'consequences' must be taken to include both the inferential powers of

---

35 For a prominent argument in favour of the contrary thesis that belief states have conceptual priority over linguistic expressions see Searle (1983).

36 in Davidson (1984), essay 11. See in particular p. 156. Compare also Wittgenstein, *Philosophical Investigations*, in particular §§ 317–318, 329–349, 540–541.

37 Brandom (1994). For a critical evaluation, see the book symposium Brandom et al. (1997) as well as Gibbard (1996). Brandom (2000) is an easily accessible exposition of the topics that are treated in his (1994).

> the statement and anything that counts as acting on the truth of the statement.[38]

The first of these factors, the conditions under which one is justified in making the statement (that is, uttering a sentence of a certain type), can be described as *appropriate circumstances of the application of a sentence of a given type*. One can consider these circumstances as the assertibility conditions. In the paradigmatic case of sentences that express perceptual beliefs such as "Here is a table" or "This is square" these conditions are non-inferential. They consist in these sorts of things – tables, square objects – being present in one's environment under normal conditions.

However, the appropriate circumstances of the application of a sentence of a given type are not sufficient to understand the sentence. Consequently, these circumstances are not sufficient for having a perceptual belief such as "This is square" either. Consider the argument against the self-sufficiency of ostensive definitions: imagine that the concept *F* is introduced by ostensive definition; someone points at something and says "This is *F*". If this gesture is not supplied with an inferential context, one does not know what sentences of the type "This is *F*" are to mean. It is not clear what they pick out – a sortal kind? a colour? a shape?, etc.[39] Hence, an ostensive definition alone does not enable a person to acquire the belief that a certain thing is *F*.

Brandom maintains that the appropriate circumstances of the application of a sentence are not sufficient to distinguish belief states from mere response mechanisms; for response mechanisms can be such that they reliably indicate circumstances in which it is appropriate to utter a sentence of a certain type. Brandom gives the following example by inventing the artificial concept "gleeb":

> You do not convey to me the content of the concept *gleeb* by supplying me with an infallible gleebness tester that lights up when and only when exposed to gleeb things. I would in that case know what things were gleeb, without knowing what I was saying about them when I called them that, what I had found out about them or committed myself to.[40]

The point is that the things that the gleebness tester selects have much more properties in common than the property of being gleeb. Consequently, being aware of that selection does not enable one to understand what it means to

---

38 Dummett (1981), p. 453. See also Kripke (1982), in particular pp. 86–87.

39 Compare Wittgenstein, *Philosophical Investigations*, in particular §§ 1, 28–33.

40 Brandom (1994), p. 122. See also pp. 87–91. See furthermore Bieri (1982), p. 5.

say that something is gleeb. So what distinguishes mastery of concepts from a disposition to respond to stimuli in a reliable and discriminatory way?

In the quotation above, Dummett proposes a second factor that determines the use of a statement: what constitutes the acceptance of the statement in question, that is, the consequences of accepting it. The consequences of accepting a sentence of a certain type can be construed in terms of the inferences that such a sentence entails and supports in a public language. I thereby follow a suggestion on which Brandom elaborates (1994, pp. 116–123) and which Davidson favours, too (1995, pp. 209–218). Hence, the use of a sentence of a certain type is the appropriate circumstances of its application plus the inferences that its application entails and supports.

A person consequently has to master these two factors if she is to have a belief that is expressed by a sentence of the type in question. Accordingly, a person has a concept *F* if and only if (a) she masters the circumstances in which it is appropriate to form beliefs or utter sentences of the type "This is *F*" and (b) she masters the inferences which a belief or sentence of this type entails and supports. Mastering the circumstances and the inferences does not mean that a person has to have beliefs about these circumstances and inferences, that is, make these circumstances and inferences the object of her beliefs. If we were to introduce such a requirement, we would face an infinite regress of beliefs. In order to have a belief of the type "This is *F*", it would be necessary to have beliefs about circumstances and inferences, and these beliefs would in turn require further higher order beliefs, etc. To master circumstances and inferences is a practical skill in the first place: it shows up in the way in which a person acquires beliefs of a certain type and connects these beliefs with other beliefs.[41] In a second step, this practical skill can be made explicit by forming beliefs about circumstances and inferences.

Based on these considerations, we can rebut the objection that inferential role semantics cannot pay tribute to perceptual beliefs such as "This is square". Mastering circumstances of application by means of a reliable discriminatory reaction capacity is not sufficient for having perceptual beliefs. One knows what it means that something is square only if one knows the inferences that a belief such as "This is square" entails and supports. The non-inferential acquisition makes them *perceptual* beliefs, i.e., beliefs that are acquired through perception and thus through certain causal connections

---

41 Compare what Pettit (1998a) calls the practical mode of belief. Pettit cites Brandom (1994) with approval in this context (p. 28 note 7).

to the world; the inferential relations make them perceptual *beliefs*, because it is these relations that provide meaning.[42]

In order to take not only perceptual beliefs into account, but also the role which beliefs have in social life, Wilfrid Sellars and Robert Brandom advance an inferential role semantics in a broad sense: over and above inferential moves from one sentence to other sentences in a language, the inferential semantics which they propose includes both non-inferential language entries such as beliefs acquired through perception as well as language departures or exits such as certain actions as appropriate practical inferences of certain beliefs.[43] For instance, if one believes that there is a tiger approaching, running away is a suitable practical inference of this belief. I adopt an inferential role semantics in this broad sense in the following.

To sum up this approach in the way I would like to accent it, something has a meaning and thus is a belief or a sentence in distinction to reliable and differentiating reactions to stimuli by virtue of inferential relations to other beliefs or sentences. These inferential relations are socially determined. But meaning in this sense that is confined to relations among beliefs or sentences requires being embedded in appropriate circumstances of application and appropriate consequences in action, that is, practical inferences. The reason for this requirement will become apparent in the next chapter when the social solution to the problem of rule-following will be considered. The use of a statement thus includes more than its meaning.

2.4.2 Inferential Role in a Public Language

Let us focus on the inferences that a belief or a sentence of a certain type entails and supports. Inferences in the sense of what Dummett calls the consequences of accepting a statement (1981, p. 453) are public. They are the inferences which the members of a linguistic community take a sentence of a certain type to entail and support. When introducing an inferential role semantics in [2.3.1], I mentioned a tripartite account of the inferential

---

42 Compare Sellars, "Empiricism and the Philosophy of Mind" in Sellars (1963), in particular §§ 16 (pp. 144–145), 19 (pp. 147–148), 35–36 (pp. 167–169) and furthermore Sellars (1979), lecture 5, §§ 26–35. See also Brandom (1997), pp. 139–144, 157–158.

43 "Some Reflections on Language Games", §§ 18–21, pp. 327–330 in Sellars (1963); Sellars (1979), lecture 4, § 31; Brandom (1994), pp. 119–121, 131–132, 233–243. Compare also Harman (1982), pp. 247–250.

context of a belief of a given type in terms of beliefs that are entailed by a given belief, that are supported by it and that are excluded by it. Note that this account is set out in terms of the inferences which beliefs of a certain *type* license, and not in terms of idiosyncratic inferences that a person is inclined to draw from certain belief tokens. Therefore, this account can simply be adopted when the focus is shifted from inferential role in the system of beliefs of a person to inferential role in a linguistic community. In [2.3.1], it was also mentioned that inferential role semantics implies a primacy of beliefs or sentences in the sense that it is beliefs or whole sentences that have meaning in the first place; accordingly, conceptual content primarily is propositional content. It was concluded that if inferential role semantics – and thus a semantic holism that bases itself on such a semantics – is to be compatible with compositionality, it has to give an account of the contribution that sub-sentential elements make to the meaning of a sentence in terms of their contribution to its inferential role.

Brandom sets out an inferential role semantics that accomplishes this task. He explains the contribution which singular terms and predicates make to the meaning of a sentence by the manner in which they function in substitution, namely the consequences that their substitution has for the inferential role of the sentences that are generated by such substitution. Consider one of Brandom's examples of a substitution of singular terms: if the inference from "Benjamin Franklin walked" to "The inventor of bifocals walked" is a good one (in the sense of a material inference that is accepted in the relevant community), then so is the inference from "The inventor of bifocals walked" to "Benjamin Franklin walked". If the substitution of singular terms results in good inferences, then these inferences are reversible. Singular terms can therefore be arranged in classes of terms that yield reversible inferences. All those singular terms whose substitution results in reversible inferences constitute one equivalence class. Brandom develops an argument why there have to be singular terms and why they enter into substitution relations that are reversible.[44]

By contrast, if the substitution of predicates results in good inferences in the material sense, then these inferences do not have to be reversible. For example, if the inference from "Benjamin Franklin walked" to "Benjamin Franklin moved" is a good one, it does not follow that the inference from "Benjamin Franklin moved" to "Benjamin Franklin walked" also is a good

---

44 Brandom (1994), pp. 360–404. The example is on p. 371. For a criticism of Brandom's inferentialist account of singular terms see Graham (1999).

one. Benjamin Franklin may have moved in another way than by walking. Predicates can be arranged according to their strength of inference. Some predicates are inferentially weaker than others: everything that follows from the applicability of "moves" also follows from the applicability of "walks", but not *vice versa*, as Brandom explains (1994, pp. 371–372).

Furthermore, Brandom develops an account of indexicals and demonstratives in terms of anaphora, that is taking the listener or the reader back to an expression already used. Pronouns have an anaphoric function. For instance, in "Kant is a famous philosopher, and he spent his entire life in Königsberg", the "he" takes us back to "Kant". Brandom maintains that indexical tokenings contribute to the inferential role of sentences by anaphoric relations to other tokenings; through these relations they are intermeshed with substitutions. According to Brandom, it is only by means of anaphoric relations that an unrepeatable indexical tokening refers to an object and is distinguished from a mere response to stimuli. Brandom therefore claims that the deictic function of indexicals is derivative from the anaphoric function of pronouns. He maintains that there can be languages which contain expressions that have a pronominal use, but which lack expressions that have a deictic use, whereas the converse is not possible (1994, Chapter 7, in particular pp. 464–466).

Brandom's work shows that an inferential role semantics can be developed in such a way that it includes an account of sub-sentential elements such as predicates and singular terms down to indexicals. Such an account can give a concrete shape to the general claim that a concept or a word has a meaning only insofar as it is used in a sentence within a whole system of sentences. This account is compatible with compositionality in the sense that the meaning of a sentence depends on its sub-sentential components and their arrangement, although this account entirely works in terms of sentences being the unit of meaning.[45]

What is the impact that these considerations have for semantic holism? We can continue to maintain that, as far as meaning is concerned, a belief is ontologically dependent on there being other beliefs together with which it constitutes the system of beliefs of a person. But the inferential role that a belief has in the system of beliefs of a person is derivative from the inferential role which it has in a linguistic community. A person has a belief of the type *p* only if she knows the inferential context in which *p* stands in a

45 See also the account of compositionality which Horwich (1998), pp. 32–37 and Chapter 7, gives in his use theory of meaning.

public language. By acquiring beliefs a person learns a public language. The semantic holism under consideration can therefore also be formulated as the thesis that in order to understand one sentence, a person has to master a whole language; for, as far as meaning is concerned, a sentence is ontologically dependent on there being other sentences together with which it constitutes a whole language.

This view of a belief system in a public language raises the problem how a language can be learned. This view excludes an account in terms of first learning one concept and then the next one, etc. According to this view, on the contrary, mastering one concept involves mastering a whole language. However, nobody completely knows the inferential context of the sentences in which a concept is used in the sense that she can produce all these sentences on request. One masters a concept by knowing a representative sample of the sentences that are a paradigmatic employment of the concept in question plus the appropriate circumstances of application of these sentences. A child learns a concept by being introduced to such a sample. The child thereby learns not only one concept, but a whole range of concepts and thus a language, however rudimentary. By enlarging its experience, a child then learns more applications of the sentences in which it uses the concepts it has and thus gets a better mastery of their inferential context by being able to intermesh them with ever more other sentences.[46]

A layman's concept of an elm and a botanist's concept of an elm are not two different concepts. The meaning of the sentence "Elms lose their leaves in autumn" uttered by a layman and the meaning of the same sentence uttered by an expert are not two different meanings. The layman has the concept of an elm. But as regards such concepts she is like the child as regards the most common concepts: she knows a number of paradigmatic sentences in which the concept of an elm is employed as well as paradigmatic circumstances of the application of sentences in which this concept is employed. However, the expert masters the inferential context of the sentences in which this concept is used in a much more fine-grained way. In doubtful cases, the layman relies on the expert. In this way, mastery of concepts comes in degrees.[47] Such cases of mastery of concepts in degrees and relying on experts in doubtful cases are known as the division of

46 Compare Sellars, "Empiricism and the Philosophy of Mind", §§ 19, 37, pp. 148, 169 in Sellars (1963).

47 See also Young (1992), pp. 317–318; Brandom (1994), pp. 120–121.

linguistic labour.[48] The proposed semantic holism includes this division of linguistic labour.

To conclude this section, consider the following summary argument for semantic holism which focuses on what it is for a person to master concepts:

1) Concepts are tied to their expression as predicates in a language. A person has a concept if and only if she knows how to employ a predicate in a language.
2) Predicates are employed in sentences. A person thus knows how to employ a predicate if and only if she is able to use the predicate in question by forming sentences in a language.
3) In order to be able to use a predicate, a person has to know the range of sentences in which the predicate in question can be employed, and she has to be able to embed a sentence in which she uses this predicate into an inferential context.
4) The inferential context that is the meaning of a sentence of the type *p* includes both linguistic stipulations and assumptions about the way the world is. These two factors cannot be separated from each other. The inferential context hence is not closed by means of those sentences which *p* entails analytically (or by another means).

This reconstruction is modelled on an argument of Peter Bieri (1982, pp. 4–5) in favour of what he calls psychological nominalism, thereby taking up Sellars' term.[49] However, Bieri does not mention the issue of there being no separation between analytic and synthetic sentences. Nonetheless, the fourth step of the argument is crucial for semantic holism: if and only if there is no closure of the inferential context, it follows that a person has to master a whole language in order to master one predicate or one concept.

2.4.3 *Inferential Semantics based on Normative Pragmatics*

The claims in the previous section are not complete as they stand. We have to spell out further what the inferential role of sentences of a certain type in a linguistic community is. If the meaning of a belief is its inferential role in a public language, it is this role that individuates the beliefs of a person, since beliefs are individuated by their meaning. However, it has to be shown how inferential role can accomplish this task, although a person knows only a representative sample of inferences for any given sentence, and this sample

48 See, for instance, Putnam, "The Meaning of 'Meaning'" (1975) in Putnam (1975), pp. 225–229; "Language and Reality" (1974) in Putnam (1975), p. 274.

49 See "Empiricism and the Philosophy of Mind", § 29, p. 160 in Sellars (1963).

may vary to a certain extent from person to person for sentences of the same type.

A use theory of meaning links up semantics, the study of meaning, with pragmatics, the study of the use of linguistic expressions. The inferential role semantics that I presented in the preceding section suggests a link with a *normative pragmatics*. I will therefore consider now a feature of inferential role in a public language that I have neglected hitherto, namely its normative feature. My claim is that "inferential role" has to be conceived as a normative notion in order to accomplish the task of individuating the beliefs of a person. That is to say: beliefs are individuated not by their relations to other beliefs which a person actually has, but by their relations to those other beliefs to which a person is committed and entitled if she has certain beliefs; these relations are determined by social interactions. Thus, if you have the belief that it will rain, this belief is individuated by its relations to those other beliefs to which you are thereby committed and entitled in the light of the members of your social, linguistic community. These norms are binding on you even if you master them only to a certain extent and rely on experts in doubtful cases. Furthermore, there are normative relations not only among beliefs, but also among beliefs and actions. For instance, there are normative relations that lead from the belief that it will rain to the rational commitment that you ought to take an umbrella when going out.[50] However, for lack of space, I shall focus only on normative relations among beliefs.

The claim that belief states are individuated by normative relations to other belief states (and other intentional states in general) is intended to be a theory of what it is to have beliefs in distinction to exhibit mere discriminatory, reliable responses to stimuli. A response is an expression of a belief if and only if (a) it can be evaluated as being correct or incorrect in terms of appropriate circumstances of application and inferences that it supports and (b) the subject that makes the response in question recognizes such an evaluation. In this vein, Sellars maintains in his seminal paper "Empiricism and the Philosophy of Mind" that expressing a belief is a move in the game of giving and asking for reasons (§ 36, pp. 168–169 in Sellars (1963)). Something is an expression of a belief if and only if reasons can be asked for it. Someone expresses a belief if and only if she recognizes that she has to give reasons for what she utters on request.

Developing this Sellarsian idea of moves in a game of giving and asking for reasons, Brandom (1994) traces an inferential role semantics back to a

50 Compare Brandom (1994), pp. 229–271, and Brandom (1998b).

normative pragmatics. He explains what I termed in [2.3.1] beliefs that can be deduced, that are supported by and that are excluded by a given belief of the type *p* in the following tripartite way:

a) *commitment*: Making a claim of the type *p* commits one to a number of other claims. For instance, if one claims that it is raining in London one is committed to the claim that London exists.
b) *entitlement*: Making a claim of the type *p* entitles one to a number of other claims. The claim that it is raining in London entitles one to the claim that the temperature will go down in London. If the latter claim is challenged, making the former claim is giving a reason for the latter claim.
c) *precluded entitlement*: Making a claim of the type *p* precludes an entitlement to a number of other claims. Claiming that it is raining in London precludes one from being entitled to claim that the sun is shining in London.

Brandom (1994, pp. 633–636) explains the inferential role of a sentence in a language in terms of those other sentences to which one is committed by endorsing a sentence of the type in question, to which one is entitled and to which entitlement is precluded according to the speakers of that language. If inferential role is conceived in these terms, there is no principled distinction within a system of beliefs or a language (such as the analytic / synthetic distinction) which separates the inferential context that determines the meaning of a belief or sentence of the type *p* from the inferential context that determines the meaning of any other type of beliefs or sentences.

Brandom develops a model of linguistic practice in terms of what he calls *deontic scorekeeping*: members of a social community act as scorekeepers of each other's commitments and entitlements in making claims (1994, Chapter 3, in particular pp. 159–161 and pp. 180–190). He adopts the idea of scorekeeping in linguistic practice from David Lewis' paper "Scorekeeping in a Language Game" (in Lewis (1983a), Chapter 13, in particular pp. 236–240). For Brandom, commitments and entitlements arise in the lights of a scorekeeper who takes a person to be committed or entitled to certain claims. Brandom considers it to be a major advantage of this normative account in comparison to causal-functional accounts of beliefs that it makes the phenomenon of inconsistent beliefs intelligible: according to him, inconsistent beliefs are to be explained in normative terms; they are incompatible commitments. That is, by committing oneself to a claim *p* entitlement to another claim *q* is precluded; one is, as it happens, however, also committed to *q* (1994, pp. 160, 196, 238, 245, 270, 332, 606).

Brandom regards this normative pragmatics as basic. Logic, for him, is making explicit relations of commitment, entitlement and precluded entitlement between claims that speakers endorse in a linguistic community.[51] For instance, he introduces the logical notion of negation or incompatibility between two claims in terms of commitment to the one claim precluding entitlement to the other claim (1994, in particular p. 160). That is why Brandom needs the third of the points indicated above, namely precluded entitlement, as a basic notion. If we presupposed the availability of negation, we could derive precluded entitlement from commitment and entitlement. However, the idea in grounding an inferential semantics on a normative pragmatics is to explain logical notions on that latter basis. Once we have introduced logical vocabulary, we can say that commitment is transitive: if the claim *p* commits one to *q* and if *q* commits one to *r*, then *p* commits one to *r*. Entitlement also is transitive: if the claim *p* entitles one to *q* and if *q* entitles one to *r*, then *p* entitles one to *r*. Furthermore, if *p* entitles one to *q* and if *q* commits one to *r*, then *p* entitles one to *r*. And if *p* commits one to *q* and if *q* entitles one to *r*, then *p* entitles one to *r*.

The claims to which one is committed by making a claim of the type *p*, to which one is entitled and to which entitlement is precluded are not fixed once and for all. Meaning as conceived in terms of commitments and entitlements is relative to time, because the commitments and entitlements that people accept in a community change. The inferential role of claims is subject to change consequent upon the acquisition of new knowledge about the way the world is, including in particular new perceptual beliefs. For instance, the inferential role of all the claims in which the concept "electron" is employed has changed in the twentieth century as a result of quantum physics. However, a change to the inferential role of a claim of a given type is not a question of the personal experience of individuals. Any suggestion which an individual makes in this respect is subject to an evaluation within a whole community. Furthermore, such a suggestion can be made only from within the practices of a community: by challenging an inferential move – even one that is shared by all the members of the community – as being incorrect one makes oneself a member of the community in question, or one makes an attempt to merge the community in question with one's home community.

---

51 Brandom (1994), pp. 104–116, 135–136. For a criticism of Brandom's claims in that respect see Levi (1996).

According to this conception, if one ascribes a belief of the type *p* to a person, one ascribes to her that she is committed to some beliefs of other types and entitled to some beliefs of other types. In short, one attributes a certain normative status to her.[52] This is not an instance of the is–ought fallacy, i.e., the attempt to deduce ought-sentences from is-sentences.[53] For the sentence that a person has a belief of the type *p* is explained in terms of the person committing herself to *p* and thereby committing herself to certain other beliefs. In short, the claim is the following: what distinguishes belief states (and intentional states in general) from other mental states (such as, for instance, being in pain) as well as merely physical states (such as, for instance, being six feet tall) is a normative feature – being in a belief state is undertaking a commitment (without, however, having beliefs about that commitment, on pain of the infinite regress described in [2.4.1]).

Consequently, if John ascribes a belief of a type *p* to Carol, he adopts a normative stance towards her: he regards her as having committed herself to certain further beliefs of other types. John does not have to undertake the commitment in question himself, because he does not have to share the belief that he ascribes to Carol;[54] the only thing he has to realize is that Carol undertakes a certain commitment. Thus, *by ascribing beliefs to persons, one describes the commitments that they undertake*. The relations of commitment, entitlement and precluded entitlement to other beliefs are the – normative – facts that constitute the meaning of a belief. Meaning is a sort of functional role, but a social functional role that can be described only in normative vocabulary.

Mark Lance and John O'Leary-Hawthrone (1997), by contrast, go further and consider meaning claims as being prescriptive in themselves.[55] Thus, there are no facts of meaning. Lance and O'Leary-Hawthrone have a point in saying that the normative relations which constitute the meaning of a belief are not explicit. Making them explicit is not simply a description of implicit practices, but also a prescription in the sense of a suggestion for a binding

---

52 Compare Gibbard (1994), pp. 99–102.

53 See, by contrast, the objection of Engel (forthcoming), section 1.

54 Gibbard (1996), pp. 710–711, and Engel (forthcoming), section 2, raise the objection that according to Brandom's position, if person *A* ascribes a commitment to person *B*, it seems that person *A* has to endorse the commitment in question herself. Compare also Heal (1989), pp. 160–166, on Kripke (1982).

55 Lance and O'Leary-Hawthrone (1997), Chapter 3, in particular pp. 208–227, 236, and furthermore pp. 2, 9–10, 374.

specification of pre-existing practices. They employ the example of children playing football and a specification of rules for playing football (pp. 222–223). However, such a specification is modelled on a pre-existing practice and tries a far as possible to encapsulate that practice. Conceding this point does not prevent us from going for a descriptive theory of meaning.

Lance and O'Leary-Hawthrone's main argument for their position of meaning claims being prescriptions is that a descriptivist view leads to social relativism (in particular pp. 208–227). However, if one adopts the position sketched above, one is free to describe the meaning of a belief in terms of the normative attitudes which the members of a community take and then go on to challenge these attitudes as being incorrect, thereby suggesting a correction of the received meaning. Thus, the position for which I shall argue is this one: Meaning is identical with actual use in a community. It is identical with social facts that consist in normative attitudes which people adopt towards their fellows – attitudes of taking people to be committed or entitled to certain claims. These social facts are describable only in normative vocabulary. The normative attitudes of taking people to be committed or entitled to certain claims are themselves subject to an assessment as being correct or incorrect. What is correct or incorrect in normative attitudes that constitute the meaning of beliefs about the world does not reduce to or supervene on social facts. The *correct* normative attitudes depend on the way the world is.

Grounding an inferential semantics on a normative pragmatics accomplishes the remaining one of the two tasks for an inferential semantics which have been outlined in [2.3.1], namely to give an account of material inferences that does not presuppose meaning. What is founded by the relations of commitment, entitlement and precluded entitlement as determined by deontic scorekeeping are material inferences which give rise to meaning and which can then be canonized in formal, logical rules for inferences as Brandom explains (1994, pp. 97–105). Furthermore, an inferential semantics that conceives the inferential role of a belief as a normative, social role avoids the two problems for an inferential role semantics which have been described in [2.3.2], i.e., the inter-personal problem of communication and the intra-personal problem of belief change.

The inferential role semantics which has been presented in this sub-chapter is committed to the position that a person can have single beliefs only in the context of a whole system of beliefs; the meaning of a single belief is relative to other beliefs within such a system, because it consists in an inferential context. However, this is not the whole story. The inferential context of a belief is the use of the sentence that expresses that belief in a linguistic community. Consequently, it is not the role in a system of beliefs which is confined to one person that determines the meaning of a belief of a person, but an external factor, namely the linguistic community. Since beliefs are individuated by their meaning, it is thus an external, social factor which individuates the beliefs of a person.

To give an account of how this social factor can determine meaning, it is reasonable to base a social inferential semantics on a normative pragmatics. From the perspective of the outlined normative pragmatics, the question is not so much what other beliefs a person actually has if she has a belief *p*, but what other beliefs she *ought* to have, that is, to what other beliefs she commits herself to by having the belief *p*. Of course, a person must actually have a number of these other beliefs, and she must endorse when questioned by far most of the beliefs to which she has committed herself. For instance, if a person claims that there is a cat on the mat, but does not endorse claims such as that there is a mammal on the mat, that there is something with a tail on the mat, etc., she does not have the belief that there is a cat on the mat. However, the meaning of a belief *p* is not determined by those other beliefs that a person actually has, but by those other beliefs to which she is committed and entitled and to which entitlement is precluded by endorsing *p*. All versions of psychologism are thereby avoided. For *what individuates a belief or a belief state are its normative relations to other beliefs or belief states*.[56] Consequently, what turns a mental state into a belief state and thus an epistemic state are normative relations to other belief states. These relations are determined by social relations such as the described ones of deontic scorekeeping in a linguistic community.

This position denies that beliefs are something in the heads or the minds of persons and which has fixed identity conditions. Instead, all there is to persons having beliefs is that they participate in the described social practices. This position therefore is not a straight solution to the inter-personal problem of communication and the intra-personal problem of belief change; it rather shows how to avoid these problems. Beliefs, thus conceived, can by their very nature be shared by an indefinite number of people; for their meaning is a role in a social practice. Change of meaning consequently is not a question of a person acquiring new beliefs, but a matter of social interaction. If a person acquires a new belief, no change ensues in

---

56 Compare Brandom (1994), pp. XVI, 16.

the normative relations among beliefs as they are determined in the linguistic community to which that person belongs. Nonetheless, as mentioned above, it would be wrong to conceive the norm that a meaning is as being fixed once and for all. Instead, it is in more or less constant change. But the point is: changes to the meaning of a claim of a certain type are achieved through proposals for a change to public norms that are subject to public examination. Communication is more like engaging in a practice of such examinations than a mutual confirmation that one shares fixed meanings.

The proposed social inferential semantics is committed to social holism: with respect to having beliefs, a person is ontologically dependent on there being other persons with whom she constitutes a social, linguistic community. The topic of the next chapter therefore is social holism. The main argument for social holism starts from the problem of rule-following. This is a problem for any semantics, independently of whether or not it employs social notions to explain meaning. As mentioned at the end of [2.3.2], my central argument for a social inferential semantics is that the rule-following problem requires us to conceive meaning in terms of social relations. Thus, the rule-following considerations open up the prospect of an argument that covers both social holism and holism about beliefs.

CHAPTER 3
FROM SOCIAL HOLISM
TO HOLISM ABOUT BELIEFS

Summary

The problem of rule-following as set out by Kripke in his interpretation of Wittgenstein's Philosophical Investigations is the main argument for social holism [3.1]. This chapter gives an account of how – and to what extent – social practices can determine meaning for our beliefs. This account builds a bridge between the natural and the normative. Meaning is determined in social practices on the basis of our biological equipment. Nonetheless, the description of meaning cannot be reduced to a naturalistic description [3.2]. However, this account is a satisfactory solution to the problem of rule-following only if it is combined with an inferential role semantics. The rule-following considerations thus lead to an argument that covers both social holism and holism about beliefs [3.3].

3.1 The Problem of Rule-Following

3.1.1 What is the Problem of Rule-Following?

Let us start by considering a concept *F*. Let *F* be a one place concept and a basic concept in the sense that it is not regarded as a function of other concepts. For instance, the concept "blond" is conceived as being derived from the concepts "hair" and "yellow", whereas these latter concepts are not conceived as being derived from other concepts. I shall consider only concepts such as the latter ones. We predicate the concept *F* of things in the world by saying "This is *F*", such as "This is yellow" or "This is a stone". According to Wittgenstein and the literature on Wittgenstein, a rule determines what constitutes correct use of a concept in a language. To avoid an unnecessary complication of the matter by distinguishing between rules and concepts, let us simplify that point: let us take a rule *F* in the first and foremost sense to be a concept *F* insofar as this concept is predicated of things (or events or whatever one takes to be there in the world). When we employ the concept *F* in sentences of the type "This is *F*", we follow a rule that tells us which things are *F* (and which things are not *F*) by telling us when it is correct to say "This is *F*". The question is: How does the concept or the rule *F* determine the use of sentences of the type "This is *F*"?

One may search for a further rule that tells us how to apply sentences of the type "This is *F*". Consider the following instruction: "Say 'This is a stone' of a thing x if and only if x is grey and hard." However, such an instruction fails to solve the problem. The search for further rules is futile. It leads to a regress of rules. For we get into the same trouble with the concepts G and H that are intended to determine the application of the concept F in the sense that "If and only if something is G and H, then it is F". What determines the application of the concept G? In the given example, what determines the correct use of the concepts "grey" or "hard" in sentences of the type "This is grey" or "This is hard"?

The contemporary discussion on the problem of rule-following goes back to Wittgenstein's *Philosophical Investigations* (1952).[1] The examples that Wittgenstein discusses can be taken to show the following: when it comes to the application of the rule F, any continuation of a given sequence of examples can be regarded as going on in the same way and thus as continuing to follow the rule F under some interpretation of what the rule F is. One of the cases that he considers is two persons transferring the rules of chess from moves on a chessboard to shouts and stamping with their feet (§ 200). Do they continue to play chess or even to play any game? Under some interpretations they do, under other interpretations they do not. Wittgenstein concludes in § 201:

> This was our paradox: no course of action could be determined by a rule, because every course of action can be made out to accord with the rule. The answer was: if everything can be made out to accord with the rule, then it can also be made out to conflict with it. And so there would be neither accord nor conflict here.[2]

The conclusion is that it cannot be interpretation in the sense of explaining the rule in question by another rule that determines the application of a rule. This is the *argument from the regress of rules*.

---

1 As regards the interpretation of the *Philosophical Investigations*, see the commentaries by Baker and Hacker (1985) and von Savigny (1994/95) as well as the papers in von Savigny (1998).

2 "Unser Paradox war dies: eine Regel könnte keine Handlungsweise bestimmen, da jede Handlungsweise mit der Regel in Übereinstimmung zu bringen sei. Die Antwort war: Ist jede mit der Regel in Übereinstimmung zu bringen, dann auch zum Widerspruch. Daher gäbe es hier weder Übereinstimmung noch Widerspruch.", *Philosophical Investigations*, § 201. Translation adopted from Wittgenstein (1958). This paragraph refers back to § 198. Compare already §§ 84–87. See also Wittgenstein (1978), part VI, § 38.

This argument has the following consequence: it is not appropriate to demand a reason for the way in which a person applies a sentence of the type "This is *F*". Giving a reason would amount to citing an interpretation of the rule and thus introducing a further rule. But the argument shows that no interpretation of a rule – and thus no further rule – can determine the application of a rule. Hence, one cannot but apply a rule without giving a justification of the manner in which one applies the rule (compare in particular §§ 211, 219, 289, 292).

So far the problem of rule-following is well known in the history of philosophy. For instance, Kant remarks in the *Critique of Pure Reason* that there cannot be a rule for subsuming cases under a given rule; searching for such a rule would lead to an infinite regress of rules. He proposes that it is the faculty of judgement which determines the application of a given rule. The faculty of judgement has to be trained in practice by working with cases (A 132 / B 171 – A 134 / B 174). This point can be traced back to the beginning of occidental philosophy: according to the interpretation of Wolfgang Wieland (1982), Plato distinguishes between several forms of knowledge in such a way that he admits a non-conceptual know-how when it comes to the application of concepts (in particular pp. 164–165, 185, 224–236, 252–263, 291–311). Hence, it is granted that the application of a rule is not determined by a further rule; instead, the ability to apply a rule is a practice, a know how. However, in all these positions, it is presupposed that the concept or the rule itself is determined; the problem is considered to be only how to apply the concept or the rule to something in the world.

The paradox of which Wittgenstein speaks in the quotation above can be seen not only as an expression of the well known problem, but also as a challenge to this latter point: what is original in Wittgenstein is that he poses the problem of rule-following in such a way that the very notion of persons having concepts with a determinate content at their disposal is challenged. At least, this is what the famous interpretation of Saul Kripke (1982) says.[3] Independently of whether or not Kripke's interpretation is a plausible reading of the *Philosophical Investigations*,[4] as far as the systematic point is

---

3 Kripke (1982), Chapter 2. See also the exposition in Stegmüller (1986), Chapter 2.

4 As regards this point, see the criticism of, for instance, Baker and Hacker (1984); McGinn (1984); Collins (1992). For recent interpretations of Wittgenstein along the lines of the thesis that a social community is indispensable for rule-following, see Bloor (1997), Wilson (1998) and Williams (1999), in particular Chapters 1, 2, 6 and 7.

concerned, Kripke offers a powerful exposition of the problem of rule-following. He focuses on this point: Which rule is a given sequence of examples meant to instantiate? Kripke shows that any finite sequence of examples satisfies infinitely many logically possible rules. Hence, whatever description of a given sequence of examples one endorses, infinitely many other descriptions would also fit the examples in question.

The case that Kripke considers is the rule of addition and the rule of quaddition which he invents. Addition is a rule for the operation "plus", quaddition is a rule for the operation "quus". Quaddition is like addition for all numbers smaller than 57, but yields the result 5 if at least one number is 57 or greater. Hence, if the rule is addition, 68 and 57 is 125; but if the rule is quaddition, 68 and 57 is 5. If a given sequence of examples includes only numbers smaller than 57, how do we know that the rule which this sequence is meant to instantiate is addition rather than quaddition?

Kripke's argument does not depend on a behaviouristic premise (pp. 15, 56–57). His argument is a point about the rule that a person means to follow in any finite sequence of actions including linguistic actions (speech acts). Since any person exists only for a finite time, any sequence of actions of hers is finite. Imagine that you are asked to add two numbers which are bigger than all the numbers that you have added together hitherto. Assume, for the sake of argument, that all the numbers which you have added together hitherto are smaller than 57. Now you are asked to add 57 and 68 together. How do you know that you should answer 125 if you intend to act in accordance with your former performances? You may say that this is determined by the rule of addition that you followed all the time. But how do you know that in your earlier actions you meant to follow the rule of addition rather than the rule of quaddition, given that you only considered numbers which are smaller than 57? Kripke's point is that there is nothing mental which determines a conceptual content. I therefore prefer talking in terms of a sequence of actions to speaking of a sequence of behaviour. On most accounts of actions, actions of a person involve states of the person which have a conceptual content, that is, states which are instances of rule-following.

The sceptical query that Kripke develops applies to any concept. Kripke argues that Wittgenstein focuses on mathematical concepts (in the *Remarks on the Foundations of Mathematics*)[5] and concepts to describe sensations (in

---

5 Part VI in the edition Wittgenstein (1978). See also part I, §§ 1–5; part IV, §§ 8–9.

the *Philosophical Investigations* from § 243 on), because these two groups of concepts are likely to be cited against the sceptic's claim that there is nothing about the mental state of a person that determines which rule the person follows in a finite sequence of actions.

Kripke's sceptical query raises two problems that should be distinguished:

- the *infinity problem*: How can a finite sequence of examples or of actions instantiate only one rule rather than infinitely many rules? The sceptical challenge is: *There are infinitely many possible ways of continuing any finite sequence of examples or of actions in any new situation. Each of these ways is in accordance with the rule which the sequence instantiates under some interpretation of what the rule is.*
- the *normativity problem*: What determines which is the correct manner to continue a finite sequence of examples or of actions in such a way that a person can follow a rule (so that she has a distinction between following the rule correctly and following it incorrectly at her disposition)?[6] The sceptical challenge is: *For any finite sequence of examples or of actions and for any new situation of continuing the sequence in question, it is not determined what is the correct way to go on.*[7]

Both these problems threaten our beliefs to be left with no meaning at all:[8] if each sequence of beliefs instantiates infinitely many rules (concepts), it has no meaning at all. For if it can mean infinitely many things, it does in fact not mean anything at all. Furthermore, if each sequence of beliefs can be correctly continued in any way, there is no such thing as correct in distinction to incorrect rule-following, and hence no following of a rule at all. That is to say: there is no employing of concepts at all. As Kripke puts it: "There can be no fact as to what I mean by 'plus', or any other word at any time." (p. 21; see also pp. 55, 62)

A solution to the problem of rule-following has to take into account three features:[9]

---

6 For an argument that there is no such normativity, see Bilgrami (1992), Chapter 3, in particular pp. 88–94.

7 See Kripke (1982), pp. 21, 55, 62.

8 McGinn (1984), pp. 144–149, by contrast, claims that Kripke's sceptical challenge applies only to the meaning of linguistic expressions in a straightforward way. See Boghassian (1989), pp. 509–517, against this claim.

9 Consider the conditions which Pettit (1996), pp. 82–83, sets out.

1) *A rule determines indefinitely many cases*.[10] A rule does not have to determine any case that one might be able to imagine. There may be some indeterminacy in a concept, and some concepts may admit of more indeterminacy than others; a paradigmatic example are common sense concepts in contrast to scientific concepts. However, a concept determines much more cases than those ones to which a person will have the chance to apply it. To put the matter in other words, a concept is identifiable independently of any one *particular* application of it or any one *particular* sequence of applications of it.[11] For instance, you can learn the concept "bed" without being introduced to the bed in which George Washington slept or the sequence of beds in which a president of the United States slept.
2) *Rules are followed directly*. Applying a rule is not interpreting the rule. Envisaging the application of a rule to be determined by another rule leads to a regress of rules. In reverse, the way in which one applies a rule cannot be justified by citing a rule.
3) Nonetheless, *rule-following is an intentional affair*, although the rule itself is not the intentional object of one's belief states (on pain of the infinite regress described in [2.4.1]). Because rule-following is an intentional affair, there is a *distinction between correct and incorrect rule-following* and *having this distinction at one's disposal is a necessary condition for following a rule.*

What determines the rule-following of a person in such a way that these three criteria are satisfied?

3.1.2 Rule-Following as Motivation for Social Holism

The problem of rule-following as presented in the last section is that there is nothing about the mental state of a person which determines in which intentional state that person is. However, there are other candidates besides something mental which could do the job of determining a conceptual content for the belief states of a person. One prominent candidate is dispositions to behaviour. Kripke, however, argues that having recourse to dispositions is no help (pp. 22–37). Dispositions are finite, too. What is more, appealing to dispositions does not settle the normative matter: Why is

10 This point is disputed by Bloor (1997), Chapter 2, in particular pp. 19–20. See also Rust (1996), pp. 115–116.

11 See Pettit (1996), p. 82.

the action that a person is disposed to do the action which she should do in order to follow the rule that she meant to follow? To illustrate this point, recall that one can have dispositions to do certain mistakes in a systematic way. For instance, one may have the disposition to do systematically a certain mistake when it comes to adding together big numbers. But nevertheless, one's answers satisfy a quus-like rule in this case. Dispositions fail to solve the normativity problem, because they do not make available for the purported rule-follower a distinction between correct and incorrect rule-following.

If we take Kripke's argument against dispositions into account, it is evident that the indeterminacies which Kripke points out as regards rule-following can be considered as a generalization of the indeterminacies which Quine sets out in *Word and Object* as regards translation on the basis of behavioural evidence.[12] Quine and Kripke agree that dispositions to behaviour are not sufficient to determine a meaning for the beliefs of a person. Kripke then makes a more general point than Quine by not limiting himself to behavioural evidence. By the way, Quine himself draws attention to the connection of his indeterminacy thesis with Wittgenstein's later philosophy,[13] and Kripke discusses Quine's work (pp. 14–15, 55–57).

Apart from the theories mentioned hitherto, there are theories of meaning which divorce the explanation of meaning from the mental and physical states including the dispositions of the beings that have beliefs. However, these theories are not a way out of the problem of rule-following either. If one assumes that there are meanings in the sense of abstract objects beyond space and time such as Platonic ideas or Fregean senses,[14] then one shifts the problem to the following queries: How can a person grasp these senses? What makes it that a finite sequence of mental acts of a person is grasping the sense *F* (say, the sense of addition) rather than the sense *G* (say, the sense of quaddition), given that the mental acts conform to both these senses?[15] Even if one assumes, as Jerrold Katz proposes, that such senses

---

12 Quine (1960), Chapter 2. For a precise, elaborate argument in favour of the indeterminacy thesis, see Massey (1992).

13 Quine (1960), § 16 note 2. As to the relation between Wittgenstein and Quine, see Heal (1989) and the papers in Arrington and Glock (1996).

14 "Der Gedanke" in Frege (1986b) (English translation in Frege (1984), pp. 351–372).

15 See Kripke (1982), pp. 53–54. Compare already Sellars, "Some Reflections on Language Games", § 10, p. 324 in Sellars (1963).

are themselves finite,[16] a finite sequence of mental acts can express more than one such sense, since no such sense is tied to any given finite sequence of mental acts. Thus, the problem remains what it is about the mental state of a person that distinguishes her grasping "addition" from her grasping "quaddition".

A similar objection applies if we shift from Platonic ideas to Aristotelian forms that are supposed to inhere in physical things: assuming natural properties that are instantiated by physical things does not solve the problem of rule-following.[17] Let us move from Kripke's example to the example for which Nelson Goodman is famous. In his "New Riddle of Induction", Goodman introduces the property of being grue (in Goodman (1983), pp. 72–81). In a slight modification of Goodman's original example, let us say that something is grue if and only if it is green up to *t* (you are invited to put in a date in the future), and blue afterwards. It seems that, if we now classify something as being green, we would also be justified in classifying it as being grue. The new riddle of induction then is the question what justifies the assumption that things which are green now will continue to be green hereafter. The relation of Goodman's example to Kripke's quaddition-rule is obvious: in both cases an artificial concept is introduced in order to challenge our justification in operating with our familiar concepts. Kripke mentions Goodman (pp. 20, 58–59).

If one assumes natural properties to solve Goodman's riddle of induction, one can exclude grue-like rules: we are justified in classifying things as green, because being green, in contrast to being grue, is a natural property. However, the problem of rule-following now appears in the following manner: assume that a person responds in a finite sequence of actions to a finite sequence of examples. Any finite sequence of examples instantiates more than one natural property. For instance, a given sequence of examples may instantiate not only the property of being a table, but also the properties of being wooden, brown, rectangular, having four legs, etc. Natural properties do not solve Kripke's problem, because they cannot determine on

---

16 Katz (1990), Chapters 2 to 4, in particular pp. 171–174, makes a case for countenancing such senses despite the criticism of Wittgenstein and Kripke. For an argument against Katz, see Rust (1996), pp. 103–108. See also Boghossian (1994), pp. 143–144.

17 See, by contrast, Lewis (1983b), pp. 375–377. But the analysis of dispositions that Lewis (1997) gives can be taken to imply that Lewis is now more favourable to a dispositionalist account of rule-following than he was in his (1983b).

which aspect a person focuses in responding to a sequence of examples. When it comes to continuing a given sequence of actions which responds to a given sequence of examples, there are many ways of going on in accordance with the natural properties that are instantiated by this sequence. The sceptical challenge reappears as the question what makes a particular way of going on correct. In other words, the problem of rule-following now is: How do we grasp one specific of indefinitely many natural properties in a finite sequence of examples?[18]

However, we do not have to follow Kripke when he claims that Wittgenstein accepts a sceptical paradox and that he proposes a sceptical solution which is analogous to Hume's solution to the problem of causation. According to Kripke, a sceptical solution provides only assertibility conditions, but not truth conditions for sentences about meaning. Thus, there are no facts that make sentences about meaning true.[19] If the problem of rule-following concerns not only the aspect of the application of a rule, but extends to the aspect of the content of the rule, it is suggestive to try the traditional solution to the former aspect out on the latter aspect too. That is to say: nothing mental, but a practice determines meaning. Since the dispositions to act of a person considered in isolation are not sufficient to determine meaning either, this practice has to be a social practice. Social practices are necessary in order to (a) determine a meaning for the beliefs of a person given the infinitely many logically possible meanings of any finite sequence and (b) enable a person to have a distinction between correct and incorrect rule-following at her disposal. The indeterminacies that Quine and Kripke point out are the main argument for a theory of meaning in terms of social practices whose description requires irreducibly normative vocabulary, i.e., cannot be reduced to a description in naturalistic terms: to the extent that meaning is determinate at all, it is determined by such social practices. Such a theory of meaning as a genuine alternative to traditional theories of meaning.[20] It is not a sceptical *Ersatz*-solution.

---

18 Thanks to Philip Pettit for making this point clear to me.

19 See Kripke (1982), pp. 66–71, 84–87, 108. For an exposition of this point, see Stegmüller (1986), Chapter 3. But see Wilson (1998), pp. 113–121, against the widespread impression that Kripke's Wittgenstein proposes a non-factualism about meaning.

20 Compare McDowell (1992), pp. 44–45.

The position that a person can have beliefs only within a social, linguistic community is known as *social holism*. Taking up the general conception of holism proposed in the first chapter, social holism can be formulated as the thesis that *with respect to having beliefs, a person is ontologically dependent on there being other persons with whom she constitutes a social, linguistic community*. Social holism is a version of the first type of the general conception of holism outlined in [1.4]: it is individual persons who have beliefs. If one says that a community has beliefs (or has a will), this is only a metaphorical use of the predicates "having beliefs" (or "having a will"). The thesis of social holism is not logically tied to humans. It can be put forward as a claim that applies to all finite thinking beings in every possible world.

The very first paragraph of the *Philosophical Investigations* can be seen as containing both a rejection of traditional theories of meaning and an allusion to the view of meaning as use in a social practice: citing Augustine, Wittgenstein repudiates the position that a meaning inheres in each word or each belief. By considering the example of someone handing in a piece of paper which says "five red apples" to the grocer, he introduces his alternative to the traditional position in terms of social practice. Later he says in § 201 immediately after formulating the problem of rule-following as quoted in the preceding section:

> It can be seen that there is a misunderstanding here from the mere fact that in the course of our argument we give one interpretation after another; as if each one contented us at least for a moment, until we thought of yet another standing behind it. What this shews is that there is a way of grasping a rule which is *not* an *interpretation*, but which is exhibited in what we call "obeying the rule" and "going against it" in actual cases.[21]

In the following paragraph, he concludes:

> And hence also 'obeying a rule' is a practice. And to *think* one is obeying a rule is not to obey a rule. Hence it is not possible to obey a rule 'privately':

---

21 "Daß da ein Mißverständnis ist, zeigt sich schon darin, daß wir in diesem Gedankengang Deutung hinter Deutung setzen; als beruhige uns eine jede wenigstens für einen Augenblick, bis wir an eine Deutung denken, die wieder hinter dieser liegt. Dadurch zeigen wir nämlich, daß es eine Auffassung einer Regel gibt, die *nicht* eine *Deutung* ist; sondern sich, von Fall zu Fall der Anwendung, in dem äußert, was wir »der Regel folgen«, und was wir »ihr entgegenhandeln« nennen.", *Philosophical Investigations*, § 201. Translation adopted from Wittgenstein (1958). See also §§ 213, 217, 219.

> otherwise thinking one was obeying a rule would be the same thing as obeying it.[22]

These statements can be read in the following manner: What the rule is and how it is applied shows up in social practice.[23] Wittgenstein speaks of a form of life (*Lebensform*) (in particular § 241 and Wittgenstein (1978), part VI, § 34) and a habit (*Gepflogenheit*) (in particular §§ 198–199). He repeats several times that a person considered in isolation cannot follow rules, because there would in this case be no distinction between believing to follow a rule and following a rule (§§ 258, 265, 380).

For a person considered in isolation, believing to follow a rule *F* would be the same as following the rule *F*, because she herself would determine the appropriate circumstances of application and the consequences of application by her behaviour. Whatever she would take herself to be committed to and entitled to is what she would be committed to and entitled to. Analogously, believing that she does not follow a rule *F* or believing that she no longer follows a rule *F* would be the same as not following the rule *F* or no longer following the rule *F*. Whatever a person considered in isolation does in order to continue a given sequence of behaviour counts as going on in the same way and as being right. Wittgenstein maintains in § 258:

> But in the present case I have no criterion of correctness. One would like to say: whatever is going to seem right to me is right. And that only means that here we can't talk about 'right'.[24]

Consequently, whatever a person considered in isolation does falls victim to the problem of rule-following as Wittgenstein formulates this problem in § 201: if there is no distinction between following a rule in a correct way and

---

22 "Darum ist >der Regel folgen< eine Praxis. Und der Regel zu folgen *glauben* ist nicht: der Regel folgen. Und darum kann man nicht der Regel >privatim< folgen, weil sonst der Regel zu folgen glauben dasselbe wäre, wie der Regel folgen.", *Philosophical Investigations*, § 202. Translation adopted from Wittgenstein (1958). See also § 199 and furthermore Wittgenstein (1978), part VI, §§ 21, 32–34, 41–43.

23 For an interpretation of Wittgenstein on rule-following in terms of social practices, which is distinct from Kripke's exposition of the problem, see von Savigny (1994/95) and von Savigny (1996), Chapters 2 to 6.

24 "Aber in unserm Falle habe ich ja kein Kriterium für die Richtigkeit. Man möchte hier sagen: richtig ist, was immer mir als richtig erscheinen wird. Und das heißt nur, daß hier von >richtig< nicht geredet werden kann.", *Philosophical Investigations*, § 258. Translation adopted from Wittgenstein (1958). I follow here the interpretation of Kripke in joining § 258 and § 202 together.

following it in an incorrect way available for the person in question, there is no following of a rule. No content of the purported rule is determined. Hence, if a person is considered in isolation, there is nothing that could determine a conceptual content for her mental states. Consequently, whatever mental states she has, these are not belief states with a determinate conceptual content. According to Kripke, Wittgenstein's reflections on the problem of rule-following lead him to develop his famous argument against private language: the point of this argument is that there can be no private rule-following (pp. 109–110).

3.2 Social Holism

3.2.1 Meaning through Social Practices: The Argument for Social Holism

On the one hand, in spite of Kripke's attack on a dispositional theory, some philosophers argue for a naturalistic account of rule-following in terms of specific dispositions (for instance Goldfarb (1985), pp. 475–480, Ginet (1992), Coates (1997). Prominent recent proposals are an account in terms of computational dispositions (Smart (1992), Miscevic (1996)) and, furthermore, a teleo–functional account in terms of biological purposes (Millikan (1990) and already McGinn (1984), pp. 168–175; Fodor (1990), pp. 71–73, and Braddon-Mitchell and Jackson (1997) present forceful counter-arguments). Although the latter approach is not an account in terms of dispositions, it can be combined with the computational approach (Miscevic (1996/97)). Naturalistic accounts claim that (a), as far as ontology is concerned, certain naturalistic features which apply to a person even considered in isolation are sufficient to determine a conceptual content for her belief states, and that (b), as far as epistemology is concerned, it is in principle possible to reduce the description of the belief states of a person to a description in the relevant naturalistic terms.

On the other hand, some philosophers are prepared to accept the social practice of rule-following as a primitive. This is, for instance, the position of John McDowell in his interpretation of Wittgenstein. According to McDowell, any attempt to develop an account of the norms that are embodied in such a practice by starting from non-normative notions is doomed to fail (1984, in particular pp. 350–351, and also McCulloch (1995), pp. 104–105, 217–220). However, we need an account of the relation between the normative and the non-normative. In the following, I attempt a conceptual reconstruction of the practice of rule-following which is intended to bridge the gap between the natural and the normative without proposing the possibility of a reduction of the normative, intentional vocabulary to a naturalistic vocabulary. I thereby follow the position of Philip Pettit to a large extent,[25] although I set the proposal out in my own terms.

Let us reconstruct an account of how social practices determine meaning for our beliefs step by step. The *first step* is: *Although it is logically possible to continue any finite sequence of examples in infinitely many different ways, for any finite thinking being there usually is one specific way in which this being is disposed to continue such a sequence.* For the sake of simplicity, let us assume that for each person there is only one disposition that determines for her what it is to continue a given sequence of examples in the same way in any real situation which she encounters. The person does not have to be aware of this disposition. To give an intuitive idea of what this step and the following ones can be like, imagine a sequence of trees in a physical environment and reactions of persons towards this sequence, in particular reactions of classifying or refusing to classify items with this sequence. This step presupposes that persons, like other living beings, have a cognitive access to their environment. This cognitive access consists in reliable and differentiating response-mechanisms that humans share with other living beings. This presupposition is unproblematic in the present context, because (a) this cognitive access does not imply conceptual content and (b) this cognitive access will not be employed in any epistemic sense in what follows (compare [4.2]).

We need further assumptions about the natural equipment of beings that are to follow rules. The *second step* is: *Persons who have the same biological equipment and who share a physical environment do not have in general bizarrely different dispositions.* If the dispositions of people on what constitutes continuing a given sequence of examples in the same way were to a large extent bizarrely different (such as in Kripke's case of addition versus quaddition), a social practice that determines conceptual content could not get off the ground. When I speak of the physical environment, everything that belongs to the non-human environment is intended, but all cultural influences are excluded.

---

25 Pettit (1996), pp. 76–108. See also Pettit (1998d) and the account of Kripke (1982), pp. 86–95. For a recent, comprehensive overview of the solutions to the problem of rule-following with an emphasis on Pettit's solution, see Haukioja (1999), part I, in particular Chapter I.7.

The *third step* is: *The dispositions of persons who have the same biological equipment and who share a physical environment include a disposition to coordinate at least parts of one's own behaviour with the behaviour of one's fellows.* This step goes beyond dispositions to respond to stimuli in a reliable and differentiating way. It brings other beings of the same kind into the focus of the being who has this disposition. This disposition is a second order disposition: it is a disposition to change some of one's dispositions and one's behaviour as a result of the behaviour of one's fellows, being directed at coordination. This change does not have to be a conscious process. Having and exercising such a disposition is a necessary and sufficient condition for behaviour to be social behaviour. An account of social behaviour can thus be given in naturalistic terms. However, as I shall argue, a social practice can only be described in and individuated by means of normative, intentional vocabulary. Countenancing a disposition to coordination still is an assumption about the natural equipment of persons.[26] This assumption says that man is a social being. It makes sociability a necessary condition for having beliefs.

On the basis of these assumptions about our natural equipment, we can now turn to social practices. The *fourth step* is: *Owing to the disposition to at least partial coordination humans react to each other's actions by applying sanctions in the sense of reinforcements or discouragements.* They reinforce actions in others that agree with their own actions, and they discourage actions in others which disagree with their own actions. By agreement or disagreement, I do not mean an intentional affair, but accord or failure of accord in the way in which a given sequence of examples is continued. Given the presupposed cognitive access to the environment such accord or failure of accord is transparent to the persons involved. Sanctions are exclusively physical reinforcements and discouragements at this stage. Once normative statuses such as permissions or prohibitions to do something are established, sanctions can consist in a change of normative statuses as well.

Sanctions are themselves liable to an assessment as being correct or incorrect. But such an assessment is available only in retrospective, once a content for a rule is determined. At this stage of the argument, sanctions enter as a primitive assessment of each other's actions as correct or incorrect. Nonetheless, sanctions thereby are the first instance in which rule-following

---

26 Compare Kripke (1982), pp. 91, 96–98, 109; Pettit (1996), pp. 86–88.

as an intentional affair becomes manifest. As long as sanctions consist in physical reinforcements and discouragements, it may be possible to give a description of them – and of the behaviour that they reinforce and discourage – in naturalistic terms. However, even if such a description is possible, it does not follow that a description of the conceptual content that becomes determined by means of sanctions can be reduced to a naturalistic description.[27] Sanctions can get a process of determining conceptual content off the ground, because they make available for a person a distinction between correct and incorrect actions by introducing an external perspective: there now is a distinction between what a person takes to be correct or incorrect and what is correct or incorrect in the light of others.

Having reached this stage of the argument, we can conceive sanctions as a means to determine conceptual content. The *fifth step* is: *Sanctions are a means to come to conditions under which persons agree in their ways of continuing a given sequence of examples.* In the case of agreement, sanctions reinforce the dispositions of persons in the way in which they react to their environment. In the case of disagreement, sanctions in the form of discouragements trigger a process of finding out in practice the obstacles in the persons or in the environment that prevent agreement. That is to say: People react to disagreement in such a way that they take disagreement as a sign that something has gone wrong and that they have to do something in order to get things right. They try to find out why they disagree. In some cases – those ones which then lead to beliefs about the way the environment is – they discover conditions under which they overcome their disagreement. These then are the normal conditions for a belief of the type in question. Sanctions thus induce a process of mutual adjustment that leads to convergence.

John Haugeland sets out a similar account in terms of conformity that is reached by censoriousness.[28] In his paper "The Intentionality All-Stars", he distinguishes between norm-abiding and rule-following (in Haugeland (1998), pp. 147–150). Rule-following, for him, is based upon an explicit formulation of the rule. By contrast, I speak of rule-following, although no

---

27 Compare Haugeland, "Truth and Rule-Following" in Haugeland (1998), pp. 355–356, note 9, contra Brandom (1994), pp. 34–36.

28 Haugeland (1982), in particular pp. 15–18, and "The Intentionality All-Stars" in Haugeland (1998), pp. 147–150, as well as "Truth and Rule-Following" in Haugeland (1998), pp. 310–313.

explicit formulation is required, because I take it to be the decisive point that a distinction between correct and incorrect is available to and recognized by the persons in question.

We need a further step that sums up the result of the described sanctions. The *sixth step* is: *As a result of the process of coming to conditions under which persons agree, the rule can be conceived as that in which the convergence of persons in their ways of continuing a given sequence of examples consists*. The rule determines what is correct and what is incorrect in continuing a given sequence of examples for the persons in question. It thereby is *eo ipso* normative, that is, binding on all the persons who take part in the assessments of each other's actions as correct or incorrect. It is not identical with the dispositions of any one of these persons. According to the way in which the terminology is used in this work, norms are rules as established in the interactions of creatures that are disposed to assess each other's actions as correct or incorrect by means of sanctions.

This reconstruction presupposes that it is transparent to the persons whether they agree or disagree in their reactions to a shared environment at a time and, in particular, the way in which they continue a given sequence of examples.[29] This presupposition is justified, because, at this stage of the argument, nothing like persons meaning different things by their reactions can come into play, since the point at issue is the reconstruction of meaning. No language is presupposed for this account. Recall the assumptions about the natural equipment of purported rule-followers: dispositions that are by and large the same, because we consider persons who have the same biological equipment and who share a physical environment. Against the background of these assumptions, it can be presupposed that agreement or disagreement in reaction to a shared environment at a time is transparent to the persons in question.

Nonetheless, one may object that the described process of reinforcements and discouragements is not apt to determine conceptual content, because agreement or disagreement in continuing a given sequence of examples may be due to different background intentions that the persons in question have. Beliefs about the physical environment are not acquired for the sake of theoretical contemplation, but are part and parcel of our practical intentions. In particular, this is emphasized by Wittgenstein's conception of a form of life. One may therefore say that a case of a disagreement in the way in which

---

29 Compare the objection of Hoffman (1985).

two persons continue a given sequence of examples may be due to different background intentions as to what they intend to do with the sequence of examples under consideration.

However, intentions to act presuppose intentional states that have a determinate conceptual content such as belief states. Invoking such intentions in an objection against the account of conceptual content under consideration therefore presupposes what this account seeks to reconstruct. Only once conceptual content is determined can a case of a disagreement in classification due to different practical intentions arise. For the sake of a reconstruction of conceptual content it is therefore warranted to consider cases of the determination of conceptual content by being exposed to sequences of examples in isolation.

The proposed account satisfies at least the last two of the three requirements for a solution to the problem of rule-following which were set out at the end of [3.1.1]. This account shows that rule-following is an intentional affair (3). In the described social practices, a public standard is established which a person can meet or fail to meet in her actions.[30] Furthermore, determining on the basis of each one's dispositions a convergence of assessments and thus establishing a content for a rule shows how it is possible to apply the rule without interpretation in one's actions (2). An interpretation of the rule does not enter this account. Rules are followed *blindly*; nonetheless, *rules* are *followed* (in contrast to mere regularities of behaviour): due to the assessments by others of the way in which a person acts on the basis of her dispositions, there is a distinction available for that person between correct and incorrect rule-following. The crucial point for this account is how it fares with respect to the first requirement, namely that a rule determines indefinitely many cases. Let us consider this point in the next section.

3.2.2 Objections to the Argument for Social Holism

There are two widespread objections to an account of rule-following in terms of social practices: (a) Given the case which Kripke's sceptic makes, does bringing in social practices really counter the attack of the sceptic? (b) Assuming that an account in terms of social practices meets Kripke's sceptical challenge, why can what is achieved by considering social practices

30 See also Williams (1999), in particular Chapter 6.

not be achieved by considering an individual in isolation? Let us tackle (a) first.

The challenge of Kripke's sceptic can be repeated on the level of the community. The interactions of a group of individuals are finite like the actions of one individual. Consequently, all the actions of addition which the members of a community have performed hitherto also satisfy infinitely many logically possible quaddition-like rules which yield divergent results only for numbers that are greater than the numbers which humans will ever consider.[31] This is right as far as a description from outside is concerned: if a Martian observes from Mars the actions of us earthlings, it is indeterminate for her whether we follow the rule of addition or a rule of quaddition or any rule at all. However, consider the conclusion which Kripke's sceptic builds on this point and which I have described as the *normativity problem* in [3.1.1]:

1) *For any finite sequence of examples or of actions and for any new situation of continuing the sequence in question, it is not determined what is the correct way to go on.*

What the account in terms of social practices achieves is to block the move to this conclusion. For this account shows that for any situation with which the people in a community deal in their ordinary practices it is determined what is the correct way to go on. Determining in the assessments of each other's actions on the basis of each one's dispositions a norm that is binding on each of the persons in question fixes what is correct to do for an indefinite number of situations. However, going beyond the ordinary practices with which the people in a given community deal, there is for any finite sequence of actions a margin conceivable beyond which it is indeterminate what is the correct way to go on. There are infinitely many possible ways imaginable how to go on beyond this margin. The norm that the social practices have established does not select one of these ways.[32] Furthermore, for any one concept, it cannot be excluded that a community will confront such a margin. That is why I speak of the norm determining indefinitely many situations with which the members of a community deal in their *ordinary* practices. In such an extraordinary situation, a further determination of the norm has to be carried out by means of the described process. (I will come back to this point in [4.3.3]). Over and above that, there is of course the

---

31 Compare Blackburn, "The Individual Strikes Back", in Blackburn (1993), p. 221.

32 Compare Pettit (1991), pp. 618–619, and Pettit (1996), pp. 349–351.

vagueness of common-sense concepts such as, for instance, "bald", which leave a number of borderline situations open.

Hence, the account in terms of social practices rejects the sceptical conclusion (1) in the following way:

2) *For any finite sequence of examples or of actions, there are indefinitely many new situations of continuing the sequence in question for which it is determined what is the correct way to go on.*

However, this account has to concede the following point:

3) *For any finite sequence of examples or of actions, there is a new situation of continuing the sequence in question conceivable for which it is not determined what is the correct way to go on.*

Conceding this point does not put conceptual content in jeopardy. No reconstruction of conceptual content in terms of the practices of finite thinking beings can achieve a determination of rules beyond the scope of these practices. Consequently, this account solves the *normativity problem* by offering a reconstruction of how persons can come to follow rules. But it does not solve the *infinity problem*, that is, the problem whether and how a finite sequence as such can determine infinitely many cases. It addresses this problem only insofar as this problem threatens our beliefs to be stripped of meaning: it shows how a finite sequence of examples or of actions can determine meaning for a community of persons within the scope of their ordinary practices.

Coming back to the connection between Kripke's Wittgenstein and Quine, when it comes to one's practice in one's mother tongue, Quine seeks in "Ontological Relativity" to avoid the indeterminacies which he has shown in the case of translation in the last resort simply by "acquiescing in our mother tongue and taking its words at face value" (in Quine (1969), p. 49). A social solution to the problem of rule-following, by contrast, shows how meaning can become determined for the practices of one's mother tongue. It thereby gives an account of how we can be justified in "acquiescing in our mother tongue and taking its words at face value".

The position under consideration is a sort of meaning finitism because of (3). But the crucial point is: it is not a meaning finitism which maintains that the extension of a concept is not determined beyond the past applications of this concept on which the members of a community agree. Such a meaning finitism, which is recently proposed by David Bloor (1997, Chapter 2, in particular pp. 19–20, 24), just states the normativity problem by claiming that for each new case a new decision is called for whether or not to admit the case under consideration to the extension of a given concept; there never is a meaning of a concept which determines this.

Consider the reply of Crispin Wright to Kripke's sceptic in his book *Realism, Meaning and Truth* (1987). He says:

> Understanding cannot be always achieved via uniquely rational extrapolation from sample uses and explanations; and is not usually. Rather the path to understanding exploits certain *natural* propensities which we have, propensities to react and judge in particular ways. The concepts which we 'exhibit' by what we count as correct, or incorrect, use of a term need not be salient to a witness who is, if I may so put it, merely rational ...(1993, p. 28)

Although I do not intend to ascribe the account sketched above to Wright, the point which he makes emphasizes a crucial feature of that account: a biological equipment is a necessary condition for conceptual content becoming determined in social interactions. The relevant equipment can be explained in teleo–functional, computational or other naturalistic terms. But however much, ontologically speaking, the behaviour of a human may be determined by what can be described in these terms and in whatever sense biological functions may be normative from the point of view of evolution, the point at issue is how a human being can follow rules in distinction to exhibiting mere regularities of behaviour. That is to say, the point is (a) how there can be determinate conceptual content and (b) how a thinking being can have from her perspective a distinction between correct and incorrect rule-following at her disposal. To describe this distinction, normative, intentional vocabulary is indispensable according to the social account sketched. Consequently, this account implies that the description of meaning cannot be reduced to a description in naturalistic terms, i.e., a description that is available for a detached observer too.[33] A description in naturalistic terms is normative in the sense explained insofar as it is an employment of concepts; but it is not in a position to attribute states with a determinate conceptual content to the objects that it describes.

Even if a detached observer is provided with a complete description of a human in naturalistic terms – and, moreover, a complete physical description of the world – she is in the situation of Kripke's sceptic, namely to be aware of no more than finite sequences of actions that satisfy infinitely many rules in such a way that it is indeterminate for any new situation what is the

---

33 For a recent elaboration on this point as a consequence of Kripke's argument, see Kober (1998a) and (1998b).

correct way to go on. For such an observer there is no such thing as rule-following in the beings which she describes and thus no such thing as people having intentional states with a determinate conceptual content. As Wright puts it, such an observer is merely rational – or, one might say, too rational.

Hence, one has to participate in the social practices of rule-following of a community in order to be in a position to describe the rule that is followed in such a way that for all cases within the scope of these practices it is determined what is correct.[34] This does not exclude the division of labour within a linguistic community. For instance, one does not have to participate in the practices of telling gold from fool's gold in order to see the gold-rule. This can be delegated to experts. The point is that the institution of experts presupposes the social practices which I reconstructed in the six steps listed above. Consequently, the proposed account of meaning meets the position that understanding of persons (*Verstehen*) is distinct from and irreducible to scientific explanation. *The other side of this irreducibility, however, is that determinate conceptual content exists only relative to those who participate in a social practice that turns some of their mental states into intentional states for them.*[35] Thus, there are facts of meaning; but (a) these facts are normative facts in the sense that they can be described only in normative vocabulary, and (b) these facts obtain only relative to those who participate in the relevant social practices. By admitting facts in this sense, the proposed solution can be seen as steering a middle course between a sceptical solution, which claims that there are no facts of meaning,[36] and a straight solution, which seeks to discover facts of meaning in the furniture of the world.

Let us now consider the second objection that is often addressed to an account in terms of social practices: Why can what is achieved by considering social practices not be achieved by considering an individual in isolation? Simon Blackburn claims in his important essay "The Individual Strikes Back" (1984, quoted according to the reprint in Blackburn (1993), essay 12) that the members of a community stand to each other as the momentary time-slices of an individual do. What applies to the members of a community therefore also applies to an individual considered in isolation

---

34 McDowell (1984) stresses this point (in particular pp. 350–351). Ebbs (1997) makes a recent, elaborate case for a participant perspective.

35 See also Stegmüller (1986), p. 86, who speaks of a contextual theory of meaning.

36 As to the difficulties with the notion of meaning facts in Kripke's sceptical solution, see Soames (1998b).

across time. An individual can correct her former actions, etc. Blackburn concludes that an individual considered in isolation can follow rules in the same way as the members of a community can (p. 224).

The premise of Blackburn's argument, namely that the members of a community stand to each other as the momentary time-slices of an individual do, is disputable. To recall, Kripke formulates his sceptical query in such a way that it applies to the time-slices of an individual: Imagine Carol being an individual considered in isolation who tries to follow the rule of addition now. How does she know that yesterday she tried to follow the rule of addition instead of a rule of quaddition? In the light of the rule which she tries to follow now she can of course judge the actions of her former self as correct or incorrect. But whatever seems to her to be correct or incorrect in these judgements now is correct or incorrect. Consequently, there is no point in speaking of judging something as correct or incorrect here.

Blackburn imagines a second type of sceptic: "So just as the original sceptic queries what it is for one person-time to be faithful to a rule adopted by a previous person-time, so the public sceptic queries what it is for one person to be faithful to the same rule as that adopted by another." (p. 224) However, as the former sceptical query shows that there is no fact of the matter whether one person-time is faithful to a rule adopted by a previous person-time, so this latter sceptical query shows that there is no fact of the matter whether one person is faithful to the rule adopted by another person in the sense that the following conception does not work: one person first adopts a rule and then a second person takes up this rule and is faithful to it. Such a conception falls victim to the objection of the sceptic that nothing determines whether the second person has the same rule in mind. But this just shows that rules can only be determined by social practices and that they have to be inter-subjective in a literal sense – all there is to rule-following is taking part in the described social practices.

Blackburn continues: "Now if the public sceptic can be by-passed by, in effect, saying that this is what we do—we see each other as mutually understanding the same rule, or dignify or compliment each other as so doing, provided the exposed practice agrees well enough—then the private sceptic can be by-passed in the same way." (p. 224) This conclusion is not warranted. The public sceptic is answered by developing an account of how a rule is determined in social practices of assessing each other's actions. But there is no such answer available as far as the private sceptic is concerned.

In an interaction of two or more persons, there can be a disagreement in the way in which these persons continue a sequence of examples in a shared environment at a time. That disagreement can trigger a process of negotiating by means of sanctions conditions under which there is convergence. By contrast, in the case of one person who is considered as being in isolation throughout her life, there is no possibility of such a process of negotiating a convergence. The present disposition of the person always has a privileged position; for the past self cannot reply and give a feedback. It cannot make available for a person a distinction between what she takes to be correct or incorrect and what is correct or incorrect. The same point applies to the simulation of another person: simulation cannot give a feedback that makes available for a person an external perspective in the sense of making available for her a distinction between taking something to be correct and something being correct.[37]

There is a second reason why there are no disagreements that can lead to a process of determining a conceptual content in the case of a person in isolation. Consider the following objection to what has been said so far: imagine that a person finds herself disposed not to apply a concept *F* to a certain object now ("This is not *F*"), whereas she was disposed to apply the concept *F* to that object some time ago and the object has not changed meanwhile. Why is this case not a disagreement with oneself which can trigger a process of negotiation with oneself that leads to convergence? The point is that it is not determined what counts as a change of a thing independently of conceptual content being determined.[38] A change of a thing is a change of its properties. In order to have epistemic access to the properties, conceptual content has to be determined. Only in the light of conceptual content being determined can we therefore speak of a thing having changed or remaining unchanged. Hence, the objection under consideration begs the question – it has to assume what is at issue, namely that there is determinate conceptual content.[39]

3.2.3 Social Practices in terms of I–Thou Relations

For the outlined conception to be a persuasive account of how normativity can be reconstructed, it is crucial to work with the notion of I–thou relations

---

37 See, by contrast, Gerrans (1998), who argues that the ability to simulate is a fact about an isolated individual that enables such an individual to follow rules (p. 583).

38 For an argument for this thesis that bases itself on the grue-green problem set out by Goodman (1983), pp. 72–81, see Gemes (1987), section 1.

39 See, in contrast to the considerations in this paragraph, Pettit (1998d), step 5.

among single persons. Kripke introduces the solution which he takes to be the one that Wittgenstein has in mind by means of the example of a child learning addition from a teacher. This is an example of an I–thou relation, albeit an asymmetric one. But Kripke goes on to consider an example of a symmetric I–thou relation, an adult judging the rule-following activities of another adult (pp. 90–91). Nonetheless, Kripke then talks in terms of a relation between a single person and a community (in particular pp. 92–95, 108–109; but see also p. 111).

If one conceives a social theory of meaning in terms of a relation between an individual and a community in the sense that the community assesses the actions of the individual according to what it takes to be fit and proper, then this theory is subject to the following objection: the *fact* of community approval or disapproval is confused with *right*, i.e., correctness and incorrectness. The arbitrariness of a person considered in isolation in the manner in which she continues a sequence of examples is simply replaced with the arbitrariness of a community in what the community takes to be the correct way of continuing the sequence in question in the same way.

An objection in this vein is sometimes addressed to Kripke.[40] Whether or not it is justified to address such an objection to Kripke, this objection hits the account of Wright in his interpretation of Wittgenstein's philosophy of mathematics (1980). In that book, Wright sets out a position which he no longer holds; he maintains there:

> None of us unilaterally can make sense of the idea of correct employment of language save by reference to the authority of securable communal assent on the matter; and for the community itself there is no authority, so no standard to meet.[41]

McDowell voices a concise objection to Wright by applying § 258 of the *Investigations* to this interpretation in the following way:

> One would like to say: whatever is going to seem right to *us* is right. And that only means that here we can't talk about 'right'.[42]

If one reconstructs norms in terms of I–thou relations, this objection does not apply. In this case, one does not work with the notion of an asymmetric relation between an individual and a community that applies its standards to

---

40 See, for instance, Davidson (1990), pp. 310–311; Hösle (1990), pp. 181–186.

41 Wright (1980), p. 220. See the whole Chapter 11. For a recent position along these lines, see Bloor (1997), in particular pp. 33, 71.

42 McDowell (1984), p. 359 note 10. Compare also Holtzman and Leich (1979), pp. 2–14.

the actions of the individual. Instead, one assumes a symmetric relation among individuals who are in principle on an equal footing and who have certain dispositions each. Of course, I–thou relations can be asymmetric too such as in the example of a teacher and a child. But we can work with symmetric I–thou relations as long as we are concerned with a *systematic* reconstruction of conceptual content – in distinction to a history of language learning and a description of the different social positions that individuals have in a society. The content of rules (and thus norms) becomes determined in what has been described as coming to conditions by means of sanctions in which there is convergence in the assessments of each other's actions. As will be shown in [5.2.1], this account makes room for the possibility that a whole community is wrong: it enables us to conceive the process of determining what is correct as being open-ended.

Consider the conception of social holism that Philip Pettit proposes. According to him, a human being is, insofar as she has thoughts and thus follows rules, dependent on there being other humans with whom she constitutes a social community. Pettit conceptualizes social holism in terms of relations among individual human beings. He contrasts social holism, thus conceived, with what he calls collectivism, namely the thesis that society exerts some sort of force of its own on its members. Pettit then argues against collectivism and in favour of social holism.[43] The work of Pettit makes clear that social holism, construed in terms of I–thou relations, is no threat to individualism in the common and garden sense of that term which is implied when, for instance, one speaks of modern Western civilization being characterized by individualism. On the contrary, far from compromising individualism, the claim is that social relations are necessary for a human being to become an individual personality. In the terms of the proposed general conception of holism, we can say that with respect to her being an individual personality a human being is ontologically dependent on there being other humans with whom she constitutes a social community by engaging in a social practice of assessing each other's actions as correct or incorrect.

---

43 Pettit (1996), Chapters 3 and 4. See also Pettit (1998d). However, the structure of the book of Pettit (1996) differs from the way of proceeding embraced in this section: Pettit (1996) first outlines a solution to the problem of rule-following in Chapter 2 and then turns to the issue of holism in Chapters 3 and 4, whereas in this work on holism I begin directly with considering an argument for social holism.

However, Pettit construes social holism as a contingent thesis about the actual practices of us human beings.[44] The account outlined above closely follows Pettit's proposal; but it is couched in such a way that we can conceive of social holism as a thesis that applies to all finite thinking beings in all possible worlds. In restricting social holism to worlds in which human practices are like those in our world, Pettit accepts the argument of Blackburn that was rejected in the preceding section.

In the whole discussion of the argument for social relations being a necessary condition for a person to have beliefs with a determinate meaning, the situation in which a person does not satisfy this condition was described as a situation of a person *being considered* in isolation. This cautious formulation is to leave open the possibility that a physically isolated person may follow rules. Although physically isolated, the person can continue to be a member of a social community. The argument for social holism does not imply that if a person chooses to live as a hermit or if she is kept in solitary confinement, she no longer has beliefs. The condition "considered in isolation" refers to the absence of any social community of which the person in question can be considered a member.

However, what about the following scenario? Imagine that all human beings but one are annihilated in a nuclear war or a natural catastrophe. Consequently, there no longer is a social community of which this physically isolated human being can be considered a member. Does this human being continue to be a person that has beliefs? The argument for social holism under consideration leaves this question open. On the one hand, one can maintain that a human being in such a condition no longer has beliefs, because there is nothing which can be considered as an external standard for a judgement of the way in which this human being continues her purported rule-following. In whatever way she goes on, that way would be right so that the predicate "right" no longer applies. On the other hand, one can argue the following: Since this human being was a member of a social community before the catastrophe occurred, she has internalized the habits (*Gepflogenheiten*) of the community. She thereby acquired the ability to judge her own actions in the light of the norms of the community even if there is no feedback from other community members any more. Having internalized the norms of the community, she is able to continue to follow these norms and to judge herself whether her actions and dispositions accord

---

44 See Pettit (1996), in particular p. 179, and Pettit (1998d), step 5.

with these norms even after the imagined catastrophe.[45] For the step from reliable and discriminatory response-mechanisms to rule-following, actual feed-back from other persons is crucial. But once that step is achieved, one may argue that actual feed-back no longer is indispensable.

If one endorses this latter option, one has to qualify the proposed general conception of holism in the following way: For some holistic systems *S*, there is the – metaphysical – possibility of a development as a result of which the constituent parts of an *S* are no longer ontologically dependent on there being other individuals with which they constitute an *S*, insofar as they instantiate the properties in question (i.e., having beliefs and thus following rules in the case of thinking beings). That is to say: there is no possible world in which only one individual that has the properties in question exists. However, if a number of such individuals exist in a possible world and constitute a holistic system *S*, it is metaphysically possible that the system dissolves in such a way that only one of its former constituents remains and that this thing continues to have the properties in question.

Finally, consider a radicalized version of the plight of Robinson Crusoe: the metaphysical possibility of a human being that spends her whole life on an isolated island without ever having contact with other humans. Kripke is liberal as regards such a case. His position is that "*if* we think of Crusoe as following rules, we are taking him into our community and applying our criteria for rule-following to him". Kripke goes on to repeat that the theory which he ascribes to Wittgenstein is one of assertibility conditions.[46]

However, since the argument for social holism which I proposed on the basis of Kripke's account does not renounce facts of meaning and thus truth conditions for rule-following, we cannot be that liberal as regards the described radicalized case of Robinson Crusoe. What is more, it is questionable whether the assertibility conditions for rule-following are satisfied in this case.[47] One can maintain that, according to the view which Kripke presents, the assertibility conditions for rule-following include direct or indirect interaction with the person to whom rule-following is attributed;

---

45 I take Rhees (1966) to defend such a position. See in particular p. 279.

46 Kripke (1982), p. 110. Peacocke (1981), pp. 93–94 note 2, defends a community view that requires nothing more than a potential community relative to which a person can be considered as a rule-follower: "The community need not be that (if any) of the rule-follower himself."

47 See also the criticism of Kripke by Bilgrami (1992), pp. 114–118.

for the point of the assertibility conditions for rule-following is a social practice of exercising a check on each other's dispositions concerning what constitutes continuing a given sequence in the same way. To count as following rules, one has to take part in such a practice. No description from outside can replace this participation.

Consequently, in the case of a human being who spends her whole life isolated from *any* interaction with other human beings, neither the truth conditions nor the assertibility conditions for rule-following are satisfied. Such a human being exhibits of course discriminatory responses to her environment. Furthermore, her behaviour including her attitudes to her own behaviour can satisfy certain patterns that we may be inclined to describe in terms of rule-following. However, according to the argument under consideration, for such a human being there is nothing which determines for her a conceptual content for some of her mental states, because there is no distinction between correct and incorrect rule-following available for her.

3.3 An Argument for Social Holism and Holism about Beliefs

Having set out the argument for social holism, let us now come back to the relation between social holism and holism about beliefs. If holism about beliefs is construed as an inferential role semantics that conceives inferential role as a social role, holism about beliefs entails social holism. For such a semantics implies that being part of a social, linguistic community is a necessary and sufficient condition for having beliefs. Does this implication also hold the other way round? If one approves of a solution to the problem of rule-following in terms of social holism, is one thereby committed to holism about beliefs? One is committed to holism about beliefs if and only if one maintains that a person cannot follow just one rule.

If one takes it to be possible for a person to follow just one rule, one regards appropriate circumstances of the application of the rule to be sufficient for the rule to have conceptual content. For if we consider one rule in isolation, what the social practices of assessing each other's actions as correct or incorrect determine are circumstances for the application of the rule in question. In this case the objection against ostensive definitions which was mentioned in [2.4.1] applies: to take up the example of Brandom cited there, the rule would in this case function as a *gleebness* tester in the sense that if one masters the rule, one knows which things are *gleeb*; but it would not tell us what it is for something to be *gleeb*. Hence, if we consider one rule in isolation, social practices give us appropriate circumstances of the application of the rule, but not meaning. Considering one rule in isolation is therefore not sufficient for the proposed account in terms of social practices to be a solution to the problem of rule-following; for it is not sufficient to give us a conceptual content for the rule.

An additional step therefore is needed for the account proposed in [3.2.1] to be a satisfactory solution to the problem of rule-following: The *seventh step* is that *assessing each other's actions by means of sanctions determines a content for one rule only together with determining a content for an open-ended number of other rules; these other rules constitute an inferential context for the beliefs that are formed by following the rule in question*. Considering assessments on what constitutes continuing a given sequence of actions in the same way in isolation merely is a move for the sake of clarity of exposition. If we intend to form an intuitive idea of how the proposed account works, we should not imagine only one sequence of examples – such as a sequence of trees – but several sequences of examples. This makes the account more realistic: in a shared physical environment, people do not react to a well-formed sequence of examples in isolation, but to a number of various things and events with which they are confronted. The seventh step claims that such a context is required for the logical reconstruction of rule-following practices, because such a context provides the material for an inferential context that is necessary for a belief to have meaning.

Thus, social practices determine the meaning of a belief of the type "This is *F*" by determining a number of moves to beliefs of other types. Having a belief consists in mastering these moves and the appropriate circumstances of application. In other words, social practices determine an open-ended number of platitudes to which people in a social community are committed and entitled if they have a belief of the type "This is *F*". A novice acquires the concept *F* by learning these platitudes (and appropriate circumstances of application).

Hence, what is assessed by other people as correct or incorrect is not one sequence of actions of going on in the same way in isolation; such a sequence is rather assessed in a context of moves to actions of other types. The mentioned sanctions determine to which other claims a person is committed and entitled by making a claim of the type "This is *F*". It is in this way that these sanctions determine a content for a rule *F*. Consequently, *the argument for social holism and holism about beliefs is that the problem of rule-following requires us to conceive the conceptual content of belief states in such a way that this content originates in social interactions which determine a conceptual content for one type of belief states by determining inferential relations to other types of belief states*. But we now also have an argument for conceiving an inferential role semantics in such a way that it includes non-inferential language entries [2.4.1]: meaning in the sense of relations among beliefs requires being embedded in appropriate circumstances of application, because meaning can become determined only by social interactions on the basis of dispositions of persons to continue finite sequences of examples in certain ways in a shared physical environment. There is no meaning detached from these interactions.

According to this account, having beliefs is tied to speaking a language, because the determination of meaning is public: whatever beliefs a person has, to the extent that these beliefs have a determinate meaning, they can be expressed in a public language. As proposed in the third step of the account of rule-following, for something to be social behaviour, having and exercising a disposition to adapt one's dispositions to the dispositions of one's fellows is necessary and sufficient. However, for something to be social *interactions* or social *practices*, an intentional element over and above this natural disposition is required. Social interactions or practices are such that (a) they can be assessed as being correct or incorrect and (b) the beings who engage in them exercise and recognize such assessments among themselves. Social practices can only be individuated by means of normative, intentional vocabulary, namely by means of the vocabulary that describes the rule which a social practice determines. Any attempt to individuate social practices by a description of biological dispositions would fall victim to the indeterminacies which the rule-following problem brings out.

Which interactions are linguistic ones? An action is *linguistic* if not only reasons can be asked for it (as for any action), but also if it can be given as a reason.[48] A linguistic action in this sense does not have to be verbal; it can also consist in other gestures. According to this conception, assertible beliefs are the basic linguistic unit, because only they have the double function that reasons can be asked for them and that they can be given as reasons. All other linguistic units presuppose assertible beliefs. For instance, the imperative "Shut the door!" presupposes several beliefs that have a content which can be asserted, such as that there is a door, that the door is open, and that the door can be shut. We now see that the restriction to assertible beliefs [2.4.1] is justified within the proposed conception.

48 Compare Brandom (1994), pp. 171–173.

As regards communication with other people in the same social, linguistic community the proposed account has the following consequence: in ordinary cases of communication such as sharing the beliefs that it is raining or that the harvest is good this year, such communication is not interpretation in the sense of the formulation of hypotheses about what other people think.[49] A person has access to the beliefs of another person without interpretation in this sense, because the content of these beliefs *eo ipso* is a public property. McDowell maintains that

> shared command of a language equips us to know one another's meaning without needing to arrive at that knowledge by interpretation, because it equips us to hear someone else's meaning in his words. ... a linguistic community is conceived as bound together, not by a match in mere externals (facts accessible to just anyone), but by a capacity for a meeting of minds.[50]

This is to say: You do not have to translate the beliefs of another person into your own beliefs, since your beliefs and the beliefs of the other person are individuated by the same external factor – norms that are binding on both of you. Nonetheless, interpretation is required in another sense, namely in the sense of navigating across different perspectives in social interactions. For instance, if John utters "I've got toothache", a competent speaker has to report this claim as "John has got toothache". In this sense, as Brandom points out, what he calls deontic scorekeeping requires interpretation.[51]

Rule-following in practice does not have to be confined to one's mother tongue and one linguistic community. One can be fluent in two or more languages, that is, master in practice the rules of two or more languages and thus be a member of more than one social, linguistic community. It is only in this way that translation is possible. In order to be able to translate, one has to participate in the rule-following practices of another linguistic community. In this context, Lance and O'Leary-Hawthrone (1997, Chapter 1) make the noteworthy suggestion to regard translation not as a descriptive business, but as a practical, normative effort to merge two or more communities into one. It is merely for the sake of convenience in setting out the position under consideration that I speak in terms of one linguistic community. The point of social holism is that one has to be a member of *at*

---

49 As to this notion of interpretation, compare Wittgenstein, *Philosophical Investigations*, § 201, last paragraph, quoted in [3.1.2].

50 McDowell (1984), pp. 350–351. See also McCulloch (1995), pp. 149–155.

51 See Brandom (1994), pp. 185, 197, 475, 477–487, 508–520, 528–529, 586–592.

least one social community in order to have beliefs with a determinate meaning. In the last resort, it may be attractive to have the vision of all linguistic communities becoming intermeshed in their practices of rule-following so that the meaning of the beliefs of a person is determined by the practices of a community that includes all human linguistic communities.

Setting out holism in the philosophy of mind in a systematic order, we should begin with social holism: the assessment of each other's actions as correct or incorrect by means of sanctions is the origin of rules, norms, and meaning. The systematic order of exposition of the proposed holism in the philosophy of mind therefore is this one:

1) *rule-following*: We begin with dispositions of each person to continue a given sequence of examples or actions in a certain way. Rules are the intersubjective norms that result from mutual assessments of each other's actions as correct or incorrect. These assessments show up in sanctions (reinforcements and discouragements).
2) *normative pragmatics*: These assessments by means of sanctions determine the material inferences to which a person is committed and entitled and to which entitlement is precluded by making a claim of the type *p*.
3) *inferential role semantics*: The claims to which a person is committed and entitled and to which entitlement is precluded by making a claim of the type *p* under appropriate circumstances are the inferential role of a claim *p* and thus its meaning.

To conclude this chapter, let us cast the argument for semantic holism at the end of [2.4.2] into the terminology of rule-following so that we have a summary of the argument for both social and semantic holism in four steps:

1) The concept *F* is employed to form beliefs of the type "This is *F*" (e.g., "This is a stone"). Concepts as employed in beliefs are rules. A person masters a concept if and only if she follows a rule.
2) A person considered in isolation cannot follow rules. The content of the rules that she follows is determined by her interactions with other persons in a social, linguistic community.
3) In order to follow a rule, a person has to know (a) the circumstances in which claiming "This is *F*" is appropriate and (b) the inferences to which she is entitled and committed by claiming "This is *F*". She consequently has to be able to embed a sentence in which she uses the concept *F* into an inferential context. This implies that she has to master a number of rules in order to follow one rule.

4) The inferential context in which the meaning of a sentence of the type p consists is not closed by means of those sentences that p entails analytically (or by another means).

Consequently, in order to be a thinking being, a person has to interact with other persons in a social, linguistic community, and she has to have a number of beliefs which are connected by inferential relations within a whole system of beliefs.

CHAPTER 4
HOLISM AS REVISION OF THE CARTESIAN TRADITION

Summary

In this chapter, I start considering the broader implications of holism about beliefs and social holism for the philosophy of mind. The outlined holism about beliefs and social holism amount to a revision of the Cartesian tradition in epistemology. Two features of this tradition are relevant in this context: (a) a representational semantics including a representational realism and (b) the thesis of internalism or epistemic self-sufficiency of belief states [4.1]. As regards (a), it is shown how the proposed holism paves the way for a direct realism in contrast to a representational realism [4.2]. As regards (b), it is examined to what extent the proposed holism implies externalism as an alternative to epistemic self-sufficiency of belief states. According to this holism, the content of belief states depends on the social and the physical environment. In particular, social holism is a social externalism. It can moreover accommodate externalism with respect to singular thoughts and a type of direct reference. However, as far as an externalism such as the one of Putnam in "The Meaning of 'Meaning'" is concerned, I argue that there is a tension between social holism and this externalism. For, according to social holism, beliefs are individuated by social practices instead of features of the physical environment. Features of the physical environment of which the speakers may be ignorant can contribute to the individuation of their beliefs only on the basis of a stipulation of the community to that effect. Consequently, the relation between beliefs and the world has to be established from within these practices [4.3].

4.1 What is the Cartesian Tradition in Philosophy of Mind?

This chapter is to prepare the ground for the examination how the proposed social holism and holism about beliefs can take into account that beliefs represent the world and go with a realism instead of leading to social relativism. I use what is known as the Cartesian tradition in modern epistemology as a background to put the proposed social holism and holism about beliefs into perspective. Two features of the Cartesian tradition are relevant to that purpose: (a) its representational semantics and representational realism, and (b) its internalism or epistemic self-sufficiency of belief states. It will be shown holism about beliefs provides an alternative to (a), and to what extent social holism is an alternative to (b).

The Cartesian tradition is that tradition in philosophy that has been inaugurated by the work of René Descartes (1596–1650). In his *Meditations on First Philosophy* (1641) in particular, Descartes seeks to establish an indubitable foundation on which philosophy in the sense of a comprehensive system of knowledge can be built. He tries to find such a foundation by means of putting all his beliefs in question. This doubt leaves one belief untouched: the belief that while doubting, he is thinking. He puts forward the proposition "I think, therefore I am" ("*Cogito, ergo sum*") as the foundation for all knowledge. On the basis of this proposition, he comes to conceiving thinking as a property of a substance that is distinct from matter.

The reflection on language, in which analytic philosophy originates, can to a certain extent be seen as a radicalization of the Cartesian reflection on the conceptual conditions for having thoughts, which is peculiar to modern philosophy. Language can be regarded as a conceptual condition for having thoughts; this condition is neglected in Descartes and his successors. However, the point of the reflection on language in analytic philosophy is not to establish a foundation for our knowledge. In those analytic philosophers whose work was considered in the preceding two chapters, reflection on what it means to have beliefs is considered to yield holism about beliefs as well as social holism. All the forms of and all the arguments for holism in the philosophy of mind can be reconstructed on the basis of a reflection on what it is for a person to have beliefs. Thus, they all can be reconstructed on the basis of a specifically modern, Cartesian starting-point. In one of his recent papers, Davidson (1995) explicitly pursues such a strategy.

My point is only that the discussion on holism in analytic philosophy of mind can be taken to retain a methodological starting-point of Descartes' reflection; no claim is made about retaining substantial theses such as, for example, authority of the first person. Nonetheless, because of that common methodological starting-point, I prefer to describe the changes which holism about beliefs and social holism induce as a revision within the Cartesian tradition instead of regarding them as amounting to Anti-Cartesianism.

In today's discussion, the notion of the Cartesian tradition and its revision is employed in such a way that it includes much more than just Descartes' work. This discussion focuses on central features that dominate modern philosophy of mind down to logical empiricism. Both Descartes' ontological dualism of physical and mental substance and traditional materialism share these features. Hence, the point at issue is not Descartes' substance dualism.

The point at issue is an epistemology that is common to Descartes' substance dualism and its materialistic opponents down to logical empiricism.[1]

What is characteristic of this epistemology? It is innocuous to say that beliefs represent what there is in the world, i.e., that they represent things and events in the world (or facts or states of affairs that obtain in the world, whatever ontology one takes to be most plausible). In the same vein, it is innocuous to say that if a person is in a belief state, she is in a state of representing things and events in the world. This is just another way of saying that beliefs and belief states are about the world. However, *one leaves what is innocuous and enters into a substantial and controversial epistemology if one conceives the representational feature of beliefs and belief states in such a way that one considers representation to be a primitive semantic feature.* When I talk about representation, I always mean representation in an intentional sense. Representation in this sense is tied to conceptual content. I do not consider pre-conceptual forms of representation.

Representation in the described sense is a semantic feature: if something represents something else, it has a meaning, and *vice versa*. Representation is a primitive semantic feature if and only if it is not possible to give an explanation of representation in terms of other semantic features. This not to say that if one regards representation as a primitive semantic feature, one is committed to accepting representation without explanation. On the contrary, if someone claims that representation is a primitive semantic feature, it is reasonable to demand an account of representation in terms of other features than semantic ones. A relation of resemblance or a relation of causation between what is represented and the representation are candidates for such an account. The point is that relations such as resemblance or causation are not semantic relations.

It is not innocuous to regard representation as a primitive semantic feature, because semantic holism challenges this position. According to semantic holism, certain relations of one belief to other beliefs are a primitive semantic feature, namely inferential relations. If one considers representation to be a primitive semantic feature, it suggests itself to seek an explanation of inferential relations in terms of representation. By contrast, if one begins with inferential relations, the task is to give an account of the representational feature of beliefs in terms of these relations.

An epistemology that considers representation to be a primitive semantic feature is characteristic of the Cartesian tradition in the philosophy of mind.

---

1 For a helpful introduction into the whole complex, see McCulloch (1995).

Three theses have to be distinguished when it comes to elaborating such an epistemology:

1) a *semantic thesis*: Beliefs and belief states represent things and events in the world in such a way that representation is a primitive semantic feature. This leads to a *representational semantics*. One can group all those semantics under this heading which begin with relations between beliefs and the world, including truth-conditional semantics and semantics that are built on the notion of reference.
2) a *thesis about the epistemic access of a person to the world*: A person has epistemic access to the world in her belief states only by means of representations. (The epistemic access is the access that a person has to the world by being in belief states). Representations unlock the world for a person so to speak. They function as an epistemic intermediary between belief states and the world: a belief state is about something in the world in virtue of being related to a mental representation. The mental representation thereby intervenes between the belief state and its intentional object in the world. This thesis can be called *weak representational realism*.
3) a *thesis about the immediate intentional objects of belief states*: mental representations are the immediate intentional objects of our belief states including in particular our perceptual belief states. It is only by being immediately about mental representations that the belief states of a person are – mediately, so to speak – about the world. Thus, for example, in a situation of an ordinary and veridical perception of a tree, what the person in question is directly aware of is in fact a mental representation of a tree. She directly perceives mental representations and indirectly perceives through these representations things and events in the physical world.[2] To bring out the way in which this thesis is related to as well as distinct from (2), this thesis can be called *strong representational realism*.

The step that is crucial to get from (1) to (2) is to conceive the representational feature of belief states in terms of representations that open up the world for a person in her thinking so to speak. The step that is crucial to get from (2) to (3) is to conceive these representations as being themselves the immediate intentional objects of the belief states of a person. It is of no importance for the epistemological issue under consideration whether these

---

2 For a statement of the distinction between direct and indirect perception see Nadler (1989), p. 11.

representations are regarded as irreducibly mental items or whether they are taken to be identical with items in the brain.

If one endorses a representational semantics (1), one can go as far as tracing the propositional content of belief states back to some sort of primary content. In this case, this primary content is in itself a representation, and the content of belief states is derivative from this content. For instance, the mental representations which Fodor assumes are such a primary content.[3] Furthermore, the ideas of which Descartes and Locke speak in their epistemologies as well as the sense data which figure prominently in positivist epistemologies are commonly regarded as such a primary content. However, if one endorses a representational semantics, one is not committed to conceiving such a primary content. This semantics can go with the thesis of the primacy of beliefs in the sense that content in an epistemic sense is in the first place the propositional content of belief states.

A representational semantics is incompatible with semantic holism, because this semantics regards the representational feature of belief states as a primitive semantic feature. Such a semantics starts with a relation between beliefs and the world, and it seeks to derive relations among beliefs from that relation. This implies that the representational feature of one belief state is independent of other belief states. With respect to its representational feature, a belief is not ontologically dependent on there being other beliefs; otherwise, the relations to other beliefs would be the primitive semantic feature. Nonetheless, the representational feature of a belief may depend on a relation of resemblance or causation to what is represented. However, even if one maintains that this dependence is an ontological one, it does not yield any form of holism, because this ontological dependence is not symmetric: what is represented is itself, with respect to all the properties that are represented, not ontologically dependent on a relation of resemblance or causation to the representation.

In the tradition of modern philosophy, a representational semantics (1) often goes with a weak representational realism (2). The reasoning that leads from (1) to (2) can be reconstructed as follows: belief states have their representational feature as a semantic primitive by having a content that is in itself a representation of something in the world. For if the content of a belief state were not in itself a representation, but would consist in the relations to other belief states, these relations – and not the representational – would be

3 See e.g. Fodor and Lepore (1992), Chapter 4; Fodor (1998), Chapter 1, in particular pp. 7–8.

the primitive semantic feature. A representational semantics thereby invites us to distinguish between a belief state and its content in the sense that the content is a representation that acts as an epistemic intermediary between the belief state and its intentional object in the world: the belief state is related to its intentional object in the world by means of being related to a representation. Mental representations then have a double function: meaning inheres in them in the sense that they are the content of belief states, and they act as an epistemic intermediary between belief states and the world.

However, this reasoning is not compulsory: if one endorses a representational semantics, one is not committed to a weak representational realism. Consider, for instance, Aristotle's theory of perception and thought in books 2 and 3 of *De Anima*: if the act of perception or thought is the reception of a sensible or an intelligible form, then it may be possible to say that this act itself represents something in the world without the received form being distinguished from this act as an epistemic intermediary. In this case, we have a representational semantics, since the intelligible form is a concept; but this position is not a representational realism. It may be that this position is tied to an ontology of there being thought-like items in the world such as, for instance, forms of things that are at the same time concepts (compare Esfeld (forthcoming a)). Nonetheless, this example shows that a representational semantics does not simply imply a weak representational realism.

Weak representational realism (2) does not imply strong representational realism (3) either. Weak representational realism implies only that the representations in question *can* be made the intentional object of one's belief states. Otherwise this epistemology would be pragmatically inconsistent: it would talk about representations that could not be made the object of talk. But weak representational realism does not imply that mental representations *are* the immediate object of one's ordinary belief states. By contrast, strong representational realism just is the stronger thesis, because it makes this latter claim. It thereby implies weak representational realism.

An influential tradition in scholarship ascribes all three of the above mentioned theses – and thus both a representational semantics and a strong representational realism – to Descartes. The monograph of Anthony Kenny, for instance, is characteristic of this tradition (1968, Chapter 5). Richard Rorty relies also on this interpretation of Descartes when he builds his case against modern epistemology (1980, pp. 45–61). However, it is in dispute whether a strong representational realism can be attributed to Descartes. The recent study by Dominik Perler (1996), for instance, is an elaborate argument against interpreting Descartes along the lines of what I term strong representational realism.

When it comes to the issue of a revision of the Cartesian tradition in today's epistemology, strong representational realism may seem to be more like a caricature than a fair characterization of this tradition. For example, it is evident that strong representational realism moves the task of accounting for the aboutness of belief states only one step back: one does not account for the aboutness of belief states by saying that they are immediately about mental items. I therefore consider only a representational semantics and a weak representational realism to be a characteristic trait of the Cartesian tradition.

There is a further characteristic trait of the Cartesian tradition in modern epistemology that is relevant to an assessment of the impact of the proposed social holism and holism about beliefs: internalism or *epistemic self-sufficiency of belief states*. Belief states are epistemically self-sufficient if and only if the content of these states is ontologically independent of what the physical world is like. That is to say: the content of these states can be the same whatever the physical world is like – or even if there is no physical world at all. Epistemic self-sufficiency is implied by Descartes' doubt exercise in the *Meditations*: Descartes takes it for granted that the meaning of belief states inheres in ideas. He keeps the ideas and their identity fixed and doubts whether there is a physical world to which his ideas correspond.[4] Over and above that, Descartes also proposes ontological self-sufficiency of belief states by conceiving them as states of a mental substance.

Epistemic self-sufficiency of belief states is the same as *content internalism*: since belief states are individuated by their content, internalism can be formulated as the thesis that only factors which are immanent to a person enter into the individuation of her belief states. All those positions which do not commit themselves to realism, that is, all those positions which do not subscribe to the existence and epistemic accessibility of a world that is external to the person who is in belief states, are committed to internalism; they cannot make room for external factors entering into the individuation of the belief states of a person. The rejection of realism thus implies internalism and hence epistemic self-sufficiency of belief states. But internalism does of course not imply a rejection of realism.

---

4 Perler, however, goes in correspondence as far as disputing that the thesis of epistemic self-sufficiency of belief states can be ascribed to Descartes. Compare also Alanen (1992).

What is the relationship between on the one hand epistemic self-sufficiency of belief states and on the other hand a representational semantics as well as weak and strong representational realism? One can endorse both these positions. But they are logically independent of each other. As regards epistemic self-sufficiency of belief states, one can combine this position with an inferential role semantics of the type considered in [2.3], namely an inferential role semantics that takes beliefs to be individuated by their inferential relations in the system of beliefs of each single person.

Furthermore, one can accept a representational semantics including a representational realism and consider representations to be externally individuated (e.g. Dretske (1995), Chapters 2 and 5): representations are individuated by the manner in which they are caused by things or events in the physical world external to the person that has these representations. In this case, one takes a certain relation of causal dependence between things and events in the world and representations to be sufficient for ontological dependence. The task then is to pick out the appropriate causal relation. If one regards the belief states of a person, in turn, to be individuated by the representations that are their content, the move to epistemic self-sufficiency of belief states is blocked.

To sum up this section, two traits of what can be called the Cartesian tradition in modern epistemology are relevant to an assessment of the impact of the proposed social holism and holism about beliefs: (a) a representational semantics including a weak representational realism; (b) internalism or epistemic self-sufficiency of belief states. The first of these traits brings the Cartesian tradition in opposition to semantic holism; for semantic holism conceives not representation, but relations of one belief to other beliefs as a primitive semantic feature. The second trait is compatible with semantic holism: one can propose an inferential role semantics that restricts itself to the system of beliefs of single persons. However, if we base semantic holism on social holism, then we cannot go for internalism; for social holism conceives the belief states of a person as being individuated by features that are external to this person. In the following, I shall introduce first the alternative to representational realism that semantic holism opens up and then go into the externalism that social holism implies.

4.2 Direct Realism versus Representational Realism

On the one hand, it is a truism to say that beliefs and belief states represent what there is in the world. This is simply another way of saying that beliefs and belief states are about the world. On the other hand, one commits oneself to a position within what is regarded as the Cartesian tradition if one conceives representation as a primitive semantic feature and endorses weak representational realism. Inferential role semantics, by contrast, does not acknowledge representation as primitive semantic feature. Instead, it works with the inferences of a belief as primitive semantic feature.

In presenting the matter this way, I do not intend to suggest that there simply is a dichotomy between a representational and an inferential role semantics. There are, of course, also other positions. However, as already mentioned at the beginning of [3.1.1], the focus is on beliefs which are non-inferentially acquired. These are beliefs of the type "This is *F*", whereby *F* is a one place concept and a basic concept in the sense that it is not regarded as a function of other concepts. Examples of such beliefs are "This is square", "This is a stone", etc. The question is whether such beliefs have their representational feature as a semantic primitive or in virtue of inferential relations to other beliefs. My rationale for presenting the matter in terms of an opposition between a representational and an inferential role semantics is (a) to bring out the task which the latter has to meet (namely to explain representation) and (b) to elaborate on the way in which the latter departs from the mainstream tradition in modern epistemology.

One can go as far as denying the representational feature of beliefs or belief states. Davidson, for instance, regards the talk of representation as being committed to a correspondence theory of truth and argues that the notion of correspondence is not intelligible (1990, pp. 303–304). But if one, pace Davidson, simply takes up intentionality as characteristic of beliefs or belief states and thus maintains that beliefs or belief states are about the world, one does not thereby subscribe to such a controversial theoretical commitment. It is reasonable to make the following demand: if an inferential role semantics is to be a genuine alternative to a representational semantics – instead of changing the subject to making a case for relativism – it has to employ its conceptual tools to account for the representational feature of beliefs. For if one refuses to accept this demand, it can be objected that one comes close to the position of Rorty who maintains that there are only social practices, but no epistemic relations to the world.[5]

---

5 See Rorty (1980). In "Robert Brandom on Social Practices and Representations" in Rorty (1998), pp. 128–130, Rorty praises Davidson for abandonning the notion of representation.

To prepare the terrain for tackling this demand in the next chapter, let us go into the relation between an inferential role semantics and realism. A *realism* about the physical world is any position according to which (a) there is a physical world that is ontologically and causally independent of our beliefs and (b) whether or not our beliefs about the physical world are true depends on the way the physical world is. A *representational realism* is the thesis that we have epistemic access to the physical world only by means of representations. Representations function as an epistemic intermediary between our belief states and the things and events in the physical world about which our belief states are.

By contrast, *direct realism* is the thesis that there are no epistemic intermediaries between belief states and the things and events in the physical world that are the intentional objects of these states. Of course there are causal intermediaries between a belief state and what the belief state is about. For instance, if one has the true perceptual belief that there is a sparrow in the tree, light, sensory stimuli, nerve signals, brain processes, etc. are causal intermediaries between the sparrow in the tree and the belief state that there is a sparrow in the tree. These factors belong to the causes of the event that a person acquires the perceptual belief that there is a sparrow in the tree. These factors make it that the belief in question is a perceptual belief, that is, a belief acquired in perception and thus non-inferentially acquired. However, when it comes to the epistemic relation of the perceptual belief being about the sparrow in the tree, there is nothing that intervenes between this belief state and the sparrow in the tree. The mentioned causal factors are a necessary (and perhaps a sufficient) condition for that epistemic relation to obtain; but they are not a *relatum* within that relation. The belief state is directly related to its intentional object, the sparrow in the tree.[6]

Whereas a representational semantics invites a representational realism, an inferential role semantics suggests a link with a direct realism. According to an inferential role semantics, the content of a belief state are its relations to other belief states. There is hence no room for a distinction between a belief state and a representation as its content. Consequently, nothing hinders that belief states are directly about things and events in the world. This is not to say that an inferential role semantics implies a direct realism. The claim only is that if one endorses such a semantics, the reason for admitting epistemic intermediaries disappears.

---

6 Compare Sellars, "Empiricism and the Philosophy of Mind", § 29, pp. 160–161 in Sellars (1963).

The major objection against a representational semantics and a representational realism is this one: Despite representation being regarded as a primitive semantic feature, an account of representation is called for in terms of other features than semantic ones. A relation of resemblance or a relation of causation between representation and what is represented are candidates for such an account. Attempts in terms of resemblance are widely acknowledged as being doomed today. Attempts in terms of causation face the following three difficulties: (a) What is the appropriate causal relation? Any causal chain between the world and a belief state comprises many links. How can an account in causal terms pick out the right link as the object of the representation? Moreover, a representation of a certain type is caused not only by the represented object, but by many other things as well. Not only cats, but also cat pictures and even dogs and indefinitely many other things may cause a cat representation. (b) How can a causal relation be that fine-grained that it is able to introduce a distinction between reference (extension) and meaning (intension)? Such a distinction, however, is indispensable: to recall Frege's classical example, the beliefs "The morning star is *F*" and "The evening star is *F*" have the same reference, but differ in meaning.[7] Given that Wittgenstein, Quine and others make a strong case against meanings as mental items or as abstract senses, it is a major argument in favour of an inferential semantics that such a semantics introduces a meaning in distinction from reference in terms of an inferential context without falling back into any sort of a mentalism about meanings. (c) How can a causal relation, however reliable it may be, make up the epistemic feature of representation at all and determine a representational content?

To further elaborate on (c), consider what Wilfrid Sellars criticizes as the "Myth of the Given" in his seminal paper "Empiricism and the Philosophy of Mind" (1956; quoted as in Sellars (1963)). (I put "Given" in capital letters whenever the specific sense in which Sellars uses this term is intended). By the Given, Sellars means anything that is on the one hand regarded as having an epistemic status, i.e., being some sort of knowledge, without on the other hand requiring mastery of concepts.[8] What is more, this knowledge is often considered to be a foundation for and ultimate justification of other knowledge. The Myth of the Given then includes the idea that there is a

---

7 "Über Sinn und Bedeutung" in Frege (1986a) (English translation in Frege (1984), pp. 157–177).

8 §§ 1–7, pp. 127–134 in Sellars (1963) and compare Brandom (1997), p. 122. See also Bonjour (1985), Chapter 4, in particular pp. 69, 75; Tuomela (1988).

foundation for our knowledge which is not in need of justification itself, i.e., for which it is never appropriate to demand a reason. Focussing on this aspect, McDowell (1996) says that "The idea of the Given is the idea that the space of reasons, the space of justifications or warrants, extends more widely than the conceptual sphere" (p. 7; see the whole lecture 1).

Sellars attacks in the first place a conception according to which sense data figure in perceptual beliefs in an epistemic sense. His main point is that the following two functions which are attributed to sense data are incompatible: On the one hand, sense data are regarded as being given without any concepts being involved. Perceptual beliefs are the result of applying concepts to sense data. On the other hand, an epistemic status is accorded to sense data: perceptual beliefs consist in applying concepts to sense data, and being in the mental state of having a sense datum can be the same thing as being in the epistemic state of knowing that one has a sense datum. According to Sellars, attributing these two functions to sense data is on a par with the naturalistic fallacy in ethics that seeks to deduce ought-sentences from is-sentences: one cannot deduce epistemic relations from the description of a causally induced mental happening (in particular § 5 (pp. 130–131) and also §§ 7 (pp. 132–133), 16 (pp. 144–145), 29 (pp. 160–161). This attack hits a representational semantics and a representational realism, if such a position includes a causal account of representation that accords an epistemic status to causally induced items by considering them as representations without mastery of concepts being required.

According to Sellars, if there are sense data at all, they are relevant to perceptual beliefs only in a causal sense. Experience as an epistemic phenomenon is conceptual all the way down. Being conceptual means for Sellars being linguistic, because being conceptual means having a position in the game of giving and asking for reasons of a social community (§§ 29 (pp. 160–161), 31 (pp. 162–164)). Sellars offers a social, inferential role semantics including a direct realism as alternative to what he considers as the Myth of the Given.[9] Sellars characterizes this position as psychological nominalism (§ 29, p. 160).

In his famous essay "On the Very Idea of a Conceptual Scheme" (1974),[10] Davidson takes up Sellars' theme of "Empiricism and the

---

9 In particular §§ 16 (pp. 144–145), 19 (p. 148), 31 (p. 164), 35–36 (pp. 167–169). See also the commentary of Brandom (1997), in particular pp. 139–144, 157–158.

10 In Davidson (1984), essay 13. For a detailed examination of what Davidson understands by scheme and content, see Child (1994a).

Philosophy of Mind" without, however, mentioning Sellars. In this essay Davidson attacks what is according to Sellars a version of the Myth of the Given: the dualism between a conceptual scheme and a given content to which the scheme is applied (whereby the content is not conceptual content); perceptual beliefs are the result of applying a conceptual scheme to a given content. A conceptual scheme being applied to a given content is an epistemic intermediary between belief states and the items in the world about which belief states are. Davidson repudiates any epistemic intermediary between our belief states and the world.[11] He considers the dualism of scheme and content to be the third dogma of empiricism.[12] Davidson maintains that, when it comes to epistemic matters, there neither is a given content nor a conceptual scheme. As in the case of Sellars' "Empiricism and the Philosophy of Mind", Davidson's rejection of the dualism of scheme and content opens the way for a direct realism.[13] It does not mean repudiating the notion of a reality that is ontologically and causally independent of our beliefs.[14] In Davidson's terms: "In giving up the dualism of scheme and world, we do not give up the world, but re-establish unmediated touch with the familiar objects whose antics make our sentences and opinions true or false."[15]

Whereas Sellars' "Empiricism and the Philosophy of Mind" can be employed to explain how perceptual beliefs fit in the proposed social holism and holism about beliefs, Davidson's main concern is the interpretation of a speaker with whom one does not share a language. The point of Davidson's essay as regards the further elaboration of the proposed social holism and holism about beliefs is this one: If belief states directly relate to the world, there is no *a priori* limit to communication.[16] No foreign conceptual scheme

---

11 "On the Very Idea of a Conceptual Scheme" in Davidson (1984), p. 198; Davidson (1983), p. 429.

12 "On the Very Idea of a Conceptual Scheme" in Davidson (1984), p. 189. On the dualism of scheme and content being a version of the Myth of the Given, see also McDowell (1996), lecture 1, as well as Rorty, "The World Well Lost" in Rorty (1982), pp. 3–5. As regards Quine's defense of what Davidson calls the third dogma, see "On the Very Idea of a Third Dogma" in Quine (1981), essay 4.

13 On reading Davidson as a direct realist, compare Thompson (1991), pp. 222–223. See also Engel (1994), Chapters 5 and 6.

14 See, by contrast, Rorty, "The World Well Lost" in Rorty (1982), pp. 3–18, as well as Hurley (1992), pp. 98–99.

15 "On the Very Idea of a Conceptual Scheme" in Davidson (1984), p. 198.

16 See also Pettit (1996), pp. 210–211.

hinders gaining access to the meaning of the beliefs of the members of another social community. These people may have dispositions that are widely different from ours, and communication with them may therefore fail. But the point is: once the Myth of the Given is rejected, there is no *a priori*, conceptual limit for getting into the practices of another community. In his theory of radical interpretation, Davidson then sets out methodological principles for gaining access to the belief systems of members of another community; charity counts prominently among these principles.[17]

4.3 Holism and Externalism

4.3.1 Social Externalism

Having considered how holism about beliefs makes available a direct realism in contrast to a representational realism, let us go into the second of the two mentioned features of the Cartesian tradition in epistemology, namely internalism or epistemic self-sufficiency of belief states. What is the link of social holism with externalism in contrast to internalism?

The externalism at issue is *meaning* or *content externalism*: The beliefs of a person are individuated by factors which are external to that person. This sort of externalism is to be distinguished from justificatory externalism. This is the claim that the beliefs of a person can be justified by factors which are external to the person and of which she is not aware (such as suitable causal relations which lead to acquiring the beliefs in question); consequently, a person can have justified beliefs without being able to justify these beliefs herself. I shall be concerned only with externalism about meaning.

Social holism implies *social externalism*. This is the thesis that the beliefs of a person are at least in part individuated by her social environment, because the meaning of the beliefs of a person depends at least in part on her social environment. In today's discussion, social externalism is often associated with the argument of Tyler Burge.[18] Burge's argument for social externalism is not as fundamental as the argument from rule-following in [3.2.1], because it is not about belief states being threatened to be left with no content at all. Burge's method is to develop thought experiments in which he keeps all the factors which are internal to a person fixed and varies the

17 "On the Very Idea of a Conceptual Scheme" in Davidson (1984), essay 13, pp. 195–197. See also essays 9 to 12 in Davidson (1984); Davidson (1983), pp. 431–438.

18 See, in particular, Burge (1979). For a recent exposition and defense, see Collin (1998). For an argument against Burge, see Antony (1993).

social environment. He thereby intends to show that the social environment takes part in determining the meaning of the beliefs of a person.

Burge develops social externalism in opposition to a position that he calls "individualism". The choice of that term is debatable. In any case, if social externalism is conceived along the lines of the argument presented in [3.2.1], it is, as mentioned in [3.2.3], not opposed to individualism in the common and garden sense. Since the argument for social holism developed in [3.2.1] is an argument for social externalism, I shall not continue to elaborate on social externalism.

4.3.2 Externalism about Singular Beliefs

When it comes to externalism in today's philosophy, the focus of attention is not only social externalism, but also externalism about the physical environment. One form of this latter externalism is the following position: the paradigmatic example of singular beliefs is beliefs that employ an indexical or a demonstrative in order to refer to a particular thing in the world. It is a necessary condition for having such a belief that one stands in a certain causal relation to the thing in question. Thus, having singular beliefs about a certain thing is ontologically dependent on the existence of the thing in question. For instance, one can have the belief that the man over there is fat only if there is a man over there. The thesis hence is that some beliefs, namely those singular beliefs in which indexicals or demonstratives are employed, are such that a person can have these beliefs only if their purported referent exists. The main argument is that if the purported referent of these beliefs did not exist, these beliefs would not be about anything at all; consequently, they would not be beliefs at all. This position owes much to Gareth Evans' *Varieties of Reference.*[19]

Externalism about singular beliefs implies the following: If I can have the belief that the man over there is fat only if there is in fact a man over there, then, for any singular belief considered in isolation, I may doubt whether I am really in the state of believing that, say, the man over there is fat or whether I am in a state which is subjectively indistinguishable from this state, but possibly not a belief state at all. As far as singular beliefs that employ indexicals or demonstratives are concerned, my belief states are hence not transparent to me.

---

19 See Evans (1982), in particular Chapters 6 and 9. See furthermore McDowell (1986) and McDowell (1998), lecture III, pp. 471–491; Burge (1986); McCulloch (1989), Chapter 6; Pettit (1996), pp. 44–45, 191.

Externalism about singular beliefs can be considered as an *externalist* alternative to the view that reference relies on a causal connection in distinction to exercising a conceptual capacity, which is commonly traced back to Kripke,[20] as well as to the view that there is a direct reference in the sense of a reference which does not depend on the conceptual content of the belief in question; David Kaplan has introduced the latter view of direct reference.[21] Externalism about singular beliefs accommodates a causal connection. But it avoids a dualism between – direct – reference as determined by a causal relation and conceptual content as representation (in the sense of a representational semantics). For it considers the conceptual content of a singular belief as being itself dependent on the existence of the referent.[22]

Therefore, this sort of externalism can be employed to vindicate a thesis to which holism about beliefs is committed: what makes something an epistemic phenomenon such as a belief is conceptual content all the way down to singular terms including indexicals and demonstratives. By being committed to this thesis, holism about beliefs is incompatible with theories that take causal relations to be sufficient for reference insofar as these theories imply that the causal relations in question are sufficient to individuate some beliefs. As described in [2.4.2], Brandom (1994) develops an account of the contribution that singular terms including indexicals and demonstratives make to the inferential role of a sentence by basing himself on relations of substitution and anaphora. Brandom then employs anaphoric relations to elaborate on externalism about singular beliefs as an alternative to the mentioned views of reference (pp. 551–552, 567–583, 616–617).

However, the proposed social holism and semantic holism is not committed to externalism about singular beliefs. That holism and this externalism do not imply one another. Basing oneself on an inferential semantics, one can say that for a singular belief to have content, relations to other beliefs are sufficient. This leaves room for some singular beliefs having content even if an appropriate referent is missing. As Akeel Bilgrami argues, such a view of singular beliefs is not committed to a description

20 Kripke (1980). See also e.g. Devitt (1985), who makes clear in what respect Evans (1982) develops an alternative to Kripke (1980) (and who then argues against this alternative).

21 See, in particular, Kaplan (1989a), pp. 483–486, and Kaplan (1989b), pp. 568–569. See also Recanati (1993).

22 See Evans (1982), pp. 60–63; McDowell (1986), pp. 161–168.

theory of reference (1992, pp. 159–177). (However, the holism and externalism for which I argue in this book is distinct from the holism and externalism that Bilgrami sets out). Even if such a view were committed to a description theory of reference, it is by no means clear that such a theory is not defensible. Jackson (1998c) develops a defense. Given the topic of this work, I have to restrict myself to these sketchy remarks on the philosophy of reference here. Let us simply note that the revision of the Cartesian tradition which the proposed social holism and holism about beliefs entails is not such that this holism is committed to externalism about singular beliefs, though it can be combined with such an externalism.

4.3.3 Externalism about Natural Kind Concepts

The form of externalism that provokes most discussion in today's philosophy is externalism with respect to beliefs in which natural kind concepts are employed. The thesis is this one: The meaning of beliefs in which concepts such as "water" or "tiger" are employed depends on the physical constitution of the things referred to. That is to say: These things have a real essence, which is independent of our beliefs about these things; their physical constitution is their real essence. The real essence of the things referred to determines the meaning of our beliefs about these things. If two things are referred to by the same statements, but do not have the same real essence, then these statements have different meanings – whatever else the two things may have in common and whether or not there is any difference in the use of the statements of these two types.

The main argument for this sort of externalism goes back to Putnam's famous thought experiment in "The Meaning of 'Meaning'" (1975): Independently of whether or not the members of a linguistic community know that water is H_2O, it is part of the meaning of their beliefs about water that water is H_2O. On the imaginary twin earth there is a stuff that is indistinguishable from water on the surface, but its microphysical composition is *XYZ*. The meaning of all beliefs about that stuff is different from the meaning of all beliefs about water. Putnam's thesis is that natural kind terms like "water" have an indexical component: By pointing to a liquid and saying "This is water", a person means that everything which is the same as the stuff pointed to also is water. That relation of sameness is determined independently of the way in which the person uses the term "water", namely by the real essence of the thing referred to; the real essence is its microphysical composition.[23] If one did not assume that there is a real essence which determines the relation of sameness, this suggestion would fall victim to the objection against ostensive definitions [2.4.1]: it would not be clear what is picked out by the ostension.[24] Putnam, however, also contemplates another lesson of this thought experiment that he repudiates: meaning (intension) fails to determine extension. That is to say, the meaning of "water" is the same on both earth and twin earth, but the extension is H_2O on earth and *XYZ* on twin earth.[25]

Whereas the proposed holism about beliefs and social holism imply social externalism and can be combined with externalism as regards singular beliefs, the externalism which Putnam sets out is a challenge to social holism. Recall that according to Kripke's reconstruction of Wittgenstein's argument, there is nothing mental – such as ideas or the grasp of abstract senses – which determines the meaning of the beliefs of a person. The solution to the resulting sceptical paradox is that a social practice determines the meaning of the beliefs of a person to the extent that meaning is determined at all. Externalism about beliefs in which natural kind concepts are employed can be seen as attempting an alternative solution to this problem: it is not anything mental which determines the meaning of the beliefs of a person, but the physical environment; physical things have a real essence, and the meaning of our beliefs about them is determined by this real essence. The person herself need not fully grasp the content of the rules which she follows. The essence of physical things can determine the content of the rules which she follows, even without the person herself being able to

---

23 See Putnam (1975), essay 12, in particular pp. 219–235. See also "Is Semantics Possible?" (1970) in Putnam (1975), essay 8, pp. 150–151. For an elaborate exposition, see McCulloch (1995), Chapters 7 and 8. I take up the term "real essence" from McCulloch. Putnam speaks of the nature of water, e.g., p. 233, and of "essential nature" in "Is Semantics Possible?" (1970) in Putnam (1975), essay 8, pp. 140–141. But see also "Possibility and Necessity" in Putnam (1983), pp. 63–64, and, later, "Is Water Necessarily H_2O?" in Putnam (1990), Chapter 4. In the terminology of McGinn (1989), pp. 7–9, Putnam's example is a case of strong externalism in distinction to weak externalism. The whole argument in this section is about strong externalism in McGinn's sense. McGinn (1989), pp. 30–36, endorses strong externalism with respect to cases such as water. As to the discussion which Putnam's claims started, see the essays in Pessin and Goldberg (1996).

24 Papineau (1979), pp. 158–162, focuses on the problem of determining a relation of sameness in this case without, however, considering the assumption of a real essence.

25 "The Meaning of 'Meaning'" in Putnam (1975), essay 12, in particular pp. 219–235.

distinguish between some different rules (such as the water-rule and the twater-rule) in her actions including her assessments of the actions of other persons.

Consequently, the proposed social holism and the externalism under consideration give different accounts of the meaning of some beliefs. According to this externalism, the meaning of the water-beliefs of a person is determined by the microphysical composition of the stuff to which she is causally related. The meaning of the respective beliefs of the people on twin earth is determined by the microphysical composition of the stuff to which they are causally related. Hence, although the social practice in question is indistinguishable on earth and on twin earth, the beliefs in question of the people on earth are different from the respective beliefs of the people on twin earth, because the hidden physical structure of the environment in which they live is different (assuming that both communities are ignorant of the microphysical composition of water and twater). This alternative to the account of rule-following as a social practice applies to beliefs about physical things in the first place. If it is to apply to other beliefs as well, it has to be shown how their objects have a real essence too that could carry out the required determination.[26]

Within the framework in which Putnam sets his argument out in "The Meaning of 'Meaning'", we consequently have to say the following: the proposed social account of meaning implies that meaning (intension) fails to determine reference (extension). For the social practices that determine meaning are the same on earth and twin earth, whereas the people on earth are causally related to water and the people on twin earth are causally related to twater. One may, however, have reservations about the view of reference that Putnam presupposes in "The Meaning of 'Meaning'".

If we maintain that real essences of physical things determine meaning via a suitable causal relation, then we cut off not only reference, but also meaning from our practices. Given Putnam's thought experiment, it is legitimate to imagine the following scenario: what applies to water applies to all beliefs of the people on earth about the physical things with which they deal in everyday life. There is a twin earth which is exactly like the earth in its macroscopic features, but entirely different from it in its microphysical composition. This is surely not a physical possibility; but that is irrelevant to the thought experiment. Both the people on earth and the people on twin

26 But see Putnam, "The Meaning of 'Meaning'" in Putnam (1975), essay 12, pp. 242–245.

earth are ignorant of the microphysical composition. Assume that a person from earth is transported to twin earth. This person becomes perfectly integrated into all the social practices of the people on twin earth. Nevertheless, according to the externalism under consideration, we have to say that the meaning of all the beliefs of this person about the physical things on twin earth is different from the meaning of the beliefs of the natives about these things.

Cutting off meaning from our practices does not answer Kripke's challenge. When assessing this challenge at the beginning of [3.1.2], it was claimed that assuming natural properties which are instantiated in physical things does not solve the problem. The point is that any finite sequence of examples instantiates more than one natural property. This objection does not apply to the externalism under consideration. One may say that not any natural property qualifies as a real essence; real essences are only those natural properties that make something a sortal in the sense of a natural kind. Consequently, it may be possible to maintain that there are finite sequences of examples which instantiate only one real essence. One may then say that the real essence which a finite sequence of examples instantiates objectively determines one correct way in which this sequence ought to be continued, and, owing to suitable causal relations, the real essence determines one correct way in which a person ought to continue a finite sequence of actions.

Nevertheless, the normativity problem remains. Recall that the problem is not simply what determines a correct way to continue a finite sequence of examples or actions, but what determines this in such a way that a person can *follow* a rule. Real essences in the world do not solve this problem: How does a person grasp a content in applying beliefs of the type "This is *F*" to each of the members of a finite sequence of examples in such a way that for her there is a distinction between correct and incorrect rule-following? Is grasping a real essence a mental act? Is it a disposition to behaviour? In both cases, Kripke's objections apply. Qua rule-following subject, a person cannot be programmed by whatever causal relations to be attuned to the real essence of reality in her intentional states.

Consequently, the externalism under consideration does not refute the following claim: to the extent that there is a solution to the problem of rule-following at all, it is a solution in terms of social practices. However, if we have to work with these practices anyhow, then we have something that is sufficient to do the job of determining the content of and individuating our belief states. We do not need the assumption of natural properties and real essences as far as the theory of the meaning and the individuation of our beliefs is concerned.

Nonetheless, beginning with the proposed social holism, we can accommodate the externalism under consideration. In a later essay, "Is Water Necessarily H_2O?", Putnam says:

> I won't insist (any more) that "it is conceivable that water may turn out not to be H_2O but it isn't logically possible that water isn't H_2O." ... I still believe that a community can stipulate that "water" is to designate *whatever has the same chemical structure* or *whatever has the same chemical behavior* as paradigms X, Y, Z ... (or as most of them, just in case a few turn out to be cuckoos in the nest) *even if it doesn't know, at the time it makes this stipulation, exactly what that chemical structure, or exactly what that lawful behavior, is.* (Putnam (1990), p. 70)

According to this quotation, the microphysical composition of water determines the meaning of the beliefs about that liquid only as a result of a stipulation of the community to that effect.[27] Consequently, it cannot be the constitution of the physical environment as such which determines the meaning and individuates the beliefs of a person. Instead, persons can stipulate in their social practices that the meaning of their beliefs about physical things is to be determined by the microphysical composition of the things referred to, whatever that composition may be. If persons make such a stipulation, they are not committed to attributing to microphysical composition the status of a real essence. Hence, by making such a stipulation, they are not committed to the claim that there are real essences.

Being H_2O thus is part of the meaning of beliefs about the liquid in question (picked out by what Putnam calls a stereotype)[28] only if the social practices are such that the members of the community in question regard their beliefs about physical things as being individuated not by the macroscopic features with which they are acquainted in their practices, but by whatever the microphysical composition of the physical things may be.. Supposing that both the people on earth and the people on twin earth are ignorant of the microphysical composition of the liquids in question, there is

27 See also the suggestions of Brandom (1994), pp. 119–120, 331–332; McCulloch (1995), pp. 171–174, 187–190, 210–211; Ebbs (1997), Chapter 7; and compare already the hint in Burge (1982), p. 119 note 8.

28 See "Is Semantics Possible?" (1970) in Putnam (1975), essay 8, in particular pp. 148, 150. See also "The Meaning of 'Meaning'" (1975) in Putnam (1975), pp. 229–230, 249–252.

a difference in the meaning of the beliefs about the respective liquids only if the members of at least one of these communities consider the meaning of their beliefs about physical things to be determined by whatever the microphysical composition of these things is. Consequently, the proposed social holism can integrate the externalism under consideration, but only by insisting that it is in fact social practices that determine the meaning and individuate the beliefs of persons.

However, one can with reason doubt whether in fact the members of any community can be expected to intend their practices to be such that not plain macroscopic features, but a hidden physical microstructure determines the meaning of their beliefs about the physical things with which they deal in everyday life.[29] It is plausible to maintain that the sketched scenario applies to a scientific community, because it is an explicit aim of scientific research to discover the microstructure of physical things; but this is not an aim of our common sense talk about the physical things with which we deal in everyday life.

This reservation suggests the following alternative account of Putnam's case: prior to the discovery of the microphysical composition of the liquid in question, it is indeterminate whether the beliefs about the liquid mean that the liquid is H_2O or that it is *XYZ*. The rules of the respective communities do not determine this, because it is beyond the scope of their practices (see [3.2.2]). As a result of the discovery of the microphysical composition of the liquid, the meaning of the beliefs about the liquid changes in such a way that being H_2O or being *XYZ* is included in the meaning of the beliefs in question. Hence, prior to the discovery of the microphysical composition, there is no difference in meaning between the beliefs of the people on earth and the beliefs of the people on twin earth about the liquid in their respective environments. Microphysical composition becomes relevant to meaning only upon discovery.[30] One may go even one step further and maintain the following: "water" is a role concept. It stands for whatever realizes the role of being a liquid that is odourless, that quenches thirst, etc. Thus, on earth water is realized as H_2O, but it can also be realized as *XYZ*.

However that may be, the thread of my argument does not hang on whether one incorporates the externalism under consideration in the proposed social holism in the sketched way or whether one employs this holism to reject that externalism. In any case, the proposed holism is at most

29 Compare, for instance, Searle (1983), pp. 203–204; Mellor (1996), §§ 3–4.

30 See, by contrast, Putnam, "A Problem about Reference" in Putnam (1981), pp. 23–25.

opposed to the specific sort of externalism considered in this section. This holism implies a revision of the Cartesian thesis of epistemic self-sufficiency of belief states. It includes not only a social, but also a physical externalism. Nonetheless, it is committed to an externalism about the physical environment only insofar as such an externalism is entailed by social externalism. Recall the assumptions about similar dispositions of people in a shared physical environment which enter into the proposed reconstruction of rule-following [3.2.1]. These assumptions imply that rule-following as a whole – and thus the beliefs of a person as a whole – ontologically depend on a physical environment. But this position is not committed to pinning this general dependence down to a dependence of specific types of beliefs on specific features of the environment, since the meaning of beliefs consists in inferential relations to other beliefs. The rule-following considerations imply that the task of content determination and belief individuation has to be accomplished from within our social practices. Consequently, externalism about the physical environment cannot discharge the proposed holism of giving from within these practices an account of how beliefs can be about the world and be answerable to the way the world is. Such an account is the topic of the next chapter.

CHAPTER 5
HOLISM AND THE RELATION BETWEEN MIND AND WORLD

Summary

This chapter reconstructs the relation between beliefs and the world from within social holism and holism about beliefs. I proceed in three steps. The first step is to employ the conceptual tools of the outlined inferential semantics in order to explain the representational dimension of beliefs. I follow Brandom's account [5.1]. The second step is to argue that the proposed theory of meaning in terms of social practices is about the way in which we gain epistemic access to a world that is independent of these practices. According to the conception of social holism in terms of I–thou relations, the assessment of normative attitudes as correct or incorrect is an open-ended process. Consequently, normative attitudes that are shared by all the members of a community can be incorrect, and having beliefs can result in commitments that outrun the commitments which the persons in question acknowledge. This account implies a sort of response-dependence of our concepts. But this is a response-dependence that encroaches neither upon the meaning nor upon the truth of our beliefs. It is a constraint only on the acquisition of concepts. The upshot is a pragmatic realism that is beyond the traditional opposition between metaphysical realism and relativism. This pragmatc realism is not committed to any idealistic consequences [5.2]. The third step is to meet the challenge that McDowell poses to an account of meaning in terms of social practices by showing how social practices are anchored in a physical environment. I argue that McDowell's position opens up a comprehensive holism that includes the physical world. However, I maintain that the described social practices are sufficient to conceive a rational constraint of the world on our beliefs [5.3]. Finally, I show that the resulting philosophy of mind is parsimonious in its ontological commitments: it is opposed to reductionism, but compatible with materialism without, however, implying materialism [5.4].

5.1 From Inference to Representation

According to an inferential semantics, given appropriate circumstances of application, the content of a belief of the type *p* are inferential relations to beliefs of the types *q*, *r*, *s*, etc. How can such a semantics account for representation on the basis of inference? Let us come back to the inferential semantics that Brandom develops by starting from a normative pragmatics [2.4]. Let us follow Brandom in his account of representation. Recall that we are allowed to work with inferential relations among sentences that express beliefs, because, as far as conceptual content is concerned, the position under consideration is committed to the view that beliefs are on an equal footing with sentences expressed in a public language. Brandom's thesis is that a belief of a certain type represents things and events in the world by virtue of its normative relations to beliefs of other types.

Brandom (1994) accounts for the contribution that singular terms and predicates make to the meaning of a sentence by the manner in which they function in substitution. Furthermore, he conceives the contribution that unrepeatable tokenings such as demonstratives and indexicals make to the meaning of a sentence in terms of their anaphoric relations to other such tokenings (Chapters 6 and 7; see [2.4.2]). Brandom explains the reference to objects as well as what is predicated of objects by means of these relations. All those singular terms whose substitution results in reversible inferences form one equivalence class. Brandom's claim is that an equivalence class of singular terms represents one individual. Furthermore, indexicals and demonstratives refer to an individual owing to their anaphoric relations. Over and above that, Brandom gives an account of "true" and "refers to" in terms of anaphora too: he suggests regarding "true" as a an operator to construct a particular kind of prosentence and "refers to" as an operator to construct a particular kind of pronoun.[1]

The argument is that a singular term taken in isolation – or a sentence in which such a term is employed taken in isolation – does not represent anything in the sense of being a representation with a determinate conceptual content, because it is in fact not more than a mere response to stimuli. Such a response may be reliable and differentiating. But that is not enough for it to be distinguished from other responses as a representation of something with a determinate conceptual content. According to Brandom, it is by virtue of relations of substitution and anaphora that singular terms including indexicals and demonstratives come to stand for an individual in the world. Since the representational feature of a belief or sentence is thus explained in terms of its inferential relations to other beliefs or sentences including the contribution that sub-sentential elements make to these inferential relations, representation is a holistic property, too: with respect to its representational feature, a belief is ontologically dependent on there being other beliefs with which it constitutes a system of beliefs.

---

1 Brandom (1994), Chapter 5. See furthermore the deflationary theory of reference that Horwich (1998), Chapter 5, proposes in conjunction with a use theory of meaning.

Let us grant that an inferential semantics has a point in arguing that inferential relations are necessary for something to be a representation with a determinate conceptual content and that such a semantics can reconstruct how representation and reference can work in terms of the inferential relations among sentences including the contribution which sub-sentential elements make to these relations. Brandom elaborates on that matter.

The problem for the proposed social holism and holism about beliefs is that inferential relations as such are not sufficient to establish a representation of and reference to particular, contingent features (or things) of our world. The pragmatics on which an inferential semantics is based according to the proposed account in the sense that inferential relations are determined by social practices has to establish the connection between specific features of our world and conceptual content that is individuated by inferential relations. That is to say: it has to be shown on the level of this pragmatics how the social practices that determine conceptual content (a) avoid an identification of what is correct with what is taken to be correct and (b) answer to the way the world is. To give an account of representation in terms of inferences is a first step in a reconstruction of how beliefs whose content consists in inferences relate to the world. Two further steps are outlined in the following two sub-chapters: the second step is to argue that the way the world is, and not social agreement, is the measure for the norms on which inferences are based. The third step is to regard social practices as being anchored in a physical environment in such a way that there is a rational constraint of the world on our beliefs.

5.2 Holism and Realism

5.2.1 Open-ended I–Thou Relations

The thesis which this sub-chapter is to make plausible is this one: *The proposed solution to the problem of rule-following in terms of social practices is about the way in which we gain epistemic access to a world that is independent of these practices*. The task hence is to combine the proposed holism with realism. Recall that, as outlined in [4.2], a realism about the physical world is any position according to which (a) there is a physical world that is ontologically and causally independent of our beliefs and (b) whether or not our beliefs about the physical world are true depends on the way the physical world is. I shall not develop an argument for realism. My strategy will be to refute arguments which may be taken to imply that a social theory of meaning is incompatible with realism about the physical world. The problem about realism is to give a philosophical account of how we can have epistemic access to an independent reality (see Willaschek (1999)).

An account of meaning in terms of social practices seems to imply social relativism instead of being compatible with realism. The content of beliefs in which the concept *F* is employed is fixed by social practices that determine (a) what constitutes going on in the same way in classifying things as being *F* and that determine (b) to which other beliefs one is committed and entitled by having a belief of the type "This is *F*". It therefore seems that we do not gain epistemic access to an independent reality; instead, it seems that we dictate in our social practices which properties things have. Consequently, it seems that an individual can be wrong in the light of the community, but that the community as a whole cannot be wrong about anything; for it seems that the practices of the community determine which things are *F* and which are not *F*, for any concept *F*. For instance, let us assume that all ancient Greeks believed that the earth is flat. According to social relativism, there is in this case no point in claiming that the community of the ancient Greeks was wrong in believing that the earth is flat. Such a claim ignores the way in which the conceptual content of the belief that the earth is flat was determined for the ancient Greeks.

When assessing whether or not social holism leads to social relativism, it is crucial to distinguish between social holism as construed in terms of I–we relations and social holism as construed in terms of I–thou relations. Social relativism ensues only if we endorse an exposition of social holism in terms of I–we relations between an individual and a community in the sense that the community assesses the actions of the individual according to its standards of what is fit and proper. Social relativism does not require us to renounce an epistemic relation between belief states and the world; but it can admit such a relation only by saying the following: there are different, incommensurable perspectives on the world in the sense of different, incommensurable ways of conceptualizing a causal input from the world; these perspectives are socially determined. This position is a version of the Myth of the Given, as is clear from Davidson's paper "On the very Idea of a Conceptual Scheme" [4.2]. Consequently, if one sets out social holism in terms of I–we relations and is thereby committed to social relativism, one can account for an epistemic relation between belief states and the world only by buying into a version of the Myth of the Given.

By contrast, if we construe social holism in terms of I–thou relations, we can combine social holism with a semantics that avoids the Myth of the

Given by replacing representationalism with inferentialism. This is another link between an inferential semantics and the proposed social holism: as an inferential semantics makes available a direct realism, so a social holism conceived in terms of I–thou relations opens up the prospect of accounting for our epistemic access to the world without being committed to incommensurable conceptual schemes that are an epistemic intermediary between belief states and the world.

In [2.4.3], I followed Brandom in basing an inferential semantics on a normative pragmatics. It was suggested that the meaning of a belief of the type *p* is fixed by the normative attitudes of the members of a community at a time. That is to say: the meaning of a belief of the type *p* are those other beliefs to which one is committed to, entitled to and to which entitlement is precluded in the light of one's fellows. This identification of meaning with normative, social facts in the sense of the normative attitudes that people adopt at a time ensures that meaning is transparent: there is no difference between what we take to be the meaning of our beliefs and what is the meaning of our beliefs. What then has to be established is the difference between meaning in the sense of taking something to be correct, as fixed by the attitudes of people, and its being correct, as fixed by the way the world is.

The purpose of I–thou relations is not to integrate individuals into a community by achieving communal agreement. These relations do not stop at any communal agreement. They are open-ended. It is always possible for a further person to join in and challenge received assessments. Over and above that, any person can come back to her former assessments and correct them from a new perspective. Such a conception of social practices does not accord any specific status to the community: we can eliminate the communal "we" in terms of "I–thous". For the purpose of a systematic reconstruction of social practices, we can start with considering a few individuals going through the process described in [3.2.1] and then imagine ever more individuals joining in so that this process is open-ended.

Conceiving a social account of meaning in terms of symmetric I–thou relations thus enables us to say the following: a person can criticize from her perspective not only beliefs of another person, but also beliefs that are shared by all the members of a community. She can claim that some of these beliefs fail to be an appropriate conceptualization of the physical environment. From the perspective of new experience in particular, beliefs that are shared by all the members of a linguistic community can turn out to be incorrect. Consequently, the attitudes of taking something to be correct or incorrect are themselves subject to an assessment as to whether or not they are in fact correct. Setting social holism out in terms of open-ended and symmetric I–thou relations therefore prevents us from having to identify at some stage what is correct with social facts in the sense of what a community takes to be correct. For any one claim that a whole community takes to be correct, it is possible that the process of determining meaning has to go on or has to be taken up again, because someone challenges that claim by giving a reason.

Hence, if a social theory of meaning is conceived in terms of I–thou relations, such a theory can be a philosophical account of how we have epistemic access to an independent world. Since I–thou relations are open-ended instead of terminating in communal agreement, we are free to conceive these relations as a ***means*** – and, if the rule-following considerations in Chapter 3 are correct, the *only* means – to gain epistemic access to the world; it then depends on the world which ones of our beliefs about the world are correct and which ones are incorrect. We can of course discover that correctness or incorrectness only within these social practices by means of the described open-ended process of assessing each other's normative attitudes. Nonetheless, we now have a conceptual distinction at our disposal between the social process of finding out whether something is correct or incorrect and its being correct or incorrect.

However, one may have reservations whether these social practices really show how we can have epistemic access to an independent world. Including a distinction between what is taken to be correct and what is correct may not be sufficient to show that we can grasp in our social practices what is correct. In other words, one may have the concern that in this account a gap arises between what we take to be correct in our social practices and what is in fact correct. I shall address that concern in a third step in [5.3.2] by proposing a pragmatic conception of rationality. The link between holism and realism will not be complete until that third step is taken into account. Let us, however, first continue with the exposition of how the proposed social theory of meaning avoids a reduction of norms of correctness to social facts.

Brandom (1994) sets out objectivity proofs: he shows that as a result of his conception of social practices in terms of I–thou relations, the normative attitude of taking something to be true can never be identified with its being true (pp. 601–607). This objectivity notwithstanding, according to Brandom's model of social practices as deontic scorekeeping, normative statuses supervene on normative attitudes (pp. 46–52, 291–297). That is to say: the statuses of a person of being correct or incorrect supervene on attitudes of taking the person in question to be correct or incorrect in certain of her claims. However, Brandom adds that these statuses do not supervene on the actual attitudes of the members of a community at a given time:

> What follows from *p* cannot be identified with how I or anyone *actually* keeps score; it is rather to be identified with a feature of *correct* scorekeeping (for it depends on what else is true, not on what anyone *takes* to be true). (p. 627)

That is to say: normative statuses supervene on normative attitudes *insofar as these attitudes are correct*. Brandom speaks of a *normative phenomenalism* about normative statuses in distinction to a phenomenalism *tout court* (pp. 626–628). This position is a phenomenalism about norms insofar as it takes normative statuses to arise from normative attitudes, namely the attitudes which persons adopt in the assessment of each other's actions. As suggested in [3.2.1], going beyond Brandom's text, we can reconstruct these normative attitudes on the basis of dispositions of persons in a shared physical environment. Such a reconstruction then yields an account not only of how normative attitudes, but also of how normative statuses can be integrated into the natural world.

Nevertheless, it is questionable whether it is appropriate to characterize such a position as a phenomenalism: saying that normative statuses supervene on the *correct* normative attitudes is not a substantial supervenience claim, unless the correct normative attitudes are at some stage identified with what is taken to be correct. However, the point of conceiving social holism in terms of I–thou relations is to avoid a commitment to such an identification. If the *correct* normative attitudes depend on what *is true* in distinction to what is *taken to be true*, the substantial supervenience claim goes the other way round: the attitude of taking something to be true, if correct, supervenes on what is true. Such a supervenience is, of course, what we need in order to combine social holism with realism. This position thus amounts to the following: there are no commitments in the world independently of social practices. But if there are social practices, what is true – and thus what are the correct normative attitudes – supervenes on the way the world is.

Taking truth to supervene on being enables this conception to provide for commitments that go beyond the commitments which people acknowledge. Brandom (1994) traces the distinction between attributions of beliefs *de dicto* and attributions of beliefs *de re* back to the different perspectives from which the same propositional content can be specified in social practices, in particular the difference between acknowledging a commitment to a claim oneself and ascribing a commitment to a claim to another person (Chapter

8). He employs this distinction to establish a difference between subjective normative attitude and objective normative status (pp. 593–597 and also pp. 52–55, 632–633). In translating beliefs *de dicto* into beliefs *de re*, a scorekeeper brings out which are the objects that the beliefs of a person refer to in the scorekeeper's perspective. According to this conception, referring to objects depends on social interactions; but this is unproblematic against the background of the proposed solution to the problem of rule-following which requires social practices for beliefs to have meaning.

The ascription of beliefs *de re* to a person can go beyond the commitments which that person is acknowledges if questioned. For instance, if Judith believes in 2000 that the president of the United States is the most powerful person in the world, a scorekeeper can ascribe to her to believe of Bill Clinton that he is the most powerful person in the world – even though it may be the case that Judith does not believe that Bill Clinton is the most powerful person in the world, because she does not know that Bill Clinton is the president of the United States. Nonetheless, by believing that the president of the United States is the most powerful person in the world, Judith is committed to believing of Bill Clinton that he is the most powerful person in the world. What is more, even if no one realized that Bill Clinton were the president of the United States, anyone who claims that the president of the United States is the most powerful person in the world would be committed to believing of Bill Clinton that he is the most powerful person in the world. Translating beliefs into beliefs *de re* brings out that Brandom's account allows for the possibility of commitments that go beyond acknowledged commitments. However, as argued above, it is reasonable to say that these commitments are part of the meaning of a claim only when they are acknowledged in normative attitudes.

To sum this section up, conceiving social holism in terms of I–thou relations vindicates the thesis that the proposed account of meaning is about the way in which we gain epistemic access to an independent world in the following manner: (a) It shows that even beliefs which are shared by a whole community are subject to an assessment as being correct or incorrect. Consequently, as regards our beliefs about the world, we never have to identify what is correct with what is taken to be correct. (b) It makes clear that what is correct and what is incorrect in our beliefs about the world supervenes on the way the world is. It thereby provides for commitments that go beyond the commitments which people acknowledge.

5.2.2 *Rule-Following and Response-Dependence*

There is a further challenge to the thesis that the proposed account of rule-following is about the way in which we gain epistemic access to a world that is independent of our practices. As Philip Pettit makes clear, we have to concede that an account of rule-following in terms of social holism – even if it is conceived in terms of I–thou relations among individuals – commits us to regarding conceptual content as being *response-dependent*.[2] I therefore intend to argue in this section that the sort of response-dependence which this account implies does not compromise the link with realism. I shall be concerned with response-dependence only insofar as response-dependence is a consequence of the proposed view of meaning in terms of social practices.

The way in which concepts of so-called secondary qualities are commonly treated in epistemology is an example of response-dependence: for instance, something is red if and only if it looks red to normal observers in normal conditions. If all concepts are response-dependent or derived from concepts that are response-dependent, then such a biconditional holds directly or indirectly for all concepts. We thus get to the thesis of *global response-dependence*. Global response-dependence applies to concepts that relate to things and events in the world in the first place. When the term "every concept" is used in the following without further qualification, I mean empirical concepts in the first place; for the purpose is only to link up holism with a realism as regards the physical realm (but not to defend a realism about, for instance, mathematical objects). Let us formulate the notion of global response-dependence in the following way:

Global response-dependence

> For every concept F which is not derived from other concepts, something is F if and only if it appears F to the members of a social community in normal conditions.

This formulation is intended to make clear that in the context of the link between rule-following and response-dependence, the claim is not an

---

2 Pettit (1991), in particular p. 606, and Pettit (1996), pp. 91, 193–213. In the latter publication, Pettit uses the term "response-authorising". The term "response-dependent" goes back to Johnston (1989), pp. 144–146; see also Johnston (1993), pp. 103–111, and Johnston (1998), pp. 8–10. Volume 81.1 (1998) of *The Monist* and volume 3 (1998) of the *European Review of Philosophy* are dedicated to a discussion of response-dependence. For a comprehensive recent discussion of the connection between rule-following and response-dependence, see Haukioja (1999), part II.

ontological one. There may be things which are *F* in a world in which there are no persons. With respect to its being *F*, a thing is not *ontologically* dependent on there being persons to whom it appears *F*.[3] But when it comes to an account of persons following the rule *F*, then we have to bring in appearing *F*. Global response-dependence is implied by a theory of conceptual content in terms of social practices as an *a priori* truth; for it is by appearing *F* to the members of a social community in normal conditions that the content of the rule *F* is determined.

If concepts in general are response-dependent, this seems to compromise our epistemic access to a world that is independent of our practices; for it seems to imply that we can know merely the way in which things behave relative to us in eliciting certain responses in us, but not what it is in the things themselves that makes them elicit these responses. Michael Smith and Daniel Stoljar (1998) claim that global response-dependence entails what they call *noumenal realism*, namely the thesis that "there is an independent reality, but the intrinsic nature of that reality is unknowable" (p. 86). Their argument begins with the statement that "According to Global Response-Dependence, the only claims we can ever make about the world are claims about the dispositions it possesses to elicit certain responses in us." They introduce the further premise that every disposition has a non-dispositional explanatory ground. They then conclude that global response-dependence has the following consequence: "the world must be a certain way in and of itself – that is, non-dispositionally – quite irrespective of the fact that we are in no position to make claims about this way it is" (p. 87; compare also Price (1998), pp. 123–124).

Smith and Stoljar give good reasons for the assumption that every disposition has a non-dispositional explanatory ground (pp. 88–108). Given this assumption, their argument from the epistemic condition that we know only the way in which things behave relative to us to noumenal realism is sound. Pettit accepts this conclusion and argues that there is no reason to baulk at the commitment to noumenal realism.[4] Let us nonetheless consider whether an account of conceptual content in terms of social practices really entails that all we can know is the way in which things elicit certain responses in us. If we can avoid this conclusion, then noumenal realism may

---

3 Pettit makes this point clear in recent papers. See Pettit (1998b), p. 114, and Pettit (1998c), pp. 59–62.

4 Pettit (1998b). But see the reservation on pp. 123–126.

be true, but the embraced theory of conceptual content does not commit us to it.

Note that global response-dependence and the reasoning which leads to noumenal realism do not imply that all the properties which we know are dispositional or relational. The properties may be non-dispositional and non-relational, i.e., intrinsic. The point is that the meaning of the concepts of these properties is conceived in terms of the relation to us, namely the disposition to elicit certain responses in us. The argument from response-dependence to a realism according to which we cannot know the intrinsic nature of things is based upon a reading of response-dependence as a *constraint on the meaning of any concept F*. Thus, Smith and Stoljar (1998) say that "Fness can be *defined* in terms of a response ... which suitable people are caused to have in suitable circumstances" (p. 86, my emphasis). Let us formulate this reading in this way:

Global response-dependence as a constraint on the meaning of concepts

> For every concept *F* which is not derived from other concepts, it is part of the meaning of the concept *F* that something is *F* if and only if it appears *F* to the members of a social community in normal conditions.

The cautious formulation "part of the meaning" is intended to express that we do not have to commit ourselves to a reductive analysis of "being *F*" in terms of "appearing *F*".

Consider the paradigmatic example of response-dependence, secondary qualities. Take the concept "red". If the concept "red" is response-dependent in the sense just defined, then the following biconditional is a constraint on the meaning of "red": something is red if and only if it appears red to normal observers in normal conditions. This biconditional leaves open the possibility that the property of being red is an intrinsic property, as long as being red includes the disposition to appear red. What this biconditional can be taken to imply is that we grasp the property of being red only under the aspect of the appearance relation to us; and this leaves open what it is in the thing itself that makes it appear red. To put the matter in functional terms, it leaves open what it is that realizes the function or the role of appearing red.[5] On this basis, one can then argue for noumenal realism.

However, an account of rule-following in terms of social practices does not require us to conceive response-dependence as a constraint on the

---

5 Compare Pettit (1998b), in particular pp. 115–118. See also Pettit (1998c), pp. 59–62.

meaning of concepts. For such an account to work it is sufficient to conceive *response-dependence as a constraint on the acquisition of any concept F* that is not derived from other concepts. Pettit himself speaks of possession conditions in the sense in which Christopher Peacocke employs that term in his book on concepts.[6] Furthermore, in a later paper, Pettit explicitly distinguishes possession conditions from claims about the content of a concept (1998b, p. 114). However, I avoid using the notion of a possession condition, because, in the sense of Peacocke (1992), a possession condition individuates a concept and states what is required for full mastery of a concept (Chapter 1, in particular p. 29). Consequently, to satisfy a possession condition in this sense, one has to know the meaning of the concept in question. Nonetheless, in another recent paper (1998c, pp. 62–65), Pettit comes close to the position which I shall endorse below, although he accepts the commitment to noumenal realism. Over and above that, the distinction I propose is related to a distinction which Huw Price (1998, in particular p. 112) makes between response-dependence as a content condition, i.e., a constraint on meaning, and response-dependence as a usage condition, i.e., a constraint on assertibility, though what I have in mind is not assertibility, but acquisition:

Global response-dependence as a constraint on the acquisition of concepts

> For every concept *F* which is not derived from other concepts, it is a necessary condition for the acquisition of the concept *F* that something is *F* if and only if being *F* includes the disposition to appear *F* to the members of a social community in normal conditions.

The formulation "includes the disposition to appear *F*" is to prevent the misunderstanding that the property of being *F* has to be dispositional.

Global response-dependence in this sense is sufficient for an account of rule-following in terms of social practices. In order for us to acquire the concept *F*, there have to be some things which are *F* in such a way that their being *F* enables a response by persons – namely the response that persons in their practices of determining a content for a rule are disposed to classify these things as being *F*. In the terms of the account in Chapter 3, response-dependence is a prerequisite for circumstances of the application of the concept *F* being determined. However, the meaning of the concept *F* does

---

6 Pettit (1991), p. 602, and Pettit (1996), pp. 103, 199–201.

not have to say anything about appearing *F*. The meaning is determined by the inferences which a sentence such as "This is *F*" supports – inferences such that if one claims "This is *F*", one is committed to the claim "This is *G*" and entitled to the claim "This is *H*".

To illustrate the difference between a constraint on acquisition and a constraint on meaning, consider the following analogy: In order to possess a house, a condition of acquisition is that one follows a certain legal procedure, e.g., appears before a notary. However, what it means to possess a house is not such a procedure, but a certain legal status – other people must not enter the house without one's permission, etc. In the same manner, in order for a person to acquire a concept in a non-inferential way, the condition of response-dependence has to be satisfied; but the meaning of a concept consists in its relations to other concepts. Consequently, if we include in the proposed solution to the problem of rule-following a theory of meaning in terms of inferential role, the relation to us does not have to enter into the meaning of the sentences in which the concept *F* is employed. The move to noumenal realism thus is blocked.[7]

Construing response-dependence as a general constraint on the acquisition of concepts enables us also to introduce the category of secondary quality concepts: secondary quality concepts are all and only those concepts for which response-dependence is not only a constraint on acquisition, but also a constraint on meaning. For instance, the concept "comfortable" may be such that it is part of the meaning of "comfortable" that things appear comfortable to persons.

Note a further consequence of this distinction between meaning and acquisition conditions: It is not necessary that all the things which belong to the extension of the concept *F* are such that they actually elicit the disposition to classify them as being *F*. In other words, we do not have to assume that our responses are sufficient to fix the extension of a concept *F*. Once the meaning of the concept *F* is determined, owing to the meaning of the concept *F* it may be determined for some things that they are *F*, although these things do not elicit the response to classify them as being *F*.

Imagine the following scenario: there are some particular things that are covered by darkness throughout their existence. This condition notwithstanding, the meaning of "red" determines that these things are red (e.g. because they have the appropriate surface quality). However, these

---

7 Thanks to Michael Smith and Daniel Stoljar for an illuminating discussion on this subject.

things do not elicit the response to classify them as being red. Nevertheless, being red includes the disposition to appear red in normal conditions: if these things were brought into daylight and thus brought into the same environment as those things to which we respond by classifying them as being red, these things would elicit in people the response to classify them as being red. (Even if bringing them into daylight would destroy these things, insofar as they are red, they have the disposition to appear red; for the factor which makes them fall under the concept "red" is not the factor that causes them to be destroyed when brought into daylight).

It may be objected that the shift from a constraint on meaning to a constraint on acquisition notwithstanding, all that we know is the effect which a thing has on us in appearing *F* to us; consequently, there is no reason for assuming that we can know the nature of the thing itself. This objection has a point insofar as a constraint on the acquisition of a concept *F* in terms of appearing *F* can be construed in such a way that appearing *F* blocks epistemic access to the thing itself. However, the theory of conceptual content advocated here is nothing like the position that has been described as representational realism in [4.1], namely that our epistemic access to the world is mediated by representations (such as sense data, for instance). As shown in [4.2], this theory of conceptual content, by contrast, can go with a direct realism. "Appearing *F*" in the formulation of response-dependence therefore does not mean that there is an *F*-appearance in persons which mediates in an epistemic sense their access to things that are *F*; it merely means that for any concept *F* which is not derived from other concepts, there are things that fall under this concept in such a way that their being *F* elicits in persons the response to classify these things as being *F*. That response may be elicited in a very indirect way: we may need sophisticated measuring instruments to gain access to the property *F*. Think of quarks and their properties. This point makes clear: the disposition to appear *F* does not have to be anything that is included in the property as such, but may depend on the means which we develop in order to gain access to the property.

In the argument that response-dependence as a constraint on the acquisition of concepts is sufficient to get an account of rule-following in terms of social practices off the ground, I have read the biconditional of response-dependence from left to right: in short, for any concept *F*, if something is *F*, it appears *F* in normal conditions (so that we can acquire the concept *F*). But this biconditional can also be read from right to left: in short, for any concept *F*, if something appears *F* in normal conditions, it is *F*. Reading the biconditional from right to left, one can object that response-dependence compromises realism by making our practices determine the truth of beliefs of the type "This is *F*", even if response-dependence is restricted to a constraint on the acquisition of concepts. For it seems that beliefs of the type "This is *F*" are true owing to the way in which the content of the concept *F* is determined in our practices.

Whether or not this charge is justified depends on the way in which we construe the notion of normal conditions. In [3.2.1], I said that through assessing our actions as correct or incorrect by means of sanctions, we enter into a process of finding out conditions under which our dispositions on what constitutes continuing a given sequence of examples in the same way converge. Thus, this process determines normal conditions. Nonetheless, we do not have to identify normal conditions with the conditions under which suitable persons agree in their assessments at a certain time.[8] Disagreements which call for a process of negotiating a convergence do not only arise among people at the same time, but also with respect to their former judgements. In the light of new evidence – and that is in the light of a further specification of what normal conditions for judgements of a certain type are – we correct our former judgements. Since, according to the account of social practices in terms of I–thou relations, the process of challenging claims with reasons is open-ended, the process of determining normal conditions is open-ended as well. Hence, in finding out conditions under which our judgements converge at a time as well as across time, we determine not only meaning, but we can also see ourselves as aiming at a determination of those conditions in which we gain access to things as they are[9] – and be it by employing sophisticated instruments. Consequently, this account of response-dependence does not compromise realism: The truth of beliefs of the type "This is *F*" is not determined by our practices, because we do not have to identify the conditions under which we agree at a time with the normal conditions.

Even if one grants the argument for linking social holism up with realism put forward so far, one can maintain that the first rules which get their content fixed in social practices are concepts that describe how things relate to us, i.e., which effects things have on us. One may think of concepts such as "pleasant", "comfortable", etc. However, even if this claim were true, it

---

8 Compare also Spohn (1997), in particular pp. 357, 376, who talks in terms of normal conditions having a hidden nature.

9 See Pettit (1999). Compare, against the background of this recent paper, also Pettit (1991), pp. 610–618, and Pettit (1996), pp. 203–212.

would not threaten the proposed link between social holism and realism. Recall that according to the position defended here the content of one rule is not fixed in isolation, but only together with the content of many other rules. In the last resort, this has the following consequence: a concept such as, say, "comfortable" gets its content fixed by its being embedded in a whole theory or a whole world view. This theory or world view contains also claims about what makes things appear comfortable. The point now is: statements that describe the effects that physical things have on us do not have a special position in this theory. In a mature physical theory, for instance, measurement is treated as one interaction among others. Hence, the world view or theory in which even common sense concepts such as "comfortable" are embedded includes an account of the nature of physical things which makes them have certain effects on us.

It may be objected that our world view as formed by contemporary physics is not an account in terms of intrinsic natures of physical things, but ascribes to physical things only properties which consist in relations to other physical things. Consider what Frank Jackson says in this respect in his recently published John Locke lectures:

> When physicists tell us about the properties they take to be fundamental, they tell us what these properties *do*. This is no accident. We know about what things are like essentially through the way they impinge on us and our measuring instruments. It does not follow from this that the fundamental properties of current physics, or of 'completed' physics, are causal cum relational ones. It may be that our terms for the fundamental properties pick out the properties they do via the causal relations the properties enter into, but that at least some of the properties so picked out are intrinsic. They have, as we might put it, relational names but intrinsic essences. (1998a, p. 23)

This statement can be seen as fitting into the argument proposed above, namely that global response-dependence as a constraint on the acquisition of concepts does not prevent us from knowing the intrinsic nature of things (if they have an intrinsic nature). Jackson goes on:

> However, it [the mentioned view of physics] does suggest the possibility that (i) there are two quite different intrinsic properties, *P* and *P**, which are exactly alike in the causal relations they enter into, (ii) sometimes one is possessed and sometimes the other, and (iii) we mistakenly think that there is just one property because the difference does not make a difference (as the point is put in information theory). An obvious extension of this possibility leads to the uncomfortable idea that we may know next to nothing about the

intrinsic nature of the world. We know only its causal cum relational nature. (1998a, pp. 23–24)

What Jackson describes as an uncomfortable idea is a possibility that global response-dependence has to concede, even if it is only a constraint on the acquisition of concepts.[10] Jackson then repudiates the view "that the nature of everything is relational cum causal, which makes a mystery of what it is that stands *in* the causal relations" (p. 24). However, if we generalize this view in such a way that it does not only concern causal relations, but all relations, there is nothing mysterious about it. As long as "relational" is not limited to the relations in which physical things stand to persons, there surely is a possible world in which physical things have only relational properties. That is to say, in such a world, any physical system apart from the whole universe is completely described by the relations in which it stands to other physical systems. Unlike dispositions, relational properties do not need in general a non-relational explanatory ground; relations do not require intrinsic properties to support them. This is evident in the case of spatio–temporal relations, and it will become the more clear in Chapter 8 when quantum holism is considered. If our best science tells us that the basic properties which are instantiated in our world are relational ones, this may be the truth of the matter. The point is that our physics accounts for what it is in physical things that makes them have the known physical effects upon us.

The upshot of this consideration is: It may be that our world has an intrinsic nature which is not taken into account in our physics and, moreover, of which we cannot gain knowledge. However, it is not necessary to go for such a metaphysics. Even if all the properties of which we gain knowledge were relational ones, nothing would hinder us to regard the resulting description of the world as being in principle a complete one. As far as the theory of conceptual content is concerned, the point is: whether or not our world is such that the physical properties are all relational properties is an empirical matter. It is not an implication of the endorsed theory of conceptual content that all physical properties are relational ones.

In this defense of a link between social holism and realism, I have spoken in terms of concepts being established in social practices, but properties being instantiated in the world independently of these practices. This way of presenting the matter may suggest the picture of concepts matching properties. This picture, in turn, may provoke the objection that such a

---

10 As to what Jackson describes as an uncomfortable idea, compare also Searle (1983), pp. 75–76, as well as Blackburn, "Filling in Space", essay 14 in Blackburn (1993).

matching relation is mysterious. However, when I say that something instantiates the property of being *F*, I do not intend more than the following: the thing is correctly described by a sentence in which the concept *F* is employed such as "This is *F*"; whether or not such a sentence is correct supervenes on the way the world is.

5.2.3 A Pragmatic Realism

The argument of this sub-chapter can be read as making a case for a direct, pragmatic realism that is beyond the traditional opposition between metaphysical realism and relativism. The way in which I shall spell out this realism in this section and the following two ones differs from Brandom's account by admitting a traditional notion of truth in terms of non-conceptual features of the world that make some of our beliefs true. My claim is: *Social holism and a social theory of meaning can go with a realism that is not committed to any idealistic consequences – neither in the Hegelian sense of the world having conceptual features in itself, nor in the Kantian sense that we can know the world only as it appears to us, but not as it is in itself.*

Metaphysical realism is a realism, because it assumes that there is a physical world which is ontologically and causally independent of the beliefs of persons. It is metaphysical, because, when it comes to our epistemic access to the physical world, it does not take our practices into account. Metaphysical realism envisages a final true theory of the way the world is. The concepts that figure in the final theory have a meaning which is fixed independently of our practices. Appropriately enough, this final theory is often referred to as the view from nowhere.[11]

Relativism objects that the notion of a final, true theory does not make sense. We cannot abstract from the conditions under which we gain epistemic access to the world. The notion of a view from nowhere therefore is not an intelligible notion. Instead, insofar as relativism grants that we have epistemic access to the world at all, it proposes that all our beliefs about the world are relative to a conceptual scheme, which acts as an epistemic intermediary between our belief states and the world. Different languages or different theories (world views) may embody different conceptual schemes. If conceptual schemes are different, they are incommensurable. That is to say, from beliefs conceived in one conceptual scheme it is not possible to access the meaning of beliefs conceived in another conceptual scheme. Since all beliefs are conceived within a conceptual scheme, there are limits to

---

11 Compare the title of Nagel (1986).

communicability: there is no meaningful communication possible across different conceptual schemes.

By contrast, the realism that can be combined with the proposed social holism is a *pragmatic realism*. It is *pragmatic*, because it conceives meaning to be determined by social practices. It is a *realism*, because it is not a pragmatic account of truth: whether or not a belief about the world is true supervenes on the way the world is. Since the proposed account of meaning is built on open-ended I–thou relations, we are free to say that this account is only about the conditions under which we gain epistemic access to the world. The account that has been put forward explains meaning without employing the concept of truth[12] – although, of course, the proposed theory of meaning puts a constraint on the theory of truth in the sense that not any theory of truth is compatible with this theory of meaning. (The same goes the other way round: the claim that the truth of our beliefs depends on the way the world is cannot go with just any theory of meaning). Advancing an account of meaning, but not an account of truth in terms of social practices is a coherent position: once social practices have established a determinate meaning for a belief, we are free to say that the truth value of a belief with that determinate meaning depends on the way the world is.

This position accords with common sense realism, because it is a direct realism and because it takes the truth of our beliefs to supervene on the way the world is. However, it does not give common sense realism priority over scientific realism. Common sense claims can turn out to be false in the process of scientific investigation. Even the common sense ontology of things with properties that endure in time may be superseded by an ontology based on science. Recall that Sellars combines a pragmatic realism with a scientific realism,[13] although this point is neglected by many of his followers today.

Refusing to accord common sense realism a privileged status and maintaining that the truth of beliefs supervenes on the way the world is has the following consequence: this position does not have a philosophical argument at its disposal to the effect that most of our beliefs are true, at least most of our common sense beliefs about the things with which we deal in everyday life. Davidson is famous for offering a philosophical argument to

---

12 Compare the description of such an account of meaning in relation to the work of the later Wittgenstein by Dummett (1993), pp. 19–20.

13 See in particular "Empiricism and the Philosophy of Mind", § 41, p. 173 in Sellars (1963) and below [5.3.2].

that effect.[14] However, he has to invoke the notion of an omniscient interpreter for his argument to be conclusive.[15] In the framework proposed here, the trouble with such an enterprise is this one: the meaning of a belief, such as "This is *F*", consists in the inferences which a belief of that type licenses. The inferences to which one is committed and entitled by claiming that something is *F* may change in a quite dramatic way consequent upon the discovery of new features of the things that are *F*. Any particular change in the inferential context of beliefs of the type "This is *F*" can only take place against a large background of inferences that remain untouched. However, a number of small changes can add up to a great change. Sellars maintains that empirical knowledge "is a self-correcting enterprise which can put *any* claim in jeopardy, though not *all* at once".[16] There is no *a priori* assurance against how far-reaching such self-correction may be. Since holism about beliefs implies that there is no principled distinction between common sense beliefs and scientific beliefs, there is no guarantee against common sense beliefs changing their meaning either. Hence, in the long run, it can turn out that a lot of the inferences which a belief of the type "This is *F*" once licensed turn out not to be true.

However, the proposed holism makes the epistemic point that any doubt is sensible only against a background of beliefs that are not called into question. For any claim to have meaning at all, a network of claims has to be built up in the described way in which beliefs get meaning. In this sense of building up a network of claims, one can be justified in making certain claims without justifying these claims.[17] Consequently, any doubt is itself in need of justification: if one claim is challenged, this requires a justification in terms of other claims why this one claim is selected for challenge.[18]

The pragmatic realism suggested in this section is beyond the traditional opposition between metaphysical realism and relativism. It is not a metaphysical realism, since it agrees with relativism in repudiating the notion of a final true theory of the world that is a view from nowhere, i.e.,

---

14 See, in particular Davidson (1983), pp. 431–437.

15 Davidson (1983), p. 435. As to the position of the omniscient interpreter in Davidson's theory, see Manning (1995).

16 "Empiricism and the Philosophy of Mind", § 38, p. 170 in Sellars (1963). See also "Some Reflections on Language Games", § 46, p. 339 in Sellars (1963).

17 See Brandom (1994), p. 204; Brandom (1997), pp. 124–125.

18 Compare Brandom (1994), pp. 176–178, 204–206.

which is formed by employing concepts whose meaning is independent of the practices of finite thinking beings. Consequently, the possibility of challenging received normative attitudes with reason can never be excluded. However, the proposed realism also overcomes relativism by repudiating the notion of a dualism of conceptual scheme and given content and thus the notion of incommensurable conceptual schemes. Since meaning is determined in social practices, one may say that our epistemic access to the world is perspectival instead of achieving a view from nowhere; but if one talks in this way, the point is that there is no *a priori* limit to integrating other perspectives. A person can in principle become integrated into the practices of another community or work herself into the practices of a past community.

5.3 Social Practices in a Physical Environment

5.3.1 A Comprehensive Holism?

One can grant that (1) an inferential semantics has the means to account for representation at its disposal and that (2) a social theory of meaning in terms of I–thou relations is about our epistemic access to the world without having to identify beliefs that are taken to be correct with their being correct. Nevertheless, one can claim that (1) and (2) come too late to ensure that our beliefs are not a spinning in the void, because the social practices as described so far do not provide for a rational constraint of the world on our beliefs. John McDowell can be seen as voicing such an objection in his influential John Locke lectures on *Mind and World* (1994) (quoted according to the second edition McDowell (1996); see the book symposium McDowell et al. (1998)). To account for the relation between our beliefs and the world within the proposed holism and to meet the challenge that McDowell poses, we therefore need a third step which shows how the social practices that determine meaning answer to the way the world is.

In the John Locke lectures, McDowell sets his conception out by arguing that we are held captive by an oscillation between the following two positions: (a) a coherentism that accepts rational relations only among beliefs themselves and (b) the Myth of the Given [4.2]. The Myth of the Given confuses a causal relation with a rational one; it leaves us, in the terms of McDowell, with an exculpation where a justification is called for (pp. 7–8). Coherentism, by contrast, leaves our beliefs spinning in the void in the sense that there is no rational constraint from the world on them. McDowell focuses on criticising Davidson in this respect (pp. 13–18, 68, 137–153), although Davidson (1983, p. 431) himself attempts to dissolve the oscillation between the two positions which McDowell attacks. McDowell's claim is not that Davidson "does not answer the question how anything can have empirical content". He claims only "that he [Davidson] does not ensure, as he wants to, that the question lacks urgency" (p. 147). McDowell extends this criticism in part to Sellars in the new introduction to *Mind and World* (1996, pp. XV–XIX).

The solution that McDowell proposes in order to escape the oscillation between coherentism and the Myth of the Given is to consider the conceptual realm as having no boundaries: the conceptual realm does not end where persons and their interactions end; instead, it encompasses the whole physical realm – all what our beliefs can be about. The content of our beliefs, if true, are the very facts that make up the world. Conceiving a boundary between what is conceptual and what is non-conceptual precludes us from acknowledging a rational constraint of the world on our beliefs (in particular lecture 2). The point of McDowell thus is not only that experience has to be conceptual in order to exercise a rational constraint on a system of beliefs. This could be interpreted as merely shifting the boundary between what is conceptual and the world in such a way that the conceptual includes experience; but the relation between experience and the world could still be a mere causal one. By contrast, when McDowell claims that experience is conceptual, his point is that the world itself is conceptual; for only in this case can there be a rational constraint of the world on our beliefs.[19]

McDowell approves of a holism about beliefs along Sellarsian lines (pp. 12–13), although he has reservations about Brandom's social inferential semantics (McDowell (1997)) as well as about Quine's confirmation holism (pp. 156–161). If we start from the proposed social holism and holism about beliefs, what McDowell's plea for considering the conceptual realm as unbounded brings into perspective is a holism in the philosophy of mind that is a *comprehensive holism* (although I do not intend to ascribe such a holism to McDowell himself): holism about beliefs encompasses the whole conceptual realm. If the conceptual realm is unbounded, then this holism encompasses in a sense the whole world: this holism includes those aspects of the physical world which make it that the physical world is, in McDowell's words, "not external to the realm of the kind of intelligibility that is proper to meaning" (p. 72). With respect to these aspects, each component of the physical world (each fact) is ontologically dependent on

19 Compare the review of Wright (1996), pp. 240–242.

there being other components (facts) which finally include the whole physical world.

What does it mean that the conceptual realm has no boundaries? McDowell elaborates by maintaining "The question is how we can take that view without offering to reinstate the idea that the movement of the planets, or the fall of a sparrow, is rightly approached in the sort of way we approach a text or an utterance or some other kind of action" (p. 72). The answer to this question involves for McDowell a view of nature that does not identify nature with what is acknowledged in modern science. He advocates what he calls a partial reenchantment of nature (in particular lecture 4). We thus have to envisage going beyond the scientific view of nature in order to grasp that aspect of the physical world which makes it that the physical world is not outside the conceptual realm. According to McDowell, science cannot capture that aspect. He regards modern science rather as an obstacle that prevents us from seeing the truism in his position of the world being composed of the very facts that are the content of our true beliefs.

Julian Dodd (1995) objects to what McDowell considers as a truism that he conflates a Fregean notion of facts as true propositions with a notion of facts as that of which the world is composed.[20] This objection can be extended to Brandom, when he proposes to identify facts with the content of true claims and then goes on to say "The world is everything that is the case, a constellation of facts" (p. 333; see also p. 622). Facts in this latter sense of what makes up the world seem to be what facts in the former sense of the contents of true claims are about – unless one goes for a Hegelian, objective idealism in the sense of the position that the world is composed of the contents of true beliefs. McDowell and Brandom may intend to make room for such a position.[21] In this case, there is no conflation of two notions of facts, although other objections then apply. However that may be, to avoid a confusion in the notion of facts, I shall continue to set out my account without employing that notion.

Nonetheless, the view proposed here may not be able to eschew the notion of facts completely. One can maintain that if we spell out the view that the truth of our beliefs supervenes on the way the world is, we have to acknowledge facts in an ontological sense: facts are what makes our beliefs

---

20 See also Lockwood (1989), pp. 147–148, 300–301.

21 As to McDowell's Hegelianism, see Sedgwick (1997). As regards Brandom, compare Habermas (1999), pp. 161–169.

true in case they are true.[22] However, even if we have to concede that point, facts in this ontological sense are distinct from the content of our beliefs. Moreover, we do not have to accept these facts as a primitive in ontology. For it is reasonable to hold that facts in this ontological sense are themselves constituted of things and properties, or events, or whatever other worldly items.[23] I therefore continue to talk in terms of things and events in the world as placeholders for whatever may be the most plausible candidates for a primitive in ontology.

Coming back to McDowell, even if the conceptual is unbounded so that the world is itself conceptual, we face the problem of rule-following and thus the problem how our beliefs can have a content: the problem then is how we can *follow* rules in our actions (compare the point against natural properties and real essences in [3.1.2] and [4.3.3] respectively). Consequently, we need the proposed social holism and holism about beliefs to solve the problem of rule-following independently of whether or not the conceptual realm has a boundary.

If we need social practices to solve the problem of rule-following anyway, then instead of committing us to an ontology of the physical world being itself somehow conceptual, we should try to account for a rational constraint of the world on our beliefs from within these practices. In the next section, I will continue to pursue my thesis that the social story of rule-following is entirely about our epistemic access to a world that is ontologically and causally independent of beliefs. I will argue that the holism which I have proposed up to now, a holism that is limited to the area of the philosophy of mind, is sufficient to meet McDowell's demand.

5.3.2 A Pragmatic Conception of Rationality

The challenge to show how the proposed holism can admit a rational constraint of the world on our beliefs has to be met on the level of the social practices that determine meaning.[24] It has to be shown that the world exerts a constraint on these practices which is not only causal, but which can also

---

22 But see the objection of Davidson in "True to the Facts" (1969) in Davidson (1984), essay 3, in particular pp. 41–42, and furthermore Dodd (1999).

23 Compare the position of Armstrong (1997), in particular Chapters 1 and 6 to 8. See also Schantz (1996), Chapter V.3, pp. 163–177.

24 Compare also Brandom (1998a), where he defends a position along the lines of Sellars and Davidson against McDowell's charge. See furthermore Brandom (1994), pp. 331–332, 632.

be regarded as rational. Such a constraint is necessary to ensure that the items whose conceptual content consists in inferential relations are beliefs about things in the world with a truth value that depends on the way the world is. My thesis is: *The physical world is not part of the conceptual content of our belief states. But it is part of the social practices in which that content is determined.*

As argued in [3.2.2], for someone who is merely rational there never is any meaning, because any finite sequence of whatever items runs into the infinity problem and the normativity problem. The solution to the problem of rule-following consists in social practices that are not conceptual themselves, but that determine a conceptual content. These practices are the origin of rationality. The proposed theory of meaning includes a *pragmatic conception of rationality*: in order to avoid a rationality that is a spinning in the void and never gets to any meaning, we have to see rationality as being tied to these practices.

To get off the ground, these practices presuppose (a) natural dispositions of the persons who take part in them as well as (b) a shared physical environment and (c) a cognitive access to this environment; that cognitive access consists in response-mechanisms which humans share with animals [3.2.1]. The social practices that develop on this basis include the physical environment; the environment literally is a part of these practices. Furthermore, focussing on these practices insofar as they determine conceptual content is an isolation of practices for the sake of concentrating on an argument that picks out social holism and holism about beliefs alone. In fact, having a belief of the type "This is *F*" involves not only beliefs of other types and appropriate circumstances of application, but also appropriate actions as practical consequences of believing of something that it is *F* [2.4.1]. Believing that something is *F* is taking a position not only in the linguistic practices of a community, but in a form of life in the Wittgensteinian sense. Consequently, holism about beliefs and social holism are aspects of a holism that encompasses all the intentional states of a person as being embedded in a physical environment.

The physical environment exerts a causal constraint on the social practices that determine conceptual content. That constraint is not a constraint from outside these practices, because (a) the physical environment is part of these practices and (b) the position that conceptual content is determined by these practices includes a pragmatic conception of rationality. On the basis of this pragmatic view of rationality, we can say the following: that very causal constraint can be regarded as rational from within these practices, because when it comes to rule-following, rationality has to open up itself in such a way that it includes the physical environment via these practices.

One may go as far as maintaining the following: the causal constraint from the environment ensures that our beliefs refer to certain items in the environment and thus that the truth value of our beliefs supervenes on the way the world is. But note that this position does not imply any dualism between reference as determined by causal relations and conceptual content as consisting in inferential relations. Instead, the conceptual content of our beliefs as consisting in inferential relations makes up for specific reference to items in the world owing to the way in which this content is determined by social practices in a physical environment.

Although the constraint from the physical environment can be regarded as rational from within these practices, it does not provide for logical relations between the physical and the content of belief states. Furthermore, the constraint from the physical environment does not go as far as ensuring that *specific* practices are ontologically dependent on *specific* features of the environment. For instance, practices of classifying things as trees are not ontologically dependent on being in touch with trees. For the inferential context can provide a content for beliefs of the type "This is a tree" even if there are no trees in the environment (compare [4.3.2]). But this conception makes having beliefs as a whole ontologically dependent on being embedded in a physical environment. This position thus is a social externalism in that it regards the specific content of our beliefs as being ontologically dependent on social practices; it is a physical externalism in that it considers our beliefs having content at all as being ontologically dependent on our being embedded in a physical environment.

It is not objectionable that this account of conceptual content presupposes the existence of a physical world, including a cognitive access to the environment that man shares with animals. Instead, this account offers something that may be construed as a transcendental argument for that presupposition: we start with rule-following. The sceptic cannot deny rule-following if her claims are to have meaning. The argument then is that having natural dispositions and sharing a physical environment with other people to which one has a cognitive access is a metaphysically necessary condition for the possibility of rule-following.

Against this background let us now return to a problem which was mentioned when discussing Quine's holism and which is a version of the problem on which McDowell elaborates. When setting out Quine's holism, I drew attention to the problem that Quine has in conceiving experience so that it can fulfill the function of inducing changes to a system of beliefs: On the one hand, experience has to be outside a system of beliefs in order to exert a constraint on the system. On the other hand, experience has to be conceptual if it is to exert a *rational* constraint on a system of beliefs. However, if experience is conceptual, it is itself part of the holistic system of beliefs [2.1.3]. Let us consider what Sellars says to answer this problem. Sellars' "Empiricism and the Philosophy of Mind" (1956) and Quine's "Two Dogmas of Empiricism" (1951) are closely related, although Sellars does not mention Quine.[25] They both attack tenets which are central to traditional empiricism: the separation between analytic and synthetic sentences, and the assumption of sense data or sense experiences that have both a causal and an epistemic function.

According to Sellars, insofar as experience is epistemic, it is integrated into the system of beliefs: it consists in perceptual beliefs. Semantic holism applies to perceptual beliefs as well as to beliefs which are acquired by inferences. Correction can hence take place only within a system of beliefs. Sellars writes, as quoted in part already in [5.2.3]:

> For empirical knowledge, like its sophisticated extension, science, is rational, not because it has a *foundation* but because it is a self-correcting enterprise which can put *any* claim in jeopardy, though not *all* at once.[26]

Although there are no joints which are both a causal ingredient for a system of beliefs and have an epistemic status, this self-correcting enterprise gets friction from the world, because of the – rational! – way in which the meaning of our beliefs is and continues to be shaped by our dealings with physical things in the described social practices. Consequently, the proposed account of meaning can conceive the self-correcting enterprise of empirical knowledge as being responsible to an independent reality without having to admit observational sentences that are not touched by semantic holism (as Quine does, see [2.1.3]).

This alternative to Quine's account of the relation between experience and a system of beliefs also avoids the instrumentalism that lurks in "Two Dogmas" (although Quine cannot be tied down to instrumentalism): Quine speaks of the objects of science as posits to predict future experience in the light of past experience (pp. 44–45). Sellars, by contrast, considers instrumentalism to be committed to a version of the Myth of the Given. He

---

25 On the systematic connection compare the interpretation of Rorty (1980), Chapter 4.

26 "Empiricism and the Philosophy of Mind", § 38, p. 170 in Sellars (1963). See also "Some Reflections on Language Games", § 46, p. 339 in Sellars (1963).

suggests regarding science as being continuous with common sense in the sense of a realistic understanding of our knowledge as a whole. All our knowledge from perceptual beliefs up to scientific theories is the self-correcting enterprise of which Sellars speaks in the quotation above.[27] This view of Sellars makes possible a notion of observation according to which even so-called theoretical entities such as elementary particles are observable by people with the appropriate training, i.e., physicists; observation here means beliefs that can be acquired in a non-inferential way.[28] As far as the proposed theory of meaning is concerned, this has the consequence that even the conceptual content of belief states about elementary particles can be seen as being shaped by direct dealings with these things – "direct" in the epistemic sense of indicating the absence of epistemic intermediaries.

What about the justification of empirical knowledge that is a central aspect of what Sellars denounces as the Myth of the Given? Giving reasons for a claim exclusively consists in making other claims: only something which has the structure of a claim can justify a claim, because only such a thing can figure as a premise in an inference of which the claim in question is the conclusion. If the claim that is challenged is an inference, justifying it consists in citing those other claims that provide entitlement to the claim in question. If the claim is a perceptual belief, the challenge can either concern the mastery of the concepts involved or the circumstances of application of a belief of the type in question; in the latter case, the challenge is that the person in question is not reliable in her disposition to make observation reports of the type in question. Accordingly, the belief is justified either by citing the inferential context or by making the appropriate circumstances of application the object of one's beliefs in the sense of formulating claims about them; one thus proves that one is reliable in one's reports of the type in question. In other words: experience qua causal happenings provides reasons for experience qua perceptual beliefs in virtue of social practices having determined circumstances under which it is appropriate to acquire beliefs or utter sentences of the type "This is *F*". Nonetheless, the circumstances of application cannot themselves justify any belief; but making them the object of one's beliefs by formulating explicit claims about

---

27 See "Empiricism and the Philosophy of Mind", §§ 39–44, pp. 170–174 in Sellars (1963).

28 See Brandom (1994), pp. 222–224, and Brandom (1997), pp. 164–166.

them can do the job. Such a move is appropriate when a perceptual belief is challenged.[29]

Thus, beliefs can only be justified by other beliefs, namely by cohering with other beliefs. We are committed to a holism about justification [2.1.3]. If we reject the Myth of the Given, there are no beliefs which are either self-evident or justified by something which is not conceptual. Unless we consider the conceptual to be unbounded, it is only other beliefs that can justify a belief. Consequently, there is no rational relation between the world and beliefs in the sense of a relation of justification. If one restricts the notion of a rational relation to a relation of justification, then indeed the world would have to be propositional for there to be a rational constraint of the world on beliefs. Nonetheless, due to the way in which beliefs can only have meaning, they are answerable to the way the world is.

To sum up this sub-chapter, it is possible to conceive holism about beliefs and social holism in such a way that the result is a comprehensive holism that encompasses the physical realm. However, the price for such a position is to countenance an aspect of the physical world that is beyond the grasp of science and that makes the physical world part of the conceptual. The concern that motivates this position can be addressed by the proposed social holism. Thus, beliefs are separated from the world in the sense that it is only them – and nothing which makes up the world – that is conceptual. But they are tied to the world in the sense that they are not epistemically self-sufficient, because the social practices which determine meaning are shaped by dealings with physical things. Consequently, the causal constraint which the physical environment exert on beliefs is not a constraint from outside these practices. From the perspective of these practices, it is a rational constraint. For rationality is at the root these social practices which include the physical environment.

We now see in a more precise way in what sense this position is a revision of the Cartesian tradition in epistemology instead of a simple rejection [4.1]. The starting point for this position is not only the Cartesian one of a reflection on what it means to have beliefs; but this position also elaborates on this point in such a way that it conceives the epistemic relation of our beliefs to the world as being established from within our spontaneity of reason, to adopt a traditional vocabulary. The world cannot take over that task either by providing causal relations that in themselves contribute to the

---

29 See also the account which Bonjour (1985), Chapter 6, puts forward. For a recent argument against such an account, see Brewer (1999), pp. 129–145.

individuation of beliefs or by being itself conceptual. However, the way in which an epistemic relation to the world is established from within our belief-forming activity then amounts to replacing (a) representational realism with direct realism and (b) epistemic self-sufficiency of beliefs with a social externalism that encompasses a physical externalism including a rational constraint of the world on our beliefs.

The three steps in which an account of the relation between our beliefs and the world has been set out in this chapter show again that holism about beliefs and social holism dovetail: the proposed holism about beliefs opens up a direct realism by rejecting a representational semantics. This direct realism removes the obstacle for conceiving the social practices that determine the meaning of beliefs as being shaped by dealings with physical things. It thereby yields a positive argument for direct realism: if the meaning of our beliefs is fixed by social practices that involve direct dealings with physical things, then it is reasonable to say that our beliefs are directly about physical things. Finally, such a conception of these social practices enables us to employ the conceptual tools of an inferential semantics to account for the representational feature of beliefs.

5.4 Holism and the Metaphysics of Mind

5.4.1 Holism and Anti-Reductionism

When a proposal for a general, substantial conception of holism was set out in the first chapter, those properties in which the arrangement of a thing with other things consists were excluded from the family of properties that make something a part of a system *S*. This requirement has the following consequence: an argument for holism in the philosophy of mind has to include an argument for the irreducibility of the intentional features in terms of which beliefs are conceived; for otherwise it would not be these features that make something a belief [1.3]. The proposed argument for social holism and holism about beliefs meets this requirement. The following two points show that the social holism and holism about beliefs advocated here are incompatible with reductionism in the philosophy of mind:

1) *Normativity*: The proposed argument for social holism and holism about beliefs, which starts from the rule-following considerations, is set out in terms of normative notions. The main reason for employing normative notions in the theory of meaning is that we need these notions to cope with the indeterminacies on which Quine and Kripke elaborate. Because of the transition to a normative level which having beliefs involves, what

makes something a belief irreducibly are intentional properties such as meaning. The description of the beliefs that are attributed to a person cannot be reduced to a description in terms of the natural sciences in the same way as ought-sentences cannot be reduced to is-sentences.[30] For a description that is available to a detached observer such as a description of the behaviour of a person in naturalistic terms is not able to describe which beliefs a person has. The argument from meaning being determined in social practices implies that meaning can only be described from within participating in the social practices of a linguistic community [3.2.2].

2) *Externalism*: The embraced account of rule-following takes us to externalism, namely to a social community of speakers and a physical environment shared with other speakers. All the arguments for an external individuation of belief states are at the same time arguments against the possibility to reduce the description of belief states to a neurophysiological description. The external factors which enter into the individuation of belief states do not enter into the individuation of brain states. Being in a certain belief state is a relational property of the person in question which involves the social and the physical environment, whereas being in a certain brain state is not such a relational property.

The argument from externalism implies also that belief states do not supervene on brain states. If one claims that belief states supervene on brain states, one maintains the following:[31] if there were two persons who are qualitatively identical as far as their brain states are concerned, then these two persons would be qualitatively identical as far as their belief states are concerned, too. There is no difference in the belief states of any two persons without there being a difference in their brain states. But there can be a difference in brain states between them without there being a difference in belief states.

If external factors enter into the individuation of belief states, then belief states do not supervene on brain states, because these external factors are irrelevant to the individuation of brain states. Nonetheless, the argument from externalism says nothing against belief states supervening on physical

---

30 See also Brandom (1994), pp. XIII, 45–46, 626.

31 On supervenience in the philosophy of mind, see, in particular, Horgan (1993), Kim (1993), and the essays in Savellos and Yalçin (1995).

states of the person in question plus her environment.[32] Consequently, this argument is compatible with claims of global supervenience of the mental including the intentional as a whole on the physical as a whole. What is more, this argument does not say anything against belief states being realized as physical states of a person in relation to a suitable environment either. As mentioned in [1.3], the ontological claim of physical realization does not imply the epistemological claim that it is possible to reduce the description of beliefs to a description in terms of the natural sciences.

What about the argument against reductionism from normativity? Does this argument clash with supervenience claims?[33] According to this argument, belief states are normative states in the sense that the belief states of a person are individuated in terms of normative relations to other such states such as relations of commitment and entitlement as determined by social practices. Whether or not this argument clashes with supervenience claims depends on the following: the argument from normativity accepts as a starting-point the indeterminacy of belief states on the basis of behavioural evidence on which Quine and Kripke elaborate. The question thus is whether this indeterminacy excludes a supervenience of belief states on the physical.[34] The claim that we have to consider is global supervenience; for the normative relations in question are determined externally by social interactions.

Global supervenience is a thesis of covariation in the following sense: two possible worlds which differ with respect to the mental including the intentional differ also with respect to the physical. Thus, if one varies the mental, one has to vary the physical as well. The converse is not the case: there are possible worlds that differ with respect to the physical without differing with respect to the mental. One can turn global supervenience into a contingent thesis by introducing suitable qualifications in the possible worlds to which it applies.[35] Global supervenience thus is an ontological

---

32 See also the position that Rudder Baker (1995) advocates as practical realism. This is a social externalism that is not opposed to global supervenience (see in particular pp. 132–134, 181, 213–217).

33 As to semantic supervenience see van Cleve (1992).

34 For an argument that Quine's thesis of the indeterminacy of translation in Quine (1960), Chapter 2, is incompatible with his physicalism, see Gemes (1991), pp. 105–107.

35 For a recent proposal of such a qualification see Jackson (1998a), pp. 10–14. The position which I set out does, however, differ from Jackson's views on supervenience;

claim. It is distinct from the epistemological claim of reduction. The thesis that *A* supervenes on *B* is compatible with, but does not imply the thesis that the description of *A* can be reduced to a description of *B*. It can therefore be misleading to explain the thesis that *A* supervenes on *B* in terms of *B* determining or fixing *A*. For determination may suggest the principal possibility of reduction, which covariation in the explained sense does not imply. To support this distinction between an ontological and an epistemological matter, one can appeal to the distinction that Kripke (1980) draws in *Naming and Necessity*: necessity, even metaphysical necessity, on the ontological level does not have to go with any sort of an *a priori* on the epistemological level. Given the sketched conceptualization of global supervenience, social holism and holism about beliefs as based on the rule-following considerations are hence compatible with global supervenience.[36]

The proposed argument for social holism and holism about beliefs implies the epistemological claim that even if one has complete physical knowledge of a possible world, one cannot know on the basis of that knowledge alone whether there are beings in this world that follow rules and which are the rules that they follow. However, nothing hinders us to go for the ontological claim that if there is a difference between the normative, social practices that determine meaning in two possible worlds, there also is some physical difference in these worlds. For these social practices have a physical implementation. Hence, we can say that the normative, social practices that determine meaning supervene on the physical as a whole.[37] This ontological claim commits us only to saying the following: assume that (a) someone participates in these social practices and thereby gains knowledge in intentional, normative vocabulary of the meanings that are determined by them and (b) that this person also has a complete physical knowledge of the world. If this person contemplates another possible world of which she knows that this new world is identical with the world under consideration as

---

for Jackson (1998a), pp. 67–84, argues in favour of identifying conceptual and metaphysical necessity.

36 Soames (1998a), pp. 211–233, in particular pp. 228–230, by contrast, construes the sceptical challenge of Kripke (1982) in such a way that it excludes even global supervenience and then argues against this challenge, employing Kripke (1980) to distinguish between determination in an ontological and determination in an epistemological sense.

37 However, Brandom refuses to commit himself to global supervenience. See, in particular, Brandom (1994), pp. 51–52, 628.

far as the physical is concerned, she knows that the meanings which are determined for the people in this world are the same as well. Owing to (a) and owing to the reference of the physical as a whole in (b), this commitment does not clash with the indeterminacies from which the proposed argument for social holism and holism about beliefs starts.

One may grant that the normative attitudes which a person adopts supervene on physical properties of the person and her environment. But what about the *correct* normative attitudes? There are no normative facts in the world, and the proposed holism is not even committed to real essences either. Nonetheless, we can say the following: once there are persons who have normative attitudes, what are the *correct* normative attitudes supervenes on the physical world as a whole, because the truth of beliefs about the world supervenes on the way the world is. Owing to the response-dependence which a social account of conceptual content implies [5.2.2], we are not in a position of a view from nowhere of the physical world as a whole [5.2.3]. But this epistemic limit only shows that there always is the possibility that any normative attitude may be challenged with reason, though not all at once.

As regards the more far reaching claim of a physical realization of belief states, nothing hinders to say that the normative attitudes of a person are realized as physical states of that person relative to a suitable environment. Consequently, properties such as belief states having a determinate content can be realized as physical properties of persons in relation to other persons and a physical environment, although they are not identical with physical properties. Thus, both the arguments from holism about beliefs and social holism against reductionism are compatible with the thesis of global supervenience of the mental as a whole on the physical as well as with the thesis that mental states are realized as physical states of persons in relation to their environment, although these arguments do not imply these theses.

5.4.2 What is the Resulting View of Ourselves?

The rule-following considerations are not only a forceful argument against a reduction of the description of intentional states to a description of physical states in today's philosophy of mind. They are also an argument against dualism in ontology. For the intentional vocabulary is irreducible to a physical vocabulary precisely because intentional states are determined only relative to those who engage in certain social practices. Conceiving the meaning of our beliefs as some sort of a mental entity over and above the physical runs into both the infinity and the normativity problem which

Kripke poses. The solution to these problems consists in focussing on the social practices of assessing each other's actions as correct or incorrect. All there is to the meaning of our beliefs is what is determined in these practices. These practices have a physical realization. Introducing anything in them which is not realized as physical states of the persons in their environment is of no help for the theory of meaning.

Note the way in which the argument against reduction works. A description that is available to a detached observer necessarily fails to notice the conceptual content that is determined for the persons it describes. Only a description that is available to a participatory observer, i.e., an observer who participates in the rule following practices of a community, can take into account the conceptual content which is determined for the members of that community. By mutually treating ourselves as thinking beings in a social practice we make it that some of our mental states are intentional states for us. As sketched in [3.2.2], the argument from social practices cannot yield more than intentional states relative to those who participate in a social practice.[38] Apart from that qualification, intentional states fall victim to the infinity problem and the normativity problem.

It may seem that one can pursue these considerations in such a way that one ends up in eliminativism. But this is not correct. (I will go into the alleged connection between holism and eliminativism in the philosophy of mind in [10.2]). The outlined normative pragmatics implies that as long as we continue to talk and thus engage in the described social practices, we have beliefs. Having beliefs is not a matter of the vocabulary in which we choose to talk about ourselves. Even if we talked in such a way that we employed only physical vocabulary (compare Rorty (1980), Chapter 2), our talk would be meaningful, and the sketched inferential semantics including social holism would apply to it. It may certainly happen that we evolve in such a way that we stop talking. But there can be no reason that we ought to do so.

The position that having beliefs consists in engaging in the described normative, social practices fits into the post-Cartesian project not to reify the mind (either as being an entity over and above the physical or as being nothing but some physical stuff among other physical stuff). The advocated normative pragmatics develops conceptual tools that are necessary to build up such a conception of the mind in positive terms. The point which is

---

38 See also the position of Sellars, "Philosophy and the Scientific Image of Man" in Sellars (1963), pp. 39–40. As to Sellars' philosophy of mind, see Bremer (1997).

necessary and sufficient to avoid a reification of the one or the other sort just is the claim that intentional states with a determinate conceptual content exist only relative to those who participate in the described normative, social practices. This position thereby opens up connections not only with the post-Cartesian project as it is pursued since Kant's *Critique of Pure Reason* at the latest, but also with existentialism, notably Heidegger's *Being and Time*. For Heidegger, in the first section of his masterpiece, sets out to develop conceptual tools for grasping ourselves as thinking and acting beings; one of his main concerns is that we cannot grasp our being minded in terms of properties of a thing.[39] The point of the account advocated here is not an ontological difference between the mental and the physical, but the normative character of intentionality. This account thereby links up this continental tradition with the argumentative tools that have been developed by analytic philosophy. Given the state of the art in the philosophy of mind, the argument from normative, social practices is a convincing way to uphold subjectivity in today's philosophy and to spell out what this subjectivity means without being committed to any sort of dualism in ontology or idealism.

5.4.3 Overview of Holism in Philosophy of Mind

This chapter and the three preceding ones have not been a neutral overview of positions that can be brought under the heading "holism in philosophy of mind". Instead, I have pursued an account of what I take to be the most plausible version of holism in philosophy of mind, which, in particular, combines holism about beliefs and social holism. To conclude the chapters on holism in philosophy of mind, it may therefore be helpful to briefly recall the main arguments for the claims that have been endorsed in these chapters:

1. The main argument against considering representation to be a primitive semantic feature is this one: we need an account of representation as a semantic and thus as an epistemic phenomenon which distinguishes representation in this sense from a mere discriminatory and reliable response to stimuli.

---

39 See *Being and Time*, in particular § 9. Compare Esfeld (forthcoming b). Haugeland (1982) goes as far as ascribing to Heidegger in *Being and Time* a social holism along the lines of the one of the later Wittgenstein (see also Brandom (1983)); but see the recent retraction in "Toward a New Existentialism" in Haugeland (1998), p. 4, and "The Intentionality All-Stars" in Haugeland (1998), p. 170.

2. The main argument for semantic holism in the form of an inferential role semantics is that this semantics gives an account of what distinguishes representation from a mere reliable response to stimuli. Consider a perceptual belief such as "This is red". What accounts for this belief having meaning is the inferential context in which it is embedded together with its appropriate circumstances of application. This is not to deny that specific causal relations are necessary for having perceptual beliefs. What is denied is that causal connections are sufficient to account for representation in an epistemic sense. Meaning as inferential context allows us to take into account the distinction between meaning (intension) and reference (extension) without falling back into any sort of a mentalism about meaning.
3. Any theory of meaning faces the problem of rule-following. Kripke's exposition of this problem in his interpretation of Wittgenstein's *Philosophical Investigations* is the main argument for (a) broadening an inferential role semantics in such a way that inferential role is conceived as a social role and (b) basing an inferential role semantics on a normative pragmatics which employs the notion of social practices: conceptual content is determined as a result of the assessment of each other's actions as correct or incorrect in a social community; these assessments show up in sanctions (reinforcements or discouragements). Holism about beliefs thus is linked up with social holism and thereby avoids the objections from shared beliefs and belief change.
4. A systematic reconstruction of holism in philosophy of mind has to start from social holism. The main argument for combining the endorsed social holism with the embraced holism about beliefs starts from the argument summed up in (2): for a belief to have meaning, appropriate circumstances of application are not sufficient. Meaning for one type of beliefs is determined only in relation to other other types of beliefs. This argument leads from the mentioned assessments of each other's actions to a normative pragmatics which conceives the meaning of a belief of the type *p* in terms of those other beliefs to which a person is committed and entitled and to which entitlement is precluded if she endorses *p*. These other beliefs are the inferential role of *p*. Hence, the meaning of a belief – or the content of a belief state – is its normative relations to other beliefs or belief states. What turns a mental state into a belief state and thus into an epistemic state are its normative relations to other such states.

5. Both the argument for social holism and the argument for holism about beliefs establish a version of these forms of holism along the lines of the first type of the proposed general conception of holism. Each part of the holistic system in question (social community, system of beliefs of a person) instantiates the property in question: each person has the property of having beliefs, and each belief has the property of meaning. However, with respect to this property, each of these things is ontologically dependent on there being other things with which it constitutes a system of the kind in question (social community, system of beliefs). This analysis shows that the proposed social holism does not clash with individualism in the common and garden sense.
6. The proposed holism about beliefs leaves no room for a distinction between a belief state and its content in the sense of its content being a representation and acting as an epistemic intermediary between the belief state and the thing or event in the world about which the belief state is. This holism about beliefs thereby makes available a direct realism: when it comes to the epistemic relation of aboutness, there is nothing which intervenes between the belief state and the thing or event in the world about which the belief state is.
7. Instead of leading to social relativism, the proposed social holism can be combined with realism: since that holism is conceived in terms of open-ended I–thou relations among individuals, it by no means identifies correct beliefs about the world with a social fact of taking something to be a correct belief about the world. Meaning is accounted for in terms of social practices; but this does not prevent us from conceiving the truth of our beliefs to supervene on the way the world is. The upshot is a direct, pragmatic realism that overcomes the opposition between metaphysical realism and relativism.
8. This pragmatic account of meaning is an argument in favour of a pragmatic conception of rationality: the described social practices are the root of rationality. The physical environment literally is part of these practices. This pragmatic view of rationality enables us to conceive a rational constraint of the world on our beliefs without having recourse to a comprehensive holism that considers the conceptual as being unbounded.
9. The proposed holism in the philosophy of mind is parsimonious in its ontological commitments: although it is incompatible with reductionism, no commitment to any sort of a dualistic ontology ensues. Its point are normative relations in distinction to mere causal ones.

CHAPTER 6
CARTESIANISM IN PHYSICS AND HOLISM ABOUT SPACE OR SPACE–TIME

Summary

To what extent does holism in the philosophy of today's physics break with the Cartesian tradition in modern philosophy of nature? To prepare the ground for an answer to this question, I show in this chapter that Cartesian philosophy of physics admits of a sort of holism. I take up the interpretation according to which Descartes and Spinoza consider matter and the continuum of physical, three-dimensional Euclidean space to be the same thing [6.1, 6.2]. Based on this identification we reach a holism which can be called "holism about matter as holism about space": the parts of matter are identical with the points or regions of space. All physical properties are properties of points or regions of space. What makes something a point or a region of space is relational properties or relations to other points or regions within the whole of space. If the physical things are the points or regions of space, then any physical property can only be instantiated if there are many material things (points or regions of space) which extend as far as the whole material realm (the whole space) [6.3]. A theory of motion and change can be included in holism about space [6.4]. Finally, I examine whether holism about space (or space–time) can be directly applied to today's physics. It is suggestive to try this holism out on general relativity. There is a programme for a further development of general relativity that amounts to a concrete elaboration of this holism, namely the geometrodynamics of Wheeler. However, this programme failed. We do not have a physics at our disposal which is not committed to physical systems over and above space–time. Even if we lay stress on the fact that Descartes' and Spinoza's philosophy of physics admits of a sort of holism, we cannot avoid a revision of Cartesianism in today's philosophy of nature [6.5].

6.1 A Basis for Holism in Descartes' Philosophy of Physics

To introduce holism in philosophy of physics, let us go back to Cartesian philosophy of physics. In contrast to widespread beliefs that regard Descartes' philosophy of physics solely as the source of the modern mechanistic view of nature, a Cartesian philosophy of physics admits of a substantial holism. By a Cartesian philosophy of physics, I do not mean only a position which can be attributed to Descartes himself, but a position which fits into the conceptual framework of that philosophy of physics for which Descartes laid the ground. That is the whole of the philosophy of classical physics down to general relativity. Accordingly, my concern here is not the exegesis of Descartes' texts.

By going into the sort of holism that Cartesian philosophy of physics incorporates, I prepare the terrain for the examination of holism in quantum physics. I have two purposes in mind: (a) I intend to show that a substantial holism in the philosophy of physics is not tied to the conceptual features of quantum theory. (b) I plan to use the holism in Cartesian philosophy of physics as the background for my evaluation of the philosophical relevance of the holism in quantum physics. I thereby intend to reach a more balanced judgement than the widespread impression that holism in quantum physics amounts to a total break with the Cartesian tradition in modern philosophy of nature.[1]

It may seem that the modern view of nature as such is a holism in the philosophy of physics. Some scholars regard the transition from the Aristotelian and medieval view of nature to the modern view of nature, to which Descartes is central, as a transition from an ontology of substances or essences to an ontology of functions (e.g. Rombach (1966)). Instead of treating non-relational, essential properties of substances, types of functional dependence among physical systems are considered in mathematical terms. Consequently, according to this tradition of scholarship, conceiving nature along the lines of modern science means conceiving nature as a network of relations. Starting from the claim that science deals exclusively with quantities, quantities are considered as relational properties in this tradition.

Let us assume for the sake of argument that the modern view of nature consists in replacing a description of essences with a description of functions. The mathematical description of the value of a physical property of a system such as, say, its length, width, and depth, or its mass can be seen as consisting in relating that value to the value of the same properties of other systems. Saying that a system is two centimetres long and weighs six grams can be taken to indicate relations of its being longer or shorter, heavier or lighter than other systems. But even if one grants that the physical description of the values of properties consists in indicating relations, it does not follow that the properties of length and mass are themselves relational. This consideration does not preclude that a system which is the only physical thing in a possible world can have a certain geometrical shape such as being spherical or being square and a certain mass. Thus, even if physics conceives

---

1 As to a comparison between Spinoza's holism and quantum holism, see also Shimony (1999).

properties in relational terms, this does not imply that these properties themselves are relational. When I claim that Cartesian philosophy of nature admits of a holism, I do not mean a consequence of the quantitative treatment of physical properties in general, but more specifically a consequence of conceiving the physical realm as extended substance.

Descartes counts among the first who propose a philosophy of nature in accordance with modern science. Modern science is opposed to the teleological view of nature that Aristotelianism suggests. Descartes not only rejects Aristotelian teleology, but also the Aristotelian version of essentialism. According to him, there is no specific form for each kind of things which constitutes the essence of each member of a kind and which is imperishable. Consequently, for Descartes, there is no hierarchical order of forms which culminates in God. All physical things are characterized merely by the attribute of extension. By extension, Descartes means spatial extension, i.e., extension in length, breadth, and depth, whereby he considers space to be Euclidean. The terms "corporeal", "material", and "extended" are identical in meaning and reference (in particular *Principles*, part II, § 4). Instead of a great number of different sorts of substances, Descartes admits only two sorts of substances – *res extensa* and *res cogitans*. All substances of both these sorts originate in God. Each human being is a cogitating substance which is linked to extended substance, i.e., a body. Substances of both these sorts interact through the brain of each human being (*Meditations* VI, § 20; *Passions of the Soul*, part I, §§ 31–32).

Descartes renounces not only Aristotelianism, but also the ontology of atomism. According to him, there are no smallest, indivisible bodies. Every corporeal thing can in principle be divided.[2] Descartes' rejection of atomism can be construed as a consequence of his complete renunciation of Aristotelian essentialism. Atomism, as well as Aristotelian essentialism, maintains that there are a number of determinate, eternal, and immutable entities which are the things that really exists. Democritus uses the terms "*οὐσία*", "*ἰδέα*", "*εἶδος*" and "*μορφή*" to refer to the atoms.[3] These are the very terms which Plato and Aristotle employ to describe the forms. Appropriately enough, the German philosopher Hans Blumenberg says that the "unchangeable and specifically formed atoms were only 'the eternal' of the one Being of Parmenides, of Plato's Ideas and Aristotle's Forms, 'in

2 *Principles*, part II, §§ 20, 34; *Météores* I (Descartes (1965–1973), Volume 6, pp. 238–239); letter 537 to More, 5 Feb. 1649 (Descartes (1965–1973), Volume 5, p. 273).

3 Diels and Kranz (1960/61), Volume 2, fragments 68 A 37, A 57, B 141.

material form'".[4] If one rejects completely Aristotelian essentialism, a philosophy which replaces this essentialism with a theory of corporeal things that have the same characteristic features as Aristotelian essences is discredited too.

Despite his opposition to Aristotelian essentialism as well as atomism, Descartes maintains that change is possible only if there is something that persists throughout all changes. In that respect, he agrees with the Aristotelians as well as the atomists. However, in Descartes' view, there are neither immaterial forms nor very small bodies that necessarily persist throughout all changes. Since every body is divisible, there is no corporeal shape which cannot fade away. Consequently, for Descartes, what necessarily persists throughout all changes can only be the corporeal realm as a whole regardless of corporeal shapes, although corporeal substance can of course not exist without some specific corporeal shapes. Descartes can thus be seen as approving of some sort of essentialism as well. But this essentialism is distinct from Aristotelianism and atomism: there is no multitude of determinate essences. There is only one essence of all corporeal things: extension in three spatial dimensions. Corporeal substance as a whole irrespective of corporeal shapes necessarily persists throughout all changes. In the *Synopsis of the Meditations*, Descartes says:

> First, we need to know that absolutely all substances ... are by their nature incorruptible and cannot ever cease to exist ... Secondly, we need to recognize that body, taken in the general sense, is a substance [*corpus quidem, in genere sumptum, esse substantiam*] so that it too never perishes. But the human body, in so far as it differs from other bodies, ... can very easily perish.[5]

No substance can cease to exist. But every particular body, such as a human body, can cease to exist. Therefore, only the corporeal realm taken as a whole is a substance. There is no multitude of corporeal substances. All corporeal things are just one substance.[6] Leibniz, by contrast, takes

---

4 Blumenberg (1988), p. 189. Translation adopted from Blumenberg (1991), pp. 167–168. See furthermore p. 241 (p. 211 in the English translation).

5 *Synopsis of the Meditations* (Descartes (1965–1973), Volume 7, pp. 13–14). Translation adopted from Cottingham, Stoothoff and Murdoch (1984–1991), Volume 2, p. 10. See furthermore *Principles*, part II, § 4, and letter 266 to Regius, Jan. 1642 (Descartes (1965–1973), Volume 3, p. 505).

6 Compare Nason (1946), pp. 455, 468; Treder (1982), p. 279; Cottingham (1986), pp. 84–85; Curley (1988), pp. 17–18, 32–33.

Descartes to mean merely that all bodies have one nature. But the sole reason which he gives for this interpretation is this one: it is evident that the world is an aggregate.[7]

Kant puts forward an argument that is similar to Descartes' one when he considers the principle of the persistence of the substance in the *Critique of Pure Reason*. He writes:

> A philosopher was asked: How much does the smoke weigh? He replied: If you take away from the weight of the smoke that was burnt the weight of the ashes that are left over, you will have the weight of the smoke. He thus assumed as incontrovertible that even in fire the matter (substance) never disappears but rather only suffers an alteration in its form.[8]

On many occasions, Kant refers to a multitude of corporeal substances, in particular in the third analogy of experience (A 211 / B 256 – A 218 / B 265). However, according to this quotation, there is only one corporeal substance – the whole of matter regardless of corporeal shapes. Solely the whole of matter persists throughout all changes.

As mentioned above, Descartes considers the terms "corporeal", "material", and "extended" to be synonymous.[9] Furthermore, he maintains that extended substance and physical space are identical.[10] If we base ourselves on the quotation from the *Synopsis of the Meditations*, the conclusion which suggests itself is that the corporeal substance is the same thing as space.[11]

However, Descartes' position is ambiguous: On the one hand, the argument that no substance can cease to exist leads to admitting only one

---

7 "*Cum Cartesius et alii dicunt omnium corporalium unam esse substantiam, intelligunt unam similem naturam nec (puto) volunt, quod omnia corpora unum vere substantiam conflent. Et sane aggregatum esse Mundum, ut grex aut machina, res ipsa ostendit.*", Leibniz, letter 31 to de Volder, 30 June 1704 (Gerhardt (1875–1890), Volume 2, p. 271, lines1–4). Compare Woolhouse (1993), p. 23.

8 Kant, *Critique of Pure Reason* A 185 / B 228. Translation according to Kant (1998).

9 Brown (1989), by contrast, maintains in the first part of his paper that there is a material substratum underlying extension. But see *Principles*, part I, § 63, and consider what, if not extension, could be the property by virtue of which this substratum is a *material* substratum.

10 *Principles*, part II, §§ 10–12. Compare Garber (1992), p. 132.

11 Compare Keeling (1934), §§ 8–9; Hesse (1961), p. 103; Graves (1971), pp. 82–83; Hartz (1989), pp. 23–24; Kobayashi (1993), p. 61. As regards opposition to this interpretation, see especially the argument by Woolhouse (1994), in particular pp. 30–33.

substance, namely the whole corporeal realm. On the other hand, in the first part of the *Principles* (§ 60), Descartes assumes that the parts of the corporeal substance are themselves substances: he maintains that there is a real distinction between the parts of the corporeal substance; a real distinction is the sort of distinction that applies to substances.

Furthermore, turning to physics, the obvious problem for attributing an identification of corporeal substance with space to Descartes is that his physics is formulated in terms of moving bodies and, in particular, a conservation of the quantity of motion in the whole universe (*Principles*, part II, §§ 36–42). Identifying matter with space, by contrast, implies that physics can be reconstructed without being committed to material things in addition to space. Moreover, regions of space may have physical properties; but they cannot move. However, let us postpone this problem to [6.4]. Before going into this problem let us take the identification of corporeal substance with space in some of Descartes' texts as a basis upon which a holism in the philosophy of physics can be built. The possibility of an identification of matter with space is the only point which I adopt from Descartes' philosophy of physics. Because there is evidence for such an identification in some of Descartes' texts, the position that will be sketched out is a holism within the framework of Cartesian philosophy of physics. Nonetheless, the point of this examination whether that holism can be attributed to Descartes himself.[12] But let us first of all broaden our basis of historical texts by briefly going into Spinoza's philosophy of physics.

6.2 Spinoza's Theory of Extension

Two of the central problems for Descartes' metaphysics can be expressed by the following questions: How can an immaterial cogitating substance be causally relevant for physical events? How can a substance be created? According to the definition of a substance which Descartes sets out in the first part of the *Principles*, the existence of a substance does in no way depend on any other thing. He concedes that, strictly speaking, only God is a substance (§ 51).

Spinoza takes up these two problems. He comes to the conclusion that there is only one substance which is identical with God and nature (*Ethics*, part I, proposition 14). The substance comprises infinitely many attributes

---

12 Kobayashi (1993), pp. 14, 98, 124–125, 127–128, makes use of the term "holism" in his interpretation of Descartes' philosophy of nature. By employing this term, he intends that all phenomena are referred to the whole system of the world.

(*Ethics*, part I, proposition 11). The attributes are the properties that the substance has as such.[13] Two of the attributes are extension and cogitation.[14] These are the only attributes known to us. Within the attributes, the substance is differentiated into modes – specific forms of, for instance, extension and cogitation. All particular things in the physical realm are modes of the one substance. They are ways which express the attribute of extension in a certain and determinate manner.[15] Nonetheless, Spinoza refers to modes as parts of the substance as well.[16] Each mode of extension parallels a mode of cogitation (and *vice versa*). Causal relations hold between modes of extension on the one hand and modes of cogitation on the other hand. There is no interaction between the corporeal and the mental realm. Corporeal and mental events are arranged in such a way that they run parallel to each other (*Ethics*, part II, propositions 6–7, and 13 scholium).

Spinoza thus introduces remarkable changes in Descartes' metaphysics. These changes also concern the ontological status of the corporeal realm: extension belongs to the divine nature.[17] Nonetheless, one can regard Spinoza as following the line of thought that leads Descartes to identify matter with space in some of his texts. Like Descartes, Spinoza considers the terms "corporeal", "material", and "extended" to be identical in meaning and reference.[18] He is also opposed to atomism.[19] Nevertheless, like Descartes,

---

13 See *Ethics*, part I, definition 4. For this interpretation, compare Bennett (1988), § 16, and the thorough discussion by Glauser (1994).

14 *Ethics*, part I, proposition 14 corollary 2; proposition 15 scholium (Spinoza (1925), Volume 2, p. 57, lines 22–23); part II, propositions 1–2.

15 See especially *Ethics*, part I, proposition 25 corollary; part II, definition 1.

16 See, for instance, *Ethics*, part I, proposition 15 scholium, and compare *Ethics*, part I, proposition 25 corollary; part II, definition 1.

17 On criticism of Descartes' notion of extension, see furthermore letter 81 to Tschirnhaus, 5 May 1676 (Spinoza (1925), Volume 4, p. 332, lines 15–21); letter 83 to Tschirnhaus, 16 July 1676 (Spinoza (1925), Volume 4, p. 334, lines 22–26).

18 See especially *Ethics*, part I, proposition 15 scholium (Spinoza (1925), Volume 2, p. 57, lines 13–14); *Ethics*, part II, definition 1; *Descartes' Principles Demonstrated in Geometrical Manner*, part II, proposition 2 and corollary. Compare Lecrivain (1986), pp. 34–39.

19 Compare *Descartes' Principles Demonstrated in Geometrical Manner*, part II, proposition 5 with *Ethics*, part I, proposition 15 scholium (Spinoza (1925), Volume 2, p. 59, line 35 – p. 60, line 5) (water example) and letter 6 to Oldenburg (Spinoza (1925), Volume 4, p. 29, lines 12–17). On Spinoza's rejection of atomism, see Rice (1975), pp. 200–201; Lecrivain (1986), pp. 37–38.

he maintains that every change requires something that persists throughout all changes; this can only be the corporeal realm as a whole regardless of specific corporeal shapes.

Spinoza sets out his theory of extension mainly in the scholium to the fifteenth proposition of part one of the *Ethics*. Adopting Descartes' terminology, he speaks of the one substance, insofar as it is extended, as corporeal or extended substance (or simply as quantity or matter).[20] In two places in that scholium, he claims that this substance can only be conceived as infinite, unique, and indivisible.[21] However, he does not intend to deny that every body is divisible. In both these places, he adds an example. In the first place, he says:

> So also others, after they feign that a line is composed of points, know how to invent many arguments, by which they show that a line cannot be divided to infinity. And indeed it is no less absurd to assert that corporeal substance is composed of bodies, *or* parts, than that a body is composed of surfaces, the surfaces of lines, and the lines, finally, of points.[22]

According to this quotation, the corporeal substance is a continuum. Every division results in something that can be divided still further. The indivisibility of the substance, which goes with the divisibility of all corporeal things, can be conceived in the same way as the indivisibility of a three-dimensional spatial continuum: focusing on the continuum, divisions are nothing but demarcations within the continuum. By claiming that the corporeal substance is indivisible, Spinoza only intends to point out that it is

---

20 *Ethics*, part I, proposition 15, scholium in Spinoza (1925), Volume 2, pp. 57–60:
substantia extensa: p. 57, line 22; p. 58, lines 22, 30–31; p. 60, line 13.
substantia corporea: p. 57, lines 25, 28; p. 58, lines 8, 12, 14–15, 18, 34–35; p. 59, lines 7, 11, 19; p. 60, line 2.
quantitas: p. 59, line 25.
materia: p. 59, lines 32–34; p. 60, line 7.

21 *Ethics*, part I, proposition 15, scholium (Spinoza (1925), Volume 2, p. 58, line 34 – p. 59, line 1 and p. 59, lines 29–30). On the substance being infinite, see *Ethics*, part I, proposition 8. On its being unique (unicus), see *Ethics*, part I, proposition 14 and corollary 2; but see also letter 50 to Jelles, 2 June 1674 (Spinoza (1925), Volume 4, p. 239, line 16 – p. 240, line 5). On its indivisibility, see *Ethics*, part I, proposition 13 and corollary. See furthermore letter 12 to Meyer, 20 April 1663 (Spinoza (1925), Volume 4, p. 56, lines 12–15).

22 *Ethics*, part I, proposition 15 scholium (Spinoza (1925), Volume 2, p. 59, lines 3–9). Translation adopted from Curley (1985), p. 423. See furthermore letter 12 to Meyer, 20 April 1663 (Spinoza (1925), Volume 4, p. 55, line 11 – p. 56, line 4).

not composed of parts which exist on their own.[23] Nonetheless, parts can be discerned within the substance.

The example which Spinoza gives in the second place confirms this interpretation. Spinoza says there:

> This will be sufficiently plain to everyone who knows how to distinguish between the intellect and the imagination—particularly if it is also noted that matter is everywhere the same, and that parts are distinguished in it only insofar as we conceive matter to be affected in different ways, so that its parts are distinguished only modally, but not really [*quod materia ubique eadem est, nec partes in eadem distinguuntur, nisi quatenus materiam diversimode affectam esse concipimus, unde ejus partes modaliter tantum distinguuntur, non autem realiter*].
>
> For example, we conceive that water is divided and its parts separated from one another—insofar as it is water, but not insofar as it is corporeal substance. For insofar as it is substance, it is neither separated nor divided. Again, water, insofar as it is water, is generated and corrupted, but insofar as it is substance, it is neither generated nor corrupted.[24]

By differentiating a modal from a real distinction, Spinoza means that the parts of matter are modes of the one substance instead of things which have a being on their own. Spinoza's philosophy of physics thereby differs from the first part of Descartes' *Principles*, where, as mentioned in the last sub-chapter, Descartes conceives a real distinction between the parts of the corporeal substance (§ 60). According to this quotation, one gallon of water can be seen as being divided up into four portions of one quart and eight portions of one pint, etc. These portions can be regarded as being separated from each other; they do not overlap. They can change from water into other stuff. However, insofar as the substance is concerned, nothing but demarcations within the substance are considered, and properties of these demarcations change.

These two quotations show that Spinoza endorses a comprehensive holism about the physical realm: no part of matter exists on its own. He writes in a letter: "If one part of matter were reduced to nothing, the whole extension

---

23 See especially *Ethics*, part I, proposition 15 scholium (Spinoza (1925), Volume 2, p. 59, lines 11–12).

24 *Ethics*, part I, proposition 15 scholium (Spinoza (1925), Volume 2, p. 59, line 32 – p. 60, line 5). Translation adopted from Curley (1985), p. 424.

would vanish instantly.'[25] Spinoza explicitly identifies extension and body with space only in his reconstruction of Descartes' *Principles*.[26] This writing is not an expresssion of Spinoza's own position in the first place.[27] Nonetheless, one can argue that Spinoza regards the corporeal substance as being identical with space. The two quoted passages support this interpretation.

Jonathan Bennett (1988) develops an interpretation according to which Spinoza identifies the physical realm with space.[28] He traces our talk of bodies in space back to properties that are predicated of regions of space. Bennett does not use the term "holism". He only speaks of monism. He sums up his position in these terms:

> It suggests that there is just one substance—namely, the whole of space—regions of which get various qualities such as impenetrability, mass, and so on, so that any proposition asserting the existence of a body reduces to one saying something about a region of space. (§ 22.1)

Hence, all propositions which refer to things in space are reduced to propositions which refer to regions of space and which attribute physical properties to regions. However, the physical properties of the regions of space are not identical with the properties that we ascribe to the bodies which we localize in these regions, although these latter properties are realized as properties of regions of space. Bennett marks the concepts of properties which are predicated of regions of space with an asterisk (*). He writes:

---

25 "... *quod si una pars materiae annihilaretur, simul etiam tota extensio evanesceret,*" letter 4 to Oldenburg (Spinoza (1925), Volume 4, p. 14, lines 19–20). See furthermore *Short Treatise on God, Man, and His Well-Being*, part I, Chapter 2, § 20 (Spinoza (1925), Volume 1, p. 25, line 19 – p. 26, line 6).

26 *Descartes' Principles Demonstrated in Geometrical Manner*, part II, definition 6 as well as proposition 2, corollary and demonstration.

27 But the reference in *Ethics*, part I, proposition 15 scholium (Spinoza (1925), Volume 2, p. 59, lines 16–17) (elaboration of the claim that there is no empty space) is apparently meant to be a reference to *Descartes' Principles Demonstrated in Geometrical Manner*, part II, proposition 3.

28 Bennett (1988), Chapter 4. Rombach (1966), pp. 9–97, especially pp. 29–33, 42–43, 88–89, and Sachs (1976), especially p. 130, are, to a certain extent, forerunners of Bennett's interpretation. See also Harris (1973), pp. 52–54. For interpretations which take up Bennett's view, see Matthews (1989), pp. 5–6; Rohs (1996), Chapter 2.

> To say that the puddle is slimy is to say that a certain region of space is slimy*—i.e., has that property of regions which we conceptualize by saying that there are slimy things in them. (§ 23.5)

Bennett thus claims that it is possible to reconstruct the properties which we attribute to bodies in space on the basis of the properties which the regions of space have according to his interpretation of Spinoza.

The programme of such a comprehensive reduction does not clash with regarding this conception as a form of holism (although Bennett himself does not employ the term "holism"). The holism concerns in this case the basic physical level. A holism about the basic physical level is compatible with the programme to reduce the description of higher physical levels to the theory of the basic level. A clash between holism and reductionism occurs only if there is holism at higher levels (such as organisms or the mental). Any such form of holism is incompatible with the claim that the description of the higher level in question can be reduced to a description of a lower level [1.3].

Bennett chooses to call the conception that he ascribes to Spinoza "field metaphysic" because, in his view, it accords with the physics of our days (§§ 20.5, 22.6, 25.1). Based on this conception, all matter can indeed be regarded as one field (although the concept of a field is not a concept of the physics of Spinoza's time, but of 19th century physics). Nonetheless, I prefer the description "holism about matter as holism about space" to "field metaphysic". One reason is that the term "field metaphysic" does not reveal the holism involved. However, a term such as "field holism" could point out the holism as well. But a field metaphysics or a field holism can be conceived in another manner than Bennett proposes. Even if one regards the physical realm as one field and even if one conceives this field as a holistic system, one is not obliged to identify this field with space. Consequently, the term "field metaphysic" fails to make the crucial point explicit: all matter is one field *which is identical with space*. Physics does not require an ontological commitment to physical systems such as fields in addition to space.

The claim that Spinoza identifies the extended substance with space – and Bennett's interpretation in particular – is in dispute.[29] I cannot go into this

29 See, for instance, the criticism of Curley (1991) and the reply of Bennett (1991). For Curley's own position, see Curley (1969), Chapter 2; Curley (1988), Chapter 1. For a criticism of the identification of matter with space in the context of relating Spinoza's

dispute here. My point is not the exegesis of Spinoza's texts, but to determine to what extent a holism can be incorporated into a Cartesian philosophy of physics.

6.3 Holism about Matter as Holism about Space

That sort of holism which Cartesian philosophy of physics includes is based upon identifying matter with physical space. Physical space is conceived as a three-dimensional, Euclidean space. If there are parts of space, regions count among the parts of space. Regions are, like space itself, three-dimensional. Regions can be arbitrarily small. Furthermore, one can maintain that dimensionless points are also parts of space. Many of our physical theories are formulated in such a way that physical properties are regarded as being instantiated at points of space or space–time. However, in metaphysics, it is in dispute whether it is plausible to admit points as parts of space and approve of a realism about points.[30] Therefore, I formulate holism about space (and holism about space–time in [6.5]) without endorsing a commitment to points as parts of space. If there are points, holism about space (and holism about space–time) applies to points of space as well as regions of space. Since being a part of space is transitive, I speak of parts of space without further qualifications and drop the terms "constituent parts" and "constituents".

According to the conception under consideration, physical space is a continuum. Whatever account of space as a continuum one accepts, if there are proper parts of space such as regions or points, there cannot be only one thing which has the properties that make something a region or a point of space. Something can have these properties only if there are other points or regions which run through the whole space. What makes something a point or a region of space is relational properties or relations to other points or regions within the whole of space if space is a continuum. There are no qualitative non-relational properties which could serve as a basis upon which these relations or relational properties could supervene. As soon as one refers to one point or region of space, one has to recognize the existence of other points or regions that extend as far as the whole space. We can therefore say: Insofar as a thing instantiates the properties that belong to the

---

philosophy of physics to today's philosophy of physics, see Paty (1986), p. 269 / Paty (1988), pp. 184–185; van Zandt (1986), p. 258.

30 For an argument that points are limits of extended parts of space and not parts of space themselves, see Hoffman and Rosenkrantz (1994), pp. 100–113, 188–193.

family of properties which make something a point or a region of space, a thing is ontologically dependent on there actually being other points and regions together with which this thing encompasses the whole space of a possible world. This ontological dependence is clearly distinct from causal dependence. It does not require any causal arrangement.

Thus far, this conception is neutral with respect to the discussion on relational versus absolute theories of space: There may be nothing more to points or regions of space than spatial relations. But points or regions of space may also be substantial entities. The only point that is needed at this stage of the argument is that relations of one point or region to other such points or regions belong to the family of properties which make something a point or a region of space. A substantialism about space and points of space in particular may not regard these relations as essential for the identity of the individual points or regions. But as long as it is acknowledged that these relations are necessary for something to be a point or a region of *space* (in distinction to being a point of some other manifold), the claim made in the preceding paragraph holds. Recall that the proposed conception of holism is not formulated in terms of the existence of the parts of a holistic systems, but in terms of the properties that make something a part of a system of the kind in question; it leaves open whether or not these properties are essential for the identity of the thing in question.

It may be regarded as philosophically uninteresting that something can have the properties that make it a point or a region of space only if there are other points or regions which run through the whole space. However, this issue turns into an interesting holism about the physical realm as soon as we identify matter with space. Identifying matter with space does not amount to the elimination of matter. It only amounts to the claim that there are no physical systems such as particles or fields in addition to space. Matter and space are the same thing. The points or regions of space are the parts of matter. They are not only the basic individuals, but also the sole individuals in the physical realm. They have physical properties, and all physical properties are realized as properties of points or regions of space. Contrary to the claim of Bennett (1988, § 23.1 and 4), bodies are not adjectival upon regions of space; they are not reduced to properties of regions of space. Bodies are regions of space having certain physical properties. An extended, material particular and a region of space are the same thing.

If the points or regions of space are the things that have physical properties, there cannot be only one thing that has physical properties, because there cannot be only one point or region of space. Since matter and space are the same thing, instantiating physical properties that make something a part of matter is the same as instantiating the properties which make something a part of space. Therefore, a thing is, insofar as it instantiates any physical property, ontologically dependent on there being other things together with which it encompasses the whole space, that is, the whole material realm. Any physical property can only be instantiated if there are a number of material things – points or regions of space which extend as far as the whole physical realm.

Thus, for matter, the identification with space means that all matter is one holistic system. For space, in turn, the identification with matter means that it is matter everywhere. Every part of space is a part of matter. There is not even a conceptual distinction between space and matter. It is therefore not possible to reconstruct this position by beginning with some sort of an empty space and then imagine instantiations of physical properties being inserted in that space. Hence, it is necessary that if something is a part of space, it instantiates physical properties of some sort. Some physical properties can be attributed to every point and every region of space. In this sense, there is no empty space.

Consequently, strictly speaking, every physical property is relational. That a point or a region of space has a physical property is relative to there being other points or regions which also have some physical properties. However, this conception alone does not contain any specific connections between property instantiations: it merely says that each thing which instantiates physical properties is ontologically dependent on there being other things which have some physical properties and which extend as far as the whole material realm; but it does not impose any restrictions upon these other physical properties. Nothing is *ontologically* dependent on there being other things that have *specific* physical properties. There are only relations of causal dependence among the parts of matter with respect to the instantiation of specific physical properties.

It could still be objected that this holism is not interesting: regarding physical properties as properties of regions of space trivially has the consequence that any physical property can only be instantiated if there are many regions of space which run through the whole space. However, this objection overlooks that the identification of matter with space does not leave space untouched, so to speak. As soon as space is considered to be the same thing as matter and different physical properties are regarded as being instantiated by its points and regions, it is seen as having an internal structure. This structure consists in the different physical properties (or the different values of these properties) which space has in different regions, i.e., different parts. The regions may even have rather sharp boundaries; in any case, there are vague boundaries that delimit regions. Conceiving matter and space as the same thing thus has two consequences:

- Identifying space with matter yields an internal structure of space in the sense of parts of space with different physical properties.
- Identifying matter with space includes the instantiation of all physical properties in the instantiation of the properties which make something a part of space. Consequently, everything which is a part of matter is, with respect to all its physical properties, ontologically dependent on there being other parts of matter together with which it encompasses the whole material realm.

It is thus both the instantiation of physical properties by regions of space and the dependence of something insofar as it is a region of space on there being other regions which run through the whole space that turns space–matter into a holistic system. This philosophy of physics can be called "holism about matter as holism about space" (or just "holism about space"):

Holism about matter as holism about space

> Matter and the continuum of physical, three-dimensional, Euclidean space is the same thing. The parts of matter are identical with the points or regions of space. All physical properties are properties of points or regions of space. Insofar as a thing instantiates the properties which belong to the family of properties that make something a part of space, a thing is ontologically dependent on there actually being other things which are also parts of space and which extend as far as the whole space. Since the instantiation of all physical properties is included in the instantiation of the properties which make something a part of space, everything is, with respect to all its physical properties, ontologically dependent on there actually being other material things (i.e., parts of space) together with which it encompasses the whole material realm (i.e., the whole space) of a possible world. Therefore, all matter is one holistic system.

Thus conceived, holism about space is a case of the first type of holism outlined in [1.4]: the parts of matter, i.e., the points and regions of space, instantiate physical properties in the first place. However, it is not mandatory to conceive holism about space in that way. Let us come back to Bennett's interpretation of Spinoza. The claim that all propositions which refer to things in space are to be reduced to propositions which attribute physical properties to regions of space is only the first step of his interpretation. In a second step, Bennett eliminates the ascription of properties to the regions of space. Space itself is the only subject of predication. All propositions are to be formulated in such a way that all properties are directly predicated of space. In order to localize the instantiations of the properties, an adverb such as the indexicals "here" or "there" has to be added. Instead of saying "Region *x* is *F*" we have to say something like "Space is *F* there". Bennett sums up the two steps of his interpretation in these words:

> To say that the puddle is slimy is to say that a certain region of space is slimy*—i.e., has that property of regions which we conceptualize by saying that there are slimy things in them. And to say that there is a slimy* region is to say that *space is slimy* locally*—where 'locally' is just my placeholder for whatever adverb would do the required job. (§ 23.5)

As a result of this second step, quantifiers are useless in Bennett's view: there is only one thing of which properties are predicated (§ 23.4).

Independently of whether or not this suggestion succeeds in reducing the reference to regions of space to the reference to space, it shows that there are two forms of holism about space: that holism can also be construed along the lines of the second type of holism. We begin with properties of the whole. Considered as properties of the whole, these properties are of course intrinsic, because there is nothing apart from the whole. Nonetheless, these properties manifest an internal structure within the whole. They are not properties such that space is *F* and *G simpliciter*. Instead, space is *F* in one region and *G* in another region. We thus get to properties of space which introduce something as a region of space among other regions of space. These properties hence make something a part of space – in a broad sense of part that does not imply independent existence and not even being itself a subject of the predication of properties. We can therefore apply the second type of the proposed conception of holism: these properties indicate the way in which the parts are related with respect to the properties that make something a part of space, because they are the very properties of space which introduce something as a part among other parts within space.

However, leaving aside the interpretation of Spinoza, there is apparently no systematic reason why we should conceive holism about space along these lines; for the physical properties at issue are instantiated locally. It therefore seems appropriate to construe holism about space as formulated above, namely in terms of physical properties of points or regions of space. Thus conceived, holism of space comes under the first type of the proposed conception of holism.

According to holism about matter as holism about space, space–matter is an absolute entity, if anything is. However, there is no absolute space in the sense of a space that is absolved from matter. Space could not exist apart from matter. Holism about matter as holism about space can therefore not be associated with the theory of an absolute space as an entity in distinction from matter. In the framework of this latter theory, one can say that matter is ontologically dependent on space. Nothing can have properties which make it a part of matter without there being absolute space. But space is not ontologically dependent on matter. There can be space without any matter that fills space. Consequently, a material thing can exist (and move) in absolute space without there being any further material thing. Objections against such a theory of space do not touch holism about matter as holism about space.

In particular, holism about space does not face the problem that the ontological status of space remains unclear. According to Democritus, who introduces the ontology of atoms in an empty, absolute space, space does not count among the things which exist: it is non-being.[31] However, being, the atoms, can only exist on the basis of space, i.e., within space. Democritus therefore comes to the paradoxical conclusion that non-being exists as well as being.[32] Newton characterizes space as the sensorium of God at one place in his *Optics*.[33] Pressed on this point by Leibniz, Clarke maintains that space is a property of God which is an immediate and necessary consequence of his existence.[34] Notwithstanding this description (or, perhaps, just because of this description), Kant's objection is plausible: the space which the atomists conceive is an absurdity (*Unding*) – something which would exist even if there were nothing real (*Critique of Pure Reason*, A 39 / B 56, B 70–71). Holism about space avoids the problem that the ontological status of space poses for a theory which admits an absolute space in distinction from matter. Space does not have a being independently of things in space. Space in identity with matter is a (the) substance.

---

31 Diels and Kranz (1960/61), Volume 2, fragments 68 A 6, A 37.

32 Diels and Kranz (1960/61), Volume 2, fragment 68 A 6.

33 Newton (1952), question 31, p. 403. See furthermore Clarke, letter 1 to Leibniz, § 3; letter 2 to Leibniz, § 3 in Gerhardt (1875–1890), Volume 7.

34 Letter 3 to Leibniz, § 3; letter 4 to Leibniz, §§ 8–10, 15, 41; letter 5 to Leibniz, §§ 36–48, 79–82.

6.4 Holism about Space and the Conception of Motion

What is the account that holism about space can give of motion and change? Consider the suggestion on which Bennett (1988) elaborates. He starts with a spatio–temporally continuous set of place–times which are similar as far as their physical properties are concerned and which are distinguished from their environment as a result of this similarity. Bennett calls each such set a string. He reduces the description of the motion of a body to a description of the physical properties of such a string. What seems to be the motion of a body through space is in fact an alteration of the basic physical properties of some regions of space: one region ceases to instantiate certain physical properties and another region in its immediate neighbourhood acquires the same or similar properties. Bennett compares this account of motion to the way thaw affects a countryside:

> When a thaw moves across a countryside, as we say, nothing really moves; there are just progressive changes in which bits of the countryside are frozen and which are melted. Analogously, Spinoza's view is that the movement of things or stuff is, deep down, the passing along of something qualitative—a change in which regions are F and which are not, for suitable values of F. (§ 22.2; see also Bennett (1994), p. 16)

In another place, Bennett speaks of "alterations in space which can be conceptualized, one level up, as movements of things in space" (§ 26.1). He characterizes these alterations as qualitative variety.[35]

As a result of this suggestion, we have to give up the assumption that there are things which move in the sense that they undergo a change in their spatial relations to other things. Instead, we have to consider a coordinated change of physical properties of neighbouring regions of space: one region of space loses a certain physical property – it changes from being *F* into being *non-F*. Immediately afterwards, a region in the immediate neighbourhood of the first region acquires that property – it changes from being *non-F* into being *F*.

This account of motion has the following consequence: there is not one individual which persists as being *F* throughout a process of motion, but there are several individuals which acquire and lose the property of being *F* successively. There is not one instantiation of the property *F*, but a number of destructions and recreations of instantiations of that property in different

---

35 Bennett (1988), § 26.3. See furthermore §§ 26.4, 49.4. Clifford (1876) and Broad (1933), p. 158, already consider such an account of motion.

individuals.[36] Consequently, nothing, not even animals and human beings, can be at different places successively and remain the same individual. The properties that are characteristic of a certain animal or a certain person are instantiated successively by different individuals, i.e., different regions of space. Whether or not one regards this account as plausible, it is possible to reconstruct the talk of the motion of bodies in space on the basis of holism about space. The internal structure of space which consists in regions that are distinguished from one another by their physical properties makes such a reconstruction possible.

6.5 Holism about Space and the Philosophy of General Relativity

6.5.1 Holism about Matter as Holism about Space–Time

Does the holism which Cartesian philosophy of physics admits have any relevance to today's philosophy of physics? When it comes to examining whether holism about space fits into current physics, it is general relativity on which we have to focus. If holism about space is applicable to the interpretation of general relativity, then general relativity admits of a substantial holism. Therefore, in this section, I consider what the application of holism about space to the philosophy of general relativity looks like. In the next section, I evaluate whether this application is a plausible option in the philosophy of relativity physics.

Special relativity shows that space and time are not independent of each other, but united: only the space–time interval between any two events and any two points in space and time, but neither the spatial nor the temporal extension between them is invariant under the transformation of reference frames. Over and above that, general relativity combines a theory of space–time with a theory of matter insofar as matter is subject to gravitation. Matter in the sense of mass determines the geometrical structure of space–time: as a result of the presence of mass, space–time is curved. The curvature of space–time, in turn, influences the path of physical systems in space–time. Gravitation (the gravitational field) and the curvature of space–time are the same thing.[37] The identity of a theory of curved

---

36 Compare, in the context of this implication, Redhead (1983), p. 77.

37 For a reliable and easily accessible introduction to the conceptual features of special and general relativity, see the papers in Audretsch and Mainzer (1988). The paper of Audretsch (1988) treats gravitation and the curvature of space–time.

space–time with a theory of the gravitational aspect of matter is the basis for an interpretation of general relativity along the lines of holism about space.

Since space and time are united in relativity physics, holism about space has to be changed into holism about space–time in order to be applicable to the philosophy of relativity physics. We have to replace the continuum of physical, three-dimensional Euclidean space with the continuum of physical, four-dimensional space–time that has the structure of a Riemannian manifold, i.e., which is curved. However, this change is nothing but an adaptation of holism about space to the contemporary state of science. We merely have to replace "space" with "space–time" in the characterization of holism about space given in [6.3]:

Holism about matter as holism about space–time

> Matter and the continuum of physical, four-dimensional Riemannian space–time are the same thing. The parts of matter are identical with points or regions of space–time. All physical properties are properties of points or regions of space–time. Insofar as a thing instantiates the properties which belong to the family of properties that make something a part of space–time, a thing is ontologically dependent on there actually being other things which are also parts of space–time and which extend as far as the whole space–time. Since the instantiation of all physical properties is included in the instantiation of the properties which make something a part of space–time, everything is, with respect to all its physical properties, ontologically dependent on there actually being other material things (i.e., parts of space–time) together with which it encompasses the whole material realm (i.e., the whole space–time) of a possible world. Therefore, all matter is one holistic system.

When it comes to examining holism about space–time in the context of contemporary philosophy of physics, it is reasonable to qualify that thesis in such a way that it refers only to all physically possible worlds. It is plausible to maintain that the laws of relativity physics are not metaphysically necessary: a world in which gravitation is not identical with the curvature of space–time is a metaphysically possible world (compare [1.3]). Consequently, if an argument for the identification of matter with space–time is to be built on relativity physics, that argument applies only to all physically possible worlds.

General relativity is a local theory.[38] All property instantiations are localized at space–time points. The connections between events at different space–time points are only relations of causal interaction. According to special and general relativity, interactions propagate contiguously with a velocity that is not higher than the velocity of light in vacuum. As regards holism about space–time, this means that everything is, with respect to all its physical properties, only ontologically dependent on there being other physical things together which with it encompasses the whole material realm, i.e., the whole space–time; but there are no relations of ontological dependence among parts of matter with respect to the instantiation of specific physical properties.[39] As in holism about space, so in holism about space–time, physical properties give rise to an internal structure in the sense that they yield regions which are distinguished from each other by the difference in the physical properties which they instantiate; the boundaries of these regions can be more or less vague. This position then amounts to a holism about matter because the instantiation of the physical properties is included in the instantiation of the properties which make something a point or a region of space–time, and there cannot be only one point or region of space–time.

Holism about space is not able to conceive one individual as the subject of what we regard as motion ([6.4]). This is a high price to pay for the replacement for our talk of motion that holism about space can offer. Does this point apply to holism about space–time as well? If we transform holism about space into holism about space–time, we can no longer endorse an ontology of points or regions of space which undergo a change of their physical properties in time. The unification of space and time enables us to give the following account of what we regard as motion of bodies in common sense: we conceive regions of space–time as subjects of predication of physical properties. We consider a continuous string of space–time regions whose physical properties are similar. We can regard this sequence of regions as one bigger region within space–time. If we abandon the notion of regions of space which undergo a change of their physical properties, we can thus conceive one individual as the basis for a reconstruction of our common sense view of motion. What we consider in common sense as the

---

38 For a definition of a local space–time theory see Earman and Norton (1987), pp. 517–518.

39 Compare in this context Shimony, "The Methodology of Synthesis: Parts and Wholes in Low-Energy Physics" (1987), in Shimony (1993), p. 209.

motion of a body through space is in fact a continuous sequence of space–time regions which have similar physical properties and which can, for that reason, be taken as one individual. Hence, there is neither change nor motion, since only regions of space–time with physical properties exist. But holism about space–time can offer a replacement for our view of motion that conceives one individual as the subject of what we regard in common sense as motion.

6.5.2 Proposed Philosophical Consequences of Relativity Physics

Is holism about space–time a plausible philosophical interpretation of general relativity? Consequent upon the physics of relativity, philosophy can no longer treat time as a separate entity from space. According to the physics of relativity, space and time are one entity, space–time. The further philosophical theses which are proposed on the basis of the physics of relativity can be divided up into three steps:

1) Existence is tenseless.
2) Physical systems have temporal parts as well as spatial parts.
3) Matter and space–time are the same thing.

The first and the second step are already suggested on the basis of special relativity. In the discussion of these theses, I limit myself to special relativity. The third step is tied to general relativity.

The *first step* is opposed to the tense view of existence according to which only what is present exists. This view faces a serious objection from relativity: temporal and spatial extension are properties of physical systems relative to a reference frame. There is no global preferred reference frame. Consequently, there is no universal present. The physics of relativity therefore is claimed to support a tenseless view of existence. Existence is not relative to a certain tense in the same way as it is not relative to a certain place. Everything exists at a space–time point or region, and all there is in space–time simply exists – although, of course, there are objective relations of some things being or occurring earlier or later than other things. This view is sometimes referred to as the conception of a *block universe*. However, that conception leaves entirely open what the content of the so-called block universe is.[40] Furthermore, it does not imply determinism: what exists at a space–time point or region does not have to be determined by what exists at other space–time points or regions. The block universe view does not imply holism about space–time. But holism about space–time is committed to the

---

40 Compare Mellor (1981), pp. 128–132. See also Butterfield (1985).

conception of a block universe; for, according to holism about space–time, what exists just is the whole of space–time with the physical properties that it has in all its regions.

The *second step* amounts to the claim that the block universe contains only *occurrents* (events), but no *continuants* (things). A continuant is something that has spatial parts, but no temporal parts, because it persists as a whole for a certain time. By contrast, something that has temporal as well as spatial parts is an occurrent. Processes, phases of processes, events, etc. are occurrents.[41] In common sense, we admit continuants as well as occurrents. An eruption of a volcano, for example, is a process that has temporal parts (phases) as well as spatial parts. An eruption can be gentle at the beginning and violent later. It extends over a certain time without existing as a whole in any part of that time. The volcano, on the contrary, has only spatial parts, but no temporal parts. It exists as a whole throughout a certain time.

Referring to the physics of relativity, eminent philosophers such as Russell and Quine endorse an ontology which admits only occurrents, but no continuants.[42] Quine says in *Word and Object*:

> ... the treating of time on a par with space is no novelty to natural science ... Physical objects, conceived thus four-dimensionally in space-time, are not to be distinguished from events or, in the concrete sense of the term, processes. Each comprises simply the content, however heterogenous, of some portion of space-time, however disconnected and gerrymandered. (1960, p. 171)

On the basis of statements such as the quoted one, Quine can be read as committing himself to the position that the physical properties are properties of regions of space–time and thus to what I call holism about space–time.[43] However, an ontology which admits only occurrents does not entail holism about space–time: One does not have to identify the content of regions of space–time with properties of these regions. One can conceive physical

---

41 As regards this terminology, compare Broad (1933), pp. 141–166, especially pp. 141–148. Lewis (1986b), pp. 202–204, uses the terms "endurance" and "perdurance" which relate to continuants and occurents respectively.

42 Russell (1954), pp. 286–288; Russell (1956), p. 329; Russell (1969), Chapter 14; Quine (1960), § 36. See furthermore Smart (1972). For a recent elaboration of such an ontology, see Heller (1990). Lewis (1986b), pp. 202–204, also endorses such an ontology; but he does not invoke the physics of relativity.

43 For instance, this is the way in which Putnam reads Quine. See, e.g., "Realism with a Human Face" in Putnam (1990), p. 26; "Is Water Necessarily H_2O?" in Putnam (1990), pp. 65–66. See also Lemmon (1967), pp. 98–99, on Davidson.

objects as processes *in* space–time without conceiving them as physical properties *of* regions of space–time. Thus, basing oneself on the physics of relativity and defending an ontology of a block universe which contains only occurrents does not imply holism about space–time.

Shall we endorse an ontology of occurrents consequent upon relativity physics? Although space and time are united in relativity physics, relativity does not treat time on a par with space.[44] According to special relativity, there is an interval between any two points of space–time which is either space-like or time-like or zero absolutely, i.e., independently of a reference frame. There is an objective difference between space-like and time-like intervals. Causal relations in the sense of physical interactions hold only among events at points with a time-like interval, i.e., events at points which lie in the past or the future light-cone of a given point. What is relative to a reference frame is only the three-dimensional spatial extension and the one-dimensional temporal extension between events at points which are separated either by a space-like or a time-like interval absolutely. There is no global preferred reference frame. Thus, no argument for an ontology of occurrents can be established by alleging that relativity physics treats time on a par with space.

Many philosophers deny that the physics of relativity requires an adaptation of our ontology in such a way that we admit only four-dimensional occurrents, but no continuants.[45] The textbook expositions of relativity physics do not merely talk of occurrents, but also of continuants. Only a continuant (such as a particle), but not an occurrent (such as a process or an event) can be at rest or in motion, and only a continuant can be accelerated. A world-line describes a continuant, although the points on a world-line refer to events. An ontology of occurrents is an option which one can endorse. But it is not simply a philosophical consequence of the physics of relativity. To be fair, I should mention that Russell and Quine have further arguments for an ontology of occurrents which do not depend on the physics of relativity.

The *third step* alone is committed to holism about space–time. At first glance, it could seem that the identification of matter with space–time is a direct consequence of physical field theories. These theories are often described in such a way that the field properties are predicated of space–time

---

44 See, for instance, Mellor (1981), pp. 66–72.

45 Compare, for instance, Strawson (1959), pp. 56–57; Geach (1972), Chapter 10.2; Mellor (1981), pp. 127–132; Hacker (1982).

points. Hence, it could seem that the field properties are properties of space–time points. However, a physical field theory says merely that the field has the properties in question *at* space–time points, but not that these are properties *of* space–time points. Therefore, holism about space–time cannot be based upon field theories alone. It depends on an argument which establishes that these theories do not require an ontological commitment to physical systems (such as fields) in addition to space–time.[46] Paul Teller maintains that, given physical field theories, we also have the option to begin with fields and to construct a relational theory of space–time on the basis of an ontology of fields.[47]

Furthermore, it is sometimes claimed that fields in general are physical systems that are holistic in character.[48] However, as far as classical fields are concerned, they can be conceived as being built up on the basis of local field sources. It is a matter of dispute to say the least whether the properties which make something a local field source are relational in the sense that their instantiation is ontologically dependent on there being other local field sources which build up a whole field. As far as quantum fields are concerned, these are holistic systems owing to the holism about quantum physics; this holism is manifest already in traditional quantum mechanics, which considers cases of a definite number of quantum particles.

Coming back to the main point, what is much more, an argument for the identification of matter with space–time cannot be based directly on general relativity. General relativity treats merely the gravitational aspect of matter. It does not consider electromagnetism and elementary particle physics. The third step is tied to there being a physics which can trace the electromagnetic field including charges as well as elementary particles back to properties of points or regions of space–time.

John Wheeler introduced some forty years ago the programme of *geometrodynamics*. This programme seeks a further development of general relativity. Geometrodynamics gives a concrete physical shape to holism about space–time. It is the only programme which sets out to construct physics in such a way that no ontological commitment to physical systems in

---

46 Compare Graves (1971), pp. 316–317.

47 Teller (1987), pp. 430–431; Teller (1991), p. 382. See furthermore Auyang (1995), § 20.

48 For instance, Suppes (1985), p. 184.

addition to space–time is required.[49] Whereas in general relativity only the theory of gravitation can be built on a geometrical description of space–time, geometrodynamics envisages constructing electromagnetism and elementary particle physics solely on the basis of a geometrical description of empty curved space–time. Empty space–time means in this context a space–time without additional physical systems. Thus, the physical properties are included in the properties which make something a part of space–time in the sense of a Riemannian manifold, because they are identical with geometrical properties such as curvature. Wheeler writes:

> Is space-time only an arena within which fields and particles move about as "physical" and "foreign" entities? Or is the four-dimensional continuum all there is? Is curved empty geometry a kind of magic building material out of which everything in the physical world is made: (1) slow curvature in one region of space describes a gravitational field; (2) a rippled geometry with a different type of curvature somewhere else describes an electromagnetic field; (3) a knotted-up region of high curvature describes a concentration of charge and mass-energy that moves like a particle? Are fields and particles foreign entities immersed *in* geometry, or are they nothing *but* geometry? (1962b, p. 361)

The theory of space–time which treats physical properties thus consists in a description of the geometrical properties of space–time such as its curvature. The ontological claim of geometrodynamics is that the physical realm is identical with the four-dimensional space–time continuum` with its geometrical – i.e., physical – properties. John Graves aptly characterizes geometrodynamics in these terms:

> Its basic goal remains the identification of matter with space, which in turn means that all the phenomena traditionally associated with matter and considered conceptually different from space must somehow be incorporated into the natural Riemannian structure of space-time, rather than being imported from outside as 'foreign' entities ... We no longer have any irreducible matter or other entities different in kind from space-time; this space-time is not a passive arena ... It is not a collection of things, but a single thing—that only thing that is really real. (1971, pp. 312, 314)

---

49 See Wheeler (1962a), in particular pp. XI–XII, 8–87, 129–130, 225–236. For an overview, see Wheeler (1962b). See furthermore Graves (1971), Chapters 4 to 5, in particular pp. 236, 312–318; Kanitscheider (1976), Chapter 10. Clifford (1876) can be considered as a precursor of Wheeler's programme.

Geometrodynamics therefore is often connected with Cartesianism and Spinoza's theory of the physical realm in particular.[50] It can be seen as the attempt to establish the Cartesian identification of matter with space in today's physics.

However, geometrodynamics failed. The problems with which it could not cope are mainly the following three ones:

- *initial value*: it has been shown by means of counterexamples that the geometrical description which geometrodynamics offered could not distinguish between some physically different initial conditions in electromagnetism.[51]
- *singularities*: geometrodynamics intended to avoid singularities.[52] However, that attempt failed. Singularities require that physical systems such as particles are acknowledged in addition to space–time, since the equations of the relevant field, which is supposed to indicate physical properties of space–time, do not apply at those points where the singularities occur.[53]
- *elementary particles*: fermions, i.e., elementary systems of spin 1/2 such as electrons, could not be treated by geometrodynamics.[54]

The first two problems concern already classical geometrodynamics, i.e., the attempt to unite gravitation with electromagnetism on the basis of an ontology of space–time without additional physical systems. The third problem concerns quantum geometrodynamics, i.e., the attempt to include quantum elementary particle physics in geometrodynamics. Especially as a result of the third problem, Wheeler repudiated geometrodynamics in the sense of a programme that sets out to build physics solely on space–time without acknowledging additional physical systems in 1973.[55]

On the one hand, a metaphysics such as holism about space or space–time is not tied to the fate of one specific programme in physics. In particular, holism about space or space–time is not committed to regarding the physical properties of space or space–time as geometrical properties such as curvature. On the other hand, if such a metaphysics is to be more than a mere philosophical speculation, it has to be able to spell out in concrete terms how

---

50 Graves (1971), pp. 79–101; Campbell (1976), pp. 100–104; Mathews (1989), pp. 5–6.

51 See Misner (1974), pp. 12–14. Compare furthermore Earman (1972), pp. 644–646.

52 See Wheeler (1962a), in particular pp. 25–31, 45–66.

53 See Stachel (1974), pp. 33–39; Kanitscheider (1976), Chapter 12.

54 See Stachel (1974), pp. 45–46; Cao (1997), pp. 110–111.

55 See Misner, Thorne and Wheeler (1973), § 44.3–4, in particular p. 1205.

a reduction of all physical properties to properties of points or regions of space–time can be carried out. If such a metaphysics is to succeed in the philosophy of physics, it has to show in concrete terms how a reconstruction of our physical theories is possible without being committed to physical systems in addition to space–time. However, we do not have a physics at our disposal that is not committed to physical systems over and above space–time, and such a physics is not in prospect as a further development of current physics either, since the programme of such a further development failed.

Shall we therefore accept an ontology according to which space–time and matter are different entities, which influence one another (matter curves space–time, curved space–time affects the path of material systems)? In this case, we would face the following problem: what could be the ontological status of space–time as an entity which exists in addition to matter? (compare the end of [6.3]). However, this philosophical problem cannot count as a sufficient reason for endorsing holism about space–time contrary to current physics. After all, there are other philosophical positions that avoid this problem such as relational theories of space–time. In his repudiation of geometrodynamics, Wheeler himself contemplates a relational theory of space–time.[56] Furthermore, as indicated above, physical field theories are not committed to an ontology of space–time as some sort of a substance. Contemporary physics requires an ontology of physical systems. This ontology cannot be reduced to an ontology that admits only space–time. But it is an open question whether contemporary physics can do simply with an ontology of material systems such as fields without a commitment to space–time as an entity in addition to matter.[57]

In conclusion, we have to acknowledge that holism about space or space–time is not a viable option in today's philosophy of physics. It is not plausible to connect general relativity with a substantial holism about matter. Consequently, even if we lay stress on the fact that Cartesian physics admits of a sort of holism, there is no prospect of circumventing a revision of Cartesianism in current philosophy of physics. However, my point is: holism about space has an indirect relevance to today's philosophy of physics. If we approach the framework which Descartes laid out for physics with

---

56 Misner, Thorne and Wheeler (1973), § 44.4.

57 In this context, the so-called hole argument is relevant. For an introduction, see Earman and Norton (1987); Earman (1989), Chapter 9; Butterfield (1989a); Nerlich (1994), Chapter 9.

hermeneutic good-will, we have to acknowledge that this framework admits of a substantial holism. Because of this holism, the changes which are called for in today's philosophy of physics and which centre around the concept of holism do not constitute a simple rejection, but a revision of Cartesianism in the philosophy of physics. Starting from the holism in Cartesianism, these changes amount to a conceptual revision of what holism in physics consists in. Keeping this background in mind, let us turn in the next chapter to that area of contemporary physics where there is convincing evidence for some sort of a holism, namely quantum physics.

CHAPTER 7
THE BASIS FOR HOLISM IN QUANTUM PHYSICS

Summary

What does holism in the philosophy of quantum theory mean? On which features of quantum physics is it based? In this chapter, the latter question is tackled. To start with, I sketch out the new conceptual features of quantum theory that are pertinent to quantum holism [7.1]. I then take up the distinction between the principles of separability and local action. Quantum theory violates separability. Einstein's argument for the incompleteness of quantum theory is reconstructed [7.2]. Going into Bell's theorem and its philosophical implications, I point out that the discussion on Einstein's objections to quantum theory can be comprehended along the lines of the Duhem–Quine thesis [7.3].

7.1 New Conceptual Features of Quantum Theory

7.1.1 Observables and States

Upon which features of quantum theory does the claim to holism in quantum physics base itself? To begin with, this sub-chapter introduces those philosophically important conceptual features of quantum theory that are new in comparison with classical physics and that are relevant to the discussion on quantum holism.[1] The exposition is restricted to non-relativistic quantum theory. It is entirely conceptual, although a few simple formulas are indicated. No familiarity with quantum theory is presupposed.

The first and foremost objects of quantum theory are the systems at the microphysical level of nature. I shall refer to these systems as quantum physical systems, or, for short, *quantum systems*. Electrons and photons, protons and neutrons including their constituents (quarks) as well as whole atoms count among such systems. I speak of quantum *systems* and not of particles, because these systems do not always behave like particles. They can also behave like waves. The term "system" is neutral to the wave–particle dualism. It does not imply any commitment on what these entities are. Recall that I use the term "system" in a broad sense according to which everything of which properties can be predicated is a system.

---

1 For an esaily accessible and reliable exposition of these features, see the papers in Audretsch and Mainzer (1990), in particular Audretsch (1990).

What about the properties of quantum systems? To come back to a distinction which was introduced at the end of [1.2.2], we have to differentiate between generic or determinable and specific or determinate properties. If a property is generic or determinable, nothing can have this property *simpliciter*, but only in a way that admits of further specification. Accordingly, if a property is specific or determinate, it admits of no further specification. Physical quantities such as position, momentum (i.e., mass times velocity), mass, charge, etc. are generic properties. Nothing can have, say, mass *simpliciter*. If something has mass, it has a specific quantity of mass, namely a certain value of mass. A certain value of mass, position, etc. is a specific property. When I talk in the following about properties without further qualification, generic properties are intended.

Within generic properties, we have to introduce a further distinction: we have to distinguish between generic properties whose value is time-independent, i.e., remains unchanged throughout the whole existence of the system, and generic properties whose value is time-dependent, i.e., can change during the existence of the system. Mass and charge of quantum systems are examples of properties whose value is time-independent. An electron, for example, always has a negative charge. If something underwent a change from a negative to a positive charge, it would no longer be an electron. By contrast, position, momentum, energy, and spin in any direction are examples of properties of quantum systems whose value is time-dependent. A system of spin 1/2 such as an electron, for example, can change its value of spin 1/2 from spin up to spin down (or *vice versa*) in a certain direction and remain an electron. Having spin 1/2 is a generic property, because it admits of further specification, and its value is time-dependent, because it can change during the existence of the system. When I speak of the spin without further qualification, I always mean the quantity of spin that is characteristic of the system under consideration such as spin 1/2, spin 1, spin 3/2, etc.

The properties of quantum systems are called "observables". Observables are not tied to measurement, although the term suggests that they are. *I use the term "observables" exclusively for those properties whose value is time-dependent.* I do not use this term when I talk about properties whose value is time-independent or when I talk about the mathematical representation of the properties of quantum systems. When I refer to the mathematical representation of a property, I employ the term "operator", because observables are usually represented by operators in an abstract mathematical space such as a Hilbert space.

Furthermore, only functionally independent observables are considered. Two observables are functionally independent if and only if they cannot be represented as functions of a third observable. By contrast, the observables A, B are functionally dependent if and only if there is an observable C to which A, B can be related by two functions α, β such that $A = \alpha(C)$ and $B = \beta(C)$. Two functionally dependent observables can thus be reduced to one observable.

The second important notion in addition to observables is the notion of the *state* of a system. In the Schrödinger representation of quantum mechanics, the state of a system changes in time. It relates to all and only those properties whose value changes in time. Therefore, I refer to the properties whose value changes in time as *state-dependent properties*. Accordingly, the properties whose value is time-independent can be referred to as *state-independent properties*. Thus, in the terminology which is adopted here, all and only observables are state-dependent properties.

As a first approximation, the state of a system at a given point of time consists in the values of the observables of this system at this time. However, this characterization is too narrow for quantum physics. A value in the first and foremost sense of the term is a definite numerical value, i.e., one number. In quantum physics, however, a system never has a definite numerical value of all its observables at once. Nonetheless, we do not have to conclude that a system does not have a state-dependent property at all if it is in a state in which it does not have a definite numerical value of the property in question. The description of the state of a system at a given point of time specifies a probability distribution for all state-dependent properties of the system at this point of time. We can therefore say that *the state of a system at a given point of time consists in the way in which the system has properties whose value is time-dependent at this time*. What this way is if it is not a definite numerical value has to be spelled out in the interpretation of quantum theory. If the system is isolated, the state is usually described by a vector in an abstract mathematical space such as a Hilbert space. A state function or a wave function is a special case of a state vector. The state vector contains all the information that is available about the observables of the system.

7.1.2 Incompatible Observables

After some terminological matters have been settled in the preceding section, let us consider the two most prominent observables, position and momentum. These observables are functionally independent. In classical physics, the value of position and the value of momentum are also independent of each other. The value of position that a system has in a given state does not impose any restriction on the value of momentum which it can have in this state. Any value of position is compatible with any value of momentum.

In quantum physics, by contrast, the value of position that a system has in a given state imposes a restriction on the value of momentum which the system can have in this state (and *vice versa*). Position and momentum are such that they never have a definite numerical value in the sense of exactly one number. They always have some dispersion. The dispersion can come arbitrarily close to zero; but it can also be very large. The mutual dependence of the values of position and momentum in quantum physics consists in the following: if the position of the system is such that it has nearly no dispersion, the momentum of the system necessarily has a great dispersion (and *vice versa*). There is no state in which the system is both localized in an arbitrarily small region of space and has an arbitrarily small dispersion of momentum. This is the content of the famous *Heisenberg inequality*:

(7.1) $$\Delta Q \Delta P \geq 1/2\,\hbar$$

In this formula, Q is the operator for position, P the operator for momentum, Δ stands for dispersion, and ∘ stands for Planck's quantum of action divided by 2π. The Heisenberg incquality says: there is no state of the system in which the product of the dispersion of position and momentum falls below a certain limit. This inequality is sometimes referred to as "indeterminacy relation". This name can be misleading: there is no indeterminacy in the sense of imprecision. The Heisenberg inequality indicates a precise lowest limit for the product of the dispersion of position and momentum. Position and momentum have a determinate, precise dispersion in each state of the system. The name "uncertainty relation" is even worse: this inequality has nothing to do with the uncertainty of observers about the position or momentum of the system.

Observables which are functionally independent, but dependent on each other in such a way that in any state of the system at most one of them has a definite numerical value (or a value with nearly no dispersion) are known as *incompatible observables*. Mathematically speaking, incompatible observables are all and only those observables that are represented by operators which do not commute:

(7.2) $$AB \neq BA$$

In this formula, A and B are operators that represent incompatible observables. In the case of position and momentum, the operators satisfy the *Heisenberg commutation relation*:

(7.3) $$QP - PQ = i\hbar 1$$

In this formula, i is the imaginary number ($\sqrt{-1}$).

Position and momentum are not the only example of incompatible observables. On the basis of the formalism of quantum theory, it is to be expected that for any observable of a quantum system, there are other observables of the same system with which this observable is incompatible. A given observable is not incompatible with all the other observables of the same system; but this does not hinder that for any observable, there are some other observables with which this observable is incompatible. A prominent example of mutually incompatible observables apart from position and momentum are the spin observables in all three orthogonal spatial directions. The observables of spin are peculiar to quantum physics; they are unknown to classical physics. The three spin observables are only mutually incompatible; each of them is compatible with position and momentum.

Let us introduce a few further notions. Consider two incompatible observables A and B and assume that their possible values are definite numerical values, i.e., a discrete sequence of numbers such as 1, 2, 3 … In this case, 1, 2, 3 … are the *eigenvalues* of the observables A and B. One says that A and B have a *discrete spectrum of eigenvalues*. The state of the system in which it has an eigenvalue of an observable is called an *eigenstate* of this observable and the state vector can be written as an *eigenvector* of this observable. Hence, if the system is in an eigenstate of A and thus has an eigenvalue of A, it cannot also be in an eigenstate of B; in this case, it cannot have an eigenvalue of B. A system can consequently be in an eigenstate and have an eigenvalue of at most one of two or more incompatible observables.

As regards position and momentum, the matter is more complicated, because these observables have a continuous spectrum of values. As explained above, they never have a value that is completely free of dispersion. One says that position and monentum merely have approximate eigenvalues and approximate eigenstates. However, I ignore the mathematical complications which are created by quantities with a continuous spectrum of values. They are not important for the philosophical issue of quantum holism. I therefore pretend that there are eigenstates and eigenvalues of position and momentum too.

What is the operational consequence of incompatible observables and the sketched dependence of their values? It is not possible to prepare a state which is an eigenstate of more than one such observables. Let A and B be two incompatible observables and assume that they have a discrete spectrum of eigenvalues and that each eigenstate corresponds only to one eigenvalue. Assume furthermore that we prepare a state which is an eigenstate of A and measure the corresponding eigenvalue. In this case, the system does not have an eigenvalue of B. Nonetheless, given the state in which the system is, the formalism of quantum theory enables us to calculate conditional probabilities for the outcome of a subsequent measurement of B: we can assign a probability to each eigenvalue which indicates how probable it is in the given state that this eigenvalue will be the outcome of a subsequent measurement of B. These probabilities add up to 1. Consequently, we can never predict the outcome of a subsequent measurement of B with certainty. If we measure first A and then prepare a measurement of B, the system is not in an eigenstate of A any more. After the measurement of B, the system is in an eigenstate of B. Given the outcome of that measurement, we can only predict probabilities for the outcome of a second measurement of A. Thus, the operations "measurement of A" and "measurement of B" do not commute: measuring first A and then B leads to a different final state of the system than first measuring B and then measuring A.

A very simple example is the spin in two orthogonal directions of a system with spin 1/2 such as an electron. Such a system has merely two eigenvalues of spin in any direction: spin up and spin down. Let us denote spin up by +1 and spin down by –1. If we measure the spin in the direction of the z-axis (spin z) and get the outcome, say, spin up, the only thing we can predict about the outcome of a subsequent measurement of the spin in the direction of the x-axis (spin x) is that the outcome "spin up" and the outcome "spin down" both have probability 0.5. Assume that we measure spin x and get the outcome, say, spin down. We cannot say in this case that the system has the value spin down in x-direction and the value spin up in z-direction. The system has no eigenvalue of spin z any more consequent upon the measurement of spin x. The only thing we can say on the operational level is that for a new measurement of spin z, the outcome "spin up" and the outcome "spin down" now both have probability 0.5.

The probabilities in quantum theory have nothing to do with an inefficiency of macroscopic measuring instruments with respect to the microscopic domain or a disturbance of quantum systems by measuring instruments.[2] They cannot be eliminated by an improvement of measuring techniques. Measuring inaccuracies are not taken into account in the calculation of these probabilities. These probabilities refer to ideal measurements. An ideal measurement is conceived as being such that its immediate repetition yields the same result. When I talk about measurements, ideal measurements are always intended, although laboratory measurements are not ideal measurements.

7.1.3 Entangled States

Let us return to the notion of a state of a system. If the state is an eigenstate of a given observable, it can be represented by an eigenvector of this observable. What is more, the eigenvectors of an observable are a basis in which any state of the system can be represented. If the state is not an eigenstate of the observable in question, it cannot be represented by one eigenvector; but it can be represented by a superposition of eigenvectors, i.e., an addition of eigenvectors each of which is multiplied by a coefficient; this coefficient serves to calculate the probability with which the corresponding eigenvalue will be the result of a measurement of the observable in question. Such a state is a *superposition* of eigenstates of the observable in question. The superposition principle in quantum theory implies that any two eigenvectors of an observable can be added (multiplied with coefficients), and the result is a vector which also represents a possible state of the system.

The superposition principle is not limited to the states of one system. The states of two or more systems can be in superposition with one another. Consider the simplest example: imagine that two systems of spin 1/2 such as two electrons or two neutrons are emitted together from a source. After the emission, their interaction dies down because they fly apart in opposite directions. Some time later, a local observable is measured on each of the two systems. A *local observable* is an observable that relates only to one of the two systems. Spin in z-direction of the one system and spin in z-direction of the other system are examples of local observables. The term "local observable" or "local property" only expresses that the property in question is a property of one elementary system; this term does not imply that the instantiation of the property has a precise localization. Assume that spin z is measured on each of the two systems. Assume furthermore that these measurements are separated by a space-like interval. Nonetheless, their outcomes are maximally correlated: if the one system has spin up, the other

2 Compare Brown and Redhead (1981).

system has spin down. Thus, there are only two possible outcomes: either the first system has spin up and the second system spin down, or the first system has spin down and the second system spin up. After the measurement, the state of the two systems is described either by

(7.4) $$\Psi = \Psi_1^+ \otimes \Psi_2^-$$

or by (7.5) $$\Psi = \Psi_1^- \otimes \Psi_2^+$$

In these formulae, Ψ stands for the joint spin state, that is the spin state of the whole, i.e., the two systems taken together; Ψ_1 and Ψ_2 stand for the spin states of the two parts; + signifies spin up, – spin down. These formulae indicate what is known as a *product state*: each of the parts is in an eigenstate of the observable under consideration, it is described by an eigenvector of this observable, and the joint state of the two systems is described by the tensor product of these two eigenvectors. The states that are attributed to the two parts hence fix the state of the whole.

The product state (7.4) or (7.5) is the result of a reduction of a superposition of product states to one of the product states in measurement. Before a measurement occurs, the joint spin state of the two systems is described as a *superposition of product states.* No matter how far apart in space the two systems are removed, their joint spin state is a superposition of the first system having spin up and the second system having spin down with the first system having spin down and the second system having spin up in any direction. This state is referred to as the *singlet state*:

(7.6) $$\Psi = 1/\sqrt{2}\left(\Psi_1^+ \otimes \Psi_2^- - \Psi_1^- \otimes \Psi_2^+\right)$$

This formula is to say: not only is each of the two systems not in an eigenstate of spin in any direction; but all the eigenstates of spin of the one system are in superposition with the respective eigenstates of the other system. There is no possibility to represent this state as a product state. When I speak of a superposition of product states or employ related notions, I always mean a superposition of product states which cannot be represented as a product state. It is only after a measurement that the state of the two systems can be represented as a product state.

If and only if two systems have at least two incompatible observables each, they can be taken together in a state which is a superposition of product states. Note that it is only observables of each of the parts which are mutually incompatible. No observables of one part are incompatible with observables of another part. In the paradigmatic example of the singlet state, the spin observables in all three directions of the one system are mutually incompatible, and the same goes for the spin observables of the other system. But no spin observable of the one system is incompatible with any spin observable of the other system.

It is not the case that for all states of a whole which is composed of two systems such as those ones in the singlet state, a given value of a spin observable of the one system imposes a restriction on the value which a spin observable of the other system can have. Nonetheless, there are states of the whole which are a superposition of product states. In any such state, there are correlations between the values which local observables of the one system and local observables of the other system can have. If in the example of the singlet state a measurement of the spin in a certain direction of the one system yields the result spin up, it is not possible that a measurement of the spin of the other system in the same direction also yields spin up; we can in this case predict with certainty that the outcome will be spin down. Two such measurements can be separated by an arbitrary distance in space–time. Thus, given that the state of the whole is a superposition of product states and that a measurement of certain observables takes place, the eigenvalue which one observable of one system acquires is at least partially determined by the eigenvalue that the same observable of another system acquires in measurement.

It is arguably the most striking and most momentous feature of quantum theory that the joint state of two or more systems can be a superposition of product states. Erwin Schrödinger said: "I would not call that *one* but rather *the* characteristic trait of quantum mechanics." (1935a, p. 555) He speaks of an *entanglement* of the state vectors of two or more systems whose joint state is a superposition of product states (1935a, p. 555; see also 1935b, p. 827). Since it makes philosophical sense to differentiate between a system and its state, it is not advisable to talk in terms of an entanglement of systems. But if we can, following Schrödinger, employ the notion of an entanglement of state vectors, we can also speak of an entanglement of the states of two or more systems.

However, if we work with the notion of an entanglement of the states or the state vectors of two or more systems, we have to bear in mind the following: only the whole, i.e., these systems taken together, is in what is known as a *pure state.* The singlet state, for example, is a pure state. It is an eigenstate of a global spin observable, namely the total spin. A *global observable* is an observable that relates to the whole. By contrast, none of the two parts is in a pure state. None of them is in an eigenstate of spin in any direction. Consequently, none of the parts can be described by a state vector: neither the state vector spin up nor the state vector spin down in any direction can be attributed to either of the two parts. Hence, entanglement means that two or more systems are related in such a way that only these systems taken together have properties of certain kinds with a definite numerical value, e.g., a property of the kind spin with a definite numerical value. This way in which two or more systems are related gives rise to the observed correlations between the outcomes of a measurement of certain local observables of each of these systems, e.g., correlations between the outcomes of a measurement of local spin observables of each of these systems.

Although only the state of these systems taken together is a pure state, we can calculate expectation values for the local observables of each of the two parts. The description by a density matrix that applies to each of the two parts contains for any local observable of this system a probability distribution for a measurement outcome. One may go as far as considering the density matrix as a state description. But one then has to work with the notion of what is known as a *mixed state*. We can also speak of a reduced state; but, in order to avoid a confusion with state reduction, I prefer to speak in the case under consideration of a mixed state. Conceptually, we have to distinguish between a *mixed state* and a *mixture of states*, although this distinction is not reflected in the density matrix. To bring this distinction to the point, Bernard d'Espagnat has introduced the terms "proper mixture" and "improper mixture" (1971, Chapter 6.3). A mixture of states (proper mixture) is an ensemble of systems which is such that the observer is ignorant of the state in which each of the systems is. Each system is in a pure state; but the observer is not in a position to specify the pure state in which each system is. However, in the case of entanglement, there is no question of a mixture of states. In the example under consideration, the observer is not ignorant of whether the first system has spin up and the second system spin down or the first system spin down and the second system spin up. Neither of the two systems is in a state in which it has either spin down or spin up.

The term "mixed state" (improper mixture) is used here exclusively to describe such a situation. A mixed state (improper mixture) is not a mixture of anything at all. What is at issue is a description that contains in the case of entanglement of the states of two or more systems all the information that is available about each of these systems if it is considered independently of the other systems. In the mentioned example of the singlet state, this description contains for any local spin observable of each of the two systems a probability distribution for a measurement outcome. But this description ignores the correlations between the possible values of these observables that can be acquired in measurement. Thus, this description ignores the disposition of each system to acquire a certain value of spin in a given direction relative to which value of spin in a given direction the other system acquires. Consequently, *the description in terms of a mixed state does not completely specify the local observables of a system.*

The entanglement thus has the consequence that the description of each of the two systems under consideration in terms of what is known as a mixed state is the same for the two systems. As far as the properties that are affected by the entanglement are concerned, the probability distribution of an observable that relates to the one system is identical with the probability distribution of the same observable that relates to the other system.[3] *The description of the parts in terms of a mixed state does hence not specify the state of the whole*, i.e., the two systems taken together. There are cases in which the mentioned description in terms of a mixed state applies to each of two systems, but these systems taken together are not in the singlet state. In the case of entanglement, *it is only the pure state of the whole*, such as the singlet state, *which completely determines the local properties of the parts and their relations* (to the extent that these properties and relations are determined at all in quantum physics).

When I speak of the entanglement of the states of two or more systems, the following is thus intended: The states of two or more systems are entangled if and only if these systems are only taken together in a pure state and can thus be described by a state vector only taken together. The entanglement of the states of two or more systems is the basis for the discussion on holism in quantum physics. Summing this sub-chapter up, we should bear in mind the following: observables can be incompatible. This implies that a system can be in an eigenstate of and consequently have an eigenvalue of at most one of two or more incompatible observables at a time. Incompatible observables are a necessary condition for the entanglement of the states of two or more systems: if two or more systems have incompatible observables each, the joint state of these systems can be a superposition of product states. In this case, the possible values of local observables of one system are correlated with the possible values of local observables of the

3 For an elaborate argument to this effect, see French and Redhead (1988). See already Margenau (1944), pp. 202–203.

other systems. Entanglement is independent of spatial or spatio–temporal distance.

7.2 The Case of Einstein, Podolsky and Rosen

7.2.1 The Principles of Separability and Local Action

The debate about the entanglement of the states of quantum systems goes back to a paper that Albert Einstein, Boris Podolsky and Nathan Rosen published in 1935. In this paper, these physicists propose the following criterion, which they regard as a sufficient, but not as a necessary condition for the reality of a physical quantity (p. 778):

> If, without in any way disturbing a system, we can predict with certainty (i.e., with probability equal to unity) the value of a physical quantity, then there exists an element of physical reality corresponding to this physical quantity. (p. 777)

The idea is: if we can predict the value of a physical quantity with certainty and if we can make this prediction without interfering with the system, then we are entitled to conclude that the system has this value independently of our prediction and our measuring operations to verify this prediction.

Over and above this sufficient condition for the reality of a physical quantity, Einstein's view of physical reality can be characterized by the principles of separability and local action. His clearest statement of these principles appears in the paper "Quanten-Mechanik und Wirklichkeit" (1948):

> If one asks what is characteristic of the realm of physical ideas independently of the quantum-theory, then above all the following attracts our attention: the concepts of physics refer to a real external world, *i.e.*, ideas are posited of things that claim a 'real existence' independent of the perceiving subject (bodies, fields, *etc.*), and these ideas are, on the other hand, brought into as secure a relationship as possible with sense impressions. Moreover, it is characteristic of these physical things that they are conceived of as being arranged in a space-time continuum. Further, it appears to be essential for this arrangement of the things introduced in physics that, at a specific time, these things claim an existence independent of one another, insofar as these things 'lie in different parts of space'. Without such an assumption of the mutually independent existence (the 'being-thus') of spatially distant things, an assumption which originates in everyday thought, physical thought in the sense familiar to us would not be possible. Nor does one see how physical laws could be formulated and tested without such a clean separation. Field

> theory has carried out this principle to the extreme, in that it localizes within infinitely small (four-dimensional) space-elements the elementary things existing independently of one another that it takes as basic, as well as the elementary laws it postulates for them.
> For the relative independence of spatially distant things (A and B), this idea is characteristic: an external influence on A has no *immediate* effect on B; this is known as the 'principle of local action', which is applied consistently only in field theory. The complete suspension of this basic principle would make impossible the idea of the existence of (quasi-) closed systems and, thereby, the establishment of empirically testable laws in the sense familiar to us.[4]

Einstein's requirement that physical things "claim an existence independent of one another, insofar as these things 'lie in different parts of space'" is known as the principle of separability. Basing himself on this and other quotations from Einstein, Don Howard formulates separability as the claim that (1) spatially separated systems possess their own, distinct physical state each and that (2) the joint state of two or more spatially separated systems is wholly determined by their separate states.[5] Taking up Howard's work, I propose the following characterization of separability:

Principle of separability

> Physical systems have a state each in the sense that (1) this state completely determines the state-dependent, local properties of the system and (2) the joint state of two or more systems supervenes on the states which each of these systems has.

Quantum theory does not conform to separability. Consider again the simplest example of entanglement: two systems of spin 1/2 that are taken together in the singlet state. One may go as far as attributing a mixed state to each of the two systems. But the mixed state does not completely determine the state-dependent, local properties of each of these systems. To recall, the mixed state does not contain the disposition of each system to acquire a certain value of spin in a given direction relative to which value of spin in a

4 Einstein (1948), pp. 321–322. Translation adopted from Howard (1985), pp. 187–188. See also Einstein (1951), p. 84; Einstein (1953), pp. 6–10, as well as the letter to Schrödinger, 19 June 1935, quoted in Howard (1985), pp. 179–180, and the letter to Born, 18 March 1948, in Born (1969), p. 223. In the letter to Schrödinger, Einstein uses the terms "Trennungsprinzip" and "Trennungshypothese".

5 Howard (1989), pp. 225–227. See also Howard (1985), pp. 173, 179; Howard (1997), p. 125.

given direction the other system acquires. It is only the singlet state which contains these correlations. Hence, the joint state of the two systems does not supervene on the states which one may envisage attributing to each of these systems.

In this characterization of separability, the condition of spatial separation is left out. Consider, for instance, the spin state of the two electrons of a helium atom in the groundstate. This state is the singlet state too. But the two electrons in a helium atom are not localized in such a way that they are spatially separated from one another. Furthermore, in quantum computation, one considers the entanglement of the states of many systems which are usually not localized in such a way that they are separated in space. It is reasonable to treat such cases as cases of quantum non-separability too. Therefore, it is sensible to explain separability without imposing the condition that this principle applies only to systems which are separated in space. *Given the proposed characterization of separability, any case of entanglement is a violation of separability*.

In the second paragraph of the quotation above, Einstein imposes a further requirement: the principle of local action. Whereas separability refers to the states of physical systems at a given time, local action concerns changes in the states of physical systems in time. The physical systems are thought of as being localized at points in space or space–time. Local action imposes a restriction on the way in which the states of physical systems can change by interaction. It thereby presupposes separability: the systems in question have states each of which completely determine their local properties. What is intended by local action is that interactions propagate contiguously from point to neighbouring point with a finite velocity. In the German original, Einstein uses the term "*Prinzip der Nahewirkung*". His general argument for local action does not establish a specific upper limit for the velocity with which interactions propagate (such as the velocity of light); it is merely required that interactions propagate with some finite velocity.[6] Nonetheless, having the discussion on whether or not quantum theory is compatible with special relativity in mind, I suggest characterizing local action in relativistic terms:

---

6 Compare Howard (1989), p. 246.

Principle of local action in relativistic terms

> Every interaction (force) propagates contiguously with a finite velocity, that is – in relativistic terms – a velocity that is not higher than the velocity of light in vacuum.

The four types of interaction that are currently acknowledged – strong and weak interaction, electromagnetism, gravitation – conform to local action. The principle of local action is a sort of locality condition. It is sometimes identified as the principle of locality without further qualification. For instance, Don Howard works with the notions of separability and locality, whereby he means by locality Einstein's principle of local action.[7] Furthermore, it is "*Nahwirkungskausalität*" (local action) which Rudolf Haag translates as "locality" and which accounts for the title of his book *Local Quantum Physics* (1992, p. 9). However, this identification of local action with locality can lead to confusion in the interpretation of quantum theory. Most philosophers of science maintain that quantum theory satisfies the principle of local action; nonetheless, quantum theory is widely seen as implying some sort of non-locality, especially if measurement is taken into account.[8] In the discussion on Bell's theorem and the subsequent experiments, it is the conjunction of separability and local action that is often referred to as locality without further qualification.

7.2.2 Einstein's Argument for the Incompleteness of Quantum Theory

Consider again the example from [7.1.3]. Two systems of spin 1/2 such as two electrons or two neutrons are emitted together from a source and then fly apart in opposite directions. Their joint spin state is the singlet state. If we measure a local observable, say spin z, of the one system, we can, on the basis of the outcome of this measurement, predict with certainty the outcome of a measurement of the spin of the other system in the same direction. If the outcome of the one measurement is, say, spin up, we can predict with certainty that the outcome of the other measurement will be spin down. This prediction holds however far apart in space or space–time the other system is. Two such measurements can be separated by a space-like interval. These

---

7 Howard (1985), pp. 173, 179; Howard (1989), pp. 226–227; Howard (1997), p. 125.

8 See the overview in Redhead (1987), Chapters 3 to 4. Instead of speaking of separability in distinction from local action, Redhead himself introduces the principles of ontological locality and environmental locality. A violation of ontological locality implies a violation of separability, and a violation of environmental locality implies a violation of local action. See Redhead (1987), Chapter 6; Redhead (1995a), pp. 48–51.

correlations among the values of local observables of two or more systems are known as *Einstein–Podolsky–Rosen correlations* (or short EPR correlations), paying tribute to the famous paper of these physicists.

However, the example of two systems of spin 1/2 in the singlet state is in fact David Bohm's version (1951, pp. 611–622) of the thought experiment which Einstein, Podolsky and Rosen propose in their paper of 1935. Einstein and his collaborators do not consider two systems whose spin states are entangled, but two systems in one-dimensional motion whose states are entangled with respect to position and momentum. An analogous argument applies in this case, although the mathematical treatment is more complicated. The point of Einstein, Podolsky and Rosen is: given the outcome of a measurement of position or momentum of the one system, we can predict with certainty the result of a measurement of the same observable of the other system.[9]

Starting from their sufficient condition for an element of reality, which was quoted at the beginning of the last section, Einstein, Podolsky and Rosen come to the conclusion that quantum theory is incomplete: We can predict the outcome of a measurement of the corresponding observable of the other system with certainty without disturbing that system. Hence, there exists an element of reality which corresponds to that physical quantity independently of whether or not the second measurement is in effect carried out. This element of reality exists even before the first measurement takes place: Einstein, Podolsky and Rosen maintain that no reasonable definition of reality could be expected to make this element of reality of the second system "depend upon the process of measurement carried out on the first system, which does not disturb the second system in any way" (1935, p. 780; see also Einstein (1948)). Since quantum theory does not reflect this element of reality of the second system before a measurement of the first system is carried out, it is incomplete (1935, p. 780, and Einstein (1948), p. 323). Einstein, Podolsky and Rosen (1935) regard it as a necessary condition for a physical theory to be complete that "every element of the physical reality must have a counterpart in the physical theory" (p. 777). Thus, what is at issue in the discusssion on whether quantum theory is complete is not the

---

9 For relativistic qualifications of this statement, see Smith and Weingard (1987). For suggestions to formulate the sufficient condition for an element of reality of Einstein, Podolsky and Rosen in a relativistic context see Ghiradi and Grassi (1994) and Redhead and La Rivière (1997) as well as the reply of Ghiradi and Weber (1997), pp. 103–104, to Redhead and La Rivière (1997).

claim that quantum theory is the final theory or that it describes everything that exists. The point is merely whether quantum theory gives a complete description of the states of microphysical systems.

Note that both the principles of separability and local action are necessary to establish the conclusion that quantum theory is incomplete. We need local action to preclude that there is an interaction between a measurement effected on the first system and an element of reality of the second system. We need separability to preclude that the determination of the local properties of a system depends on anything else than the state in which this system is. Note furthermore that the sketched reconstruction of the argument for the incompleteness of quantum theory requires only the consideration of one pair of local observables – position, momentum, or the spin in an arbitrary direction of both systems.

Einstein, Podolsky and Rosen claim, in fact, to have established a stronger conclusion: not only is quantum theory incomplete in the sense that there are elements of reality of quantum systems which it does not acknowledge; but also, more precisely, quantum systems have a definite numerical value of two or more incompatible observables at once (1935, p. 780, and Einstein (1948), p. 323). In order to establish this conclusion, we have to consider at least two pairs of incompatible observables, such as position and momentum in the same direction or spin z and spin x of each of the two systems. The point of Einstein, Podolsky and Rosen can then be reconstructed as follows: An experimenter or a random generator chooses only well after the emission of the two systems from the source which observable is to be measured on the one system. Separability and local action imply that this choice of the experimenter or the random generator is irrelevant to the physical reality of the other system. Nonetheless, whatever choice the experimenter or the random generator makes, once we have carried out the measurement of the chosen observable of the one system and given the outcome of that measurement, we can predict with certainty the outcome of a measurement of the corresponding observable of the other system. Therefore, Einstein, Podolsky and Rosen conclude that both systems in fact have a definite numerical value of all the observables among which the experimenter or the random generator chooses already at their emission from the source. This reconstruction of the argument of Einstein and his collaborators does not have recourse to any controversial counterfactual propositions.[10]

---

10 See Butterfield (1990) for a detailed analysis of the arguments with one pair and two pairs of observables.

Note that Einstein does not consider this conclusion to be self-evident. It is established by an argument which is built on separability, local action, and his sufficient condition for the reality of a physical quantity.[11] Consequently, according to Einstein, a measurement merely discovers values which the system in question has independently of whether or not a measurement is carried out. We have to envisage a more fundamental theory than quantum theory which describes these values. Nonetheless, such a theory has to match the predictions of quantum theory, which are confirmed by experiments.

7.3 Bell's Theorem and the Philosophical Implications

7.3.1 Bell's Theorem and the Bell Experiments

The argument of Einstein, Podolsky and Rosen for quantum theory being incomplete triggered a debate about adding hidden variables to quantum theory. It is, however, in dispute whether Einstein himself can be associated with the programme of hidden variables. He may well have had a deeper change in mind – a replacement of quantum theory with a new theory.[12] On the other hand, it can be argued that any addition of hidden variables to quantum theory amounts to replacing quantum theory with a new theory of microphysical systems. The upshot of the discussion on hidden variables is this one: *it is not possible to add hidden variables to quantum theory that satisfy separability as well as local action and retain the predictions of measurement outcomes which quantum theory yields* – unless one is prepared to give up a presupposition of experimental science, namely that the setting of the parameters which are to be measured on a system is causally independent of the state of the system that is prepared at the source of the experimental set-up.

In 1967, Simon Kochen and Ernst Specker published a theorem which implies that the hidden variables have to be contextual if the predictions of quantum theory are to be reproduced.[13] Their theorem has the consequence that it is impossible to add hidden variables to quantum theory to the effect that quantum systems have a definite numerical value of incompatible observables at once independently of which measurements are carried out on

---

11 See Howard (1985), sections 1 and 2.

12 See Jammer (1974), pp. 254–255; Cushing (1994), Chapter 9.1.2.

13 Kochen and Specker (1967). For a non-technical exposition of the point at issue, see Specker (1960). See furthermore the exposition in Redhead (1987), Chapter 5. The theorem of Kochen and Specker builds on the theorem of Gleason (1957).

these systems. The definite numerical value which can be attributed to one observable by hidden variables depends on which other observables are measured together with this observable. The only exception is the spin of spin 1/2 systems, whose mathematical representation merely requires a two-dimensional Hilbert space.

Whereas the theorem of Kochen and Specker establishes that hidden variables have to be contextual if they are to match the predictions of quantum theory, the theorem of John Bell (1964) shows that no hidden variables whatsoever which satisfy separability and local action can come up to the predictions of quantum theory (unless one gives up the above mentioned presupposition of experimental science). Consider again the example of two systems of spin 1/2 in the singlet state. The outcomes of a measurement of the spin in the same direction on both systems are maximally correlated: if the one outcome is spin up, the other outcome is spin down. Measuring the spin in the same direction on both systems means that the same local observable is measured on both systems: there is no difference in the two parameters that are measured. If the parameters differ by an angle, there is no longer a maximal correlation between the outcomes of both measurements, but only a probabilistic one. If we know the outcome of a measurement of the one system, we can no longer predict the outcome of a measurement of the other system with certainty. But given the outcome of a measurement of the one system, the probability for the outcome spin up and the outcome spin down of the other system is not 0.5 each either. One outcome is more probable than the other one. Such a probability can be calculated for each difference of the angles of the parameters that are measured on both systems. Only if that difference is orthogonal so that, for example, spin z on the one and spin x on the other system are measured, there are no correlations between the outcomes any more.

Bell's theorem imposes an upper limit on the correlations between the outcomes of such measurements. The original version of this theorem consists in an inequality that sets an upper limit on such correlations. Subsequent to Bell's paper of 1964, different versions of such an inequality have been developed. Unless one goes for hidden variables that contradict the above mentioned presupposition of experimental science, all the hidden variables that conform to separability and local action satisfy Bell's theorem. Quantum theory, by contrast, violates Bell's theorem. For certain differences between the angle of the parameters that are measured on two systems in the singlet state, quantum theory predicts probability correlations between the outcomes of both measurements that are higher than Bell's theorem permits.

This result can be generalized: For every case of entanglement of the states of two or more systems there are some local observables whose measurement on these systems leads to higher correlations between the outcomes than Bell's theorem allows.[14] Consequently, any addition of hidden variables to quantum theory that are in accordance with separability and local action entails a contradiction with the predictions of quantum theory. It is of no importance for that matter whether such hidden variables are deterministic or stochastic.[15] *The violation of Bell's theorem implies for a case like an experiment on two systems of spin 1/2 in the singlet state that we cannot consider the emission of the two systems from the source as a common cause of the correlations that screens off the one measurement event from the other one.*

At the end of the eighties, a further development of Bell's theorem has been achieved. If one considers a whole which is composed of at least three instead of merely two systems of spin 1/2, it is possible to derive a contradiction between the predictions of quantum theory and all theories with hidden variables that satisfy separability and local action by working with maximal correlations alone: given the outcomes of measurements of two systems, one can predict the outcome of the measurement of the third system with certainty, and that correlation contradicts all the theories that base themselves on separability and local action. All these measurements can be space-like separated. No inequality is needed to derive this result.[16] Furthermore, this result enables an elegant unification of Bell's theorem and the theorem of Kochen and Specker.[17] Since it has thus been demonstrated that there are maximal quantum correlations which cannot be taken into account by any theory based on separability and local action, I will ignore the difference between maximal and merely probabilistic correlations in my discussion of philosophical options. For the sake of simplicity, I will limit myself to Einstein–Podolsky–Rosen correlations between the outcomes of measurements on two systems.

To be more precise, the assumption which Bell makes in the proof of his theorem is *factorizability*. I adopt the formula of Abner Shimony in his paper

---

14 For a general proof of that proposition, see Gisin (1991) and Popescu and Rohrlich (1992).

15 Compare, for instance, Hellman (1987); Redhead (1987), pp. 98–107.

16 Greenberger et al. (1990). See also Clifton, Redhead and Butterfield (1991).

17 Mermin (1990). See already Heywood and Redhead (1983).

"Events and Processes in the Quantum World" (1986) (quoted from Shinomy (1993), p. 147) with slight modifications:

(7.7) $$p^{12}_{\lambda}(x_a, x_b \mid a, b) = p^{1}_{\lambda}(x_a \mid a) \cdot p^{2}_{\lambda}(x_b \mid b)$$

In this formula, p is probability, and λ denotes the state of the whole system, i.e., the singlet state in the case under consideration; 1 and 2 refer to the two parts, i.e., the two systems of spin 1/2; x_a and x_b stand for the outcomes of a measurement of part 1 and part 2 respectively ($x_a = \pm 1$, $x_b = \pm 1$); a denotes the parameter which is measured on part 1, and b denotes the parameter which is measured on part 2. Factorizability means: given the state of the whole system, the probability for the outcome of a measurement of part 1 depends only on the parameter which is measured on part 1; and the probability for the outcome of a measurement of part 2 depends only on the parameter which is measured on part 2. The probability for both outcomes, given both parameters, is the product of these probabilities. Hence, factorizability is the precise mathematical formulation of the assumption which Bell sets out in the introduction to his paper, namely that "the result of a measurement on one system be unaffected by operations on a distant system with which it has interacted in the past" (1964, p. 195).

Jon Jarrett (1984) showed that factorizability, to which he refers as strong locality, can be analyzed into two conditions: it is the conjunction of what he calls "locality" and "completeness". Shimony (1993, pp. 146–147) introduced the terms "parameter independence" and "outcome independence" for Jarrett's locality and Jarrett's completeness respectively. Shimony's terminology is more appropriate: it has no misleading connotations for the interpretation of quantum theory. I adopt Shimony's formulae for *parameter independence* with a slight simplification:

(7.8) $$p^{1}_{\lambda}(x_a \mid a, b) = p^{1}_{\lambda}(x_a \mid a)$$

(7.9) $$p^{2}_{\lambda}(x_b \mid a, b) = p^{2}_{\lambda}(x_b \mid b)$$

These formulae say: in the state λ of the whole system, the probability for the outcome of a measurement of part 1 is independent of the parameter that is measured on part 2 (and *vice versa*). *Outcome independence* states the following (again, I adopt Shimony's formulae):

(7.10) $$p^{1}_{\lambda}(x_a \mid a, b) = p^{1}_{\lambda}(x_a \mid a, b, x_b)$$

(7.11) $$p^{2}_{\lambda}(x_b \mid a, b) = p^{2}_{\lambda}(x_b \mid a, b, x_a)$$

These formulae say: in the state λ of the whole system and given the parameters that are measured on both parts, the outcome of a measurement of part 1 is independent of the outcome of a measurement of part 2. If the outcome of a measurement of part 2 is given, the probability for the outcome of a measurement of part 1 is not changed (and *vice versa*).

Quantum theory conforms to parameter independence: if the parameter that is to be measured on the one system is given, the probabilities for the outcome of a measurement of the other system are not changed. However, quantum theory violates outcome independence: if the outcome of a measurement of the one system and the parameter measured are given, the probabilities for the outcome of a measurement of the other system are changed. In the spin case under consideration, the probability is no longer 0.5 for spin up and 0.5 for spin down, but one result is more probable than the other one (unless the spin is measured on both systems in orthogonal directions). In the following, I therefore speak in terms of quantum theory implying *outcome dependence*.

The work of Jarrett shows that a violation of factorizability and hence a violation of Bell's theorem can be accounted for in two different ways: by a violation of parameter independence or by a violation of outcome independence. Quantum theory satisfies parameter independence, but it violates outcome independence. Parameter independence and outcome independence are distinct from separability and local action. First of all, no consideration of measurement and probabilities enters into the formulation of separability. As far as local action is concerned, a violation of local action can be conceived as a violation of parameter independence or as a violation of outcome independence. Taking quantum theory to be incomplete and endorsing separability, one can assume a signal which is sent with superluminal velocity at the time when the parameter is adjusted so that parameter independence is violated. But one can also imagine a signal which instantaneously transmits the outcome of the measurement of the one system to the other system. Thus, one can imagine outcome independence to be violated without separability being violated. Therefore, outcome independence and separability are not equivalent.[18] Furthermore, if the outcomes of measurements were under the control of the experimenter, a

---

18 For a claim to the contrary, see Howard (1992); Howard (1997), pp. 126–128. For a criticism of Howard's claim, see Maudlin (1994), pp. 97–98; Laudisa (1995), pp. 314–322; Belousek (1999).

violation of outcome independence could enable the experimenter to send superluminal signals.[19]

However, outcome dependence in quantum theory does not contradict special or general relativity, at least not on the operational level. Outcome dependence does not make superluminal signalling possible, because we cannot manipulate the outcome: in the mentioned example, the result "spin up" and the result "spin down" are equally probable. Moreover, for a local observer, i.e., an observer who does not know which measurements are carried out at a distance, both these results are always equally probable. It is only *post festum* by comparing the outcomes of a sequence of measurements on the one side with the outcomes of the corresponding sequence of measurements on the other side that the correlations can be discovered.[20] We can, however, manipulate the parameter that is to be measured. But, since quantum theory satisfies parameter independence, we cannot change the probabilities for the outcome of a measurement in a space-like separated region by manipulating the parameter of an apparatus. Shimony therefore speaks of a peaceful coexistence between quantum theory and relativity physics.[21]

Subsequent to Bell's theorem, a number of experiments have been carried out, which confirm the predictions of quantum theory. All these experiments are referred to as *Bell experiments*. The most significant type of these experiments goes back to the experiment of Alain Aspect and his team which was performed in Paris at the beginning of the eighties (Aspect, Dalibard and Roger (1982)). This experiment, like most other experiments, consists in correlation measurements of the polarization of two photons that are emitted together from a source and then fly apart in opposite directions. Although photons are not systems of spin 1/2, the polarization state of such pairs of photons is the singlet state. Aspect's experiment is very significant, because two automatic switches adjust the parameter that is to be measured on each photon only after the photons have been emitted from the source; the switching event and the detection event on the one side of the source are

---

19 Compare Kronz (1990), pp. 431–441; Butterfield (1989b), pp. 131–135; Butterfield (1992a), pp. 68–77; Jones and Clifton (1993); Berkovitz (1998a). These papers offer a criticism of the claims of Jarrett (1984), pp. 573–578, and Shimony, "Controllable and Uncontrollable Non-Locality" (1984) in Shimony (1993), Chapter 10.

20 But see Mittelstaedt (1998b).

21 See, for example, Shimony (1989), p. 29; "Controllable and Uncontrollable Non-Locality" (1984) in Shimony (1993), Chapter 10, p. 133.

separated by a space-like interval from the switching event and the detection event on the other side of the source. This experiment yields a violation of Bell's theorem.[22] It conforms to the predictions of quantum theory. In recent years more experiments have been carried out which show Einstein–Podolsky–Rosen correlations between the outcomes of space-like separated measurements.[23]

The correlations between the measurement outcomes in the Bell experiments cannot be dismissed as coincidence. These correlations have been confirmed in a lot of experiments. In each of these experiments, measurements have been carried out on a great number of pairs of photons. To dismiss these correlations as coincidence would therefore amount to a dismissal of any experimental evidence. However, as in every experiment, loopholes can be found in the Bell experiments too. The following three objections can be raised:

- Almost all the experiments carried out hitherto are experiments with photons.[24] Photons have zero rest mass. This implies that, for the photons, no proper time passes between their emission from the source and the measurement.
- In the experiment of Aspect, there is no question any more of a signal which transmits the parameter that is to be measured on the one side to the other side of the experimental arrangement in one measurement (unless one settles for superluminal signals). However, the switches do not adjust randomly the parameter that is to be measured.[25]
- In all experiments carried out up to now, the detectors are inefficient: a considerable number of photons are not detected.

The third objection is more important than the other ones. Building on the inefficiency of the detectors, it is possible to construct models of hidden variables which account for the results of the Bell experiments that are performed hitherto without assuming a violation of Bell's theorem.[26]

---

22 The version of a Bell inequality which Aspect uses is the one of Clauser et al. (1969).

23 See, in particular, the recent experiment by Tittel et al. (1998) and the experiment planned by Weihs, Weinfurter and Zeilinger (1997).

24 But consider the experiment recently reported by Hagley et al. (1997).

25 Compare the critical remark in Aspect, Dalibard and Roger (1982), p. 1807; Aspect and Grangier (1985), p. 69.

26 See already Fine (1982), and, subsequent to the experiment of Aspect, see Marshall, Santos and Selleri (1985); Pascazio (1988); Uchiyama (1995). For a calculation of the

However, one can retort that it would amount to conspiracy if hidden variables of each of the systems in accordance with separability and local action influenced the detection of each of the photons just in such a way that the result is a confirmation of the predictions of quantum theory.[27] Despite the mentioned loopholes, by far the most physicists and philosophers of science acknowledge that a violation of Bell's theorem has been shown by experiment. The point at issue then is which one is the most plausible option to account for a violation of Bell's theorem.

7.3.2 Non-Locality and the Bell Experiments

Even if the outcome dependence in quantum theory does not make superluminal signalling possible, one may wonder whether the Bell experiments bring quantum theory into conflict with special relativity. For the violation of Bell's theorem by experiment manifests some sort of non-locality: two space-like separated measurement outcomes are correlated without the preparation event, i.e., the emission of the two systems from the source, acting as a common cause that screens the one measurement outcome off from the other measurement outcome.[28] Does this non-locality mean that there is superluminal causation between the two wings of a Bell experiment?

Jeremy Butterfield and Tim Maudlin argue that if we endorse a counterfactual criterion of causal dependence such as the suggestion of David Lewis (1986a, Chapter 21, in particular pp. 175–179), we have to conclude that there is causal dependence and hence causation between two space-like separated events.[29] Maudlin (1994) speaks of a superluminal causal connection between two space-like separated measuring events in a Bell experiment and of a superluminal transmission of information (in particular Chapter 5). Given a measurement outcome x_a in one wing of such an experiment, the probability for a specific outcome x_b in the other wing is

---

efficiency rate that the detectors would have to achieve in order to exclude these models see Mermin (1986).

27 Compare Mermin (1986), p. 422. See also Maudlin (1994), pp. 175–186.

28 For an argument to the effect that we should accept these correlations without explanation, see Fine (1989); van Fraassen (1991), Chapter 10.

29 Butterfield (1992b). For an examination under which concepts of causality there is causation between the two wings in such an experiment, see Skryms (1984). As to applying a counterfactual conception of causation to the EPR-correlations, see also Laudisa (1999).

raised. If there had not been the outcome x_a in the one wing, the outcome x_b in the other wing would have been less probable. Causation in Lewis' sense of counterfactual dependence is a consequence of outcome dependence in quantum theory (although Lewis does not formulate his criterion of counterfactual dependence in terms of conditional probabilities).[30]

There are two strategies to avoid this argument, as Butterfield makes clear (1992b, pp. 41–42): (a) One can dispute that Lewis' criterion for causation applies by arguing that there are not two distinct events. However, this strategy is not convincing: Let us focus on the changes in the states of the two measuring instruments that register the outcomes. These changes are clearly two distinct events, and these events are correlated. (b) Basing oneself on Lewis' theory of causation in terms of counterfactuals, one can introduce further conditions for a relation of causal dependence. This is the strategy of Michael Redhead. As a necessary condition for a causal relation, he introduces a condition of robustness against differences in the way in which the cause comes into existence. On the basis of this condition, Redhead maintains that there is no causal link between the two wings in a Bell experiment[31] and endorses Shimony's thesis of a peaceful coexistence between quantum theory and special relativity [7.3.1] (for instance Redhead (1987), pp. 168–169).

My point is: Even if we accept the argument of Butterfield and Maudlin for non-locality in the sense of a causal connection between the two space-like separated events in which the changes in the states of the measuring instruments consist, we are free to side with Shimony and Redhead in maintaining that there is peaceful coexistence between quantum theory and special relativity. Speaking of causation on the basis of a counterfactual criterion does not tie us down to admitting causation in the sense of a superluminal interaction on the quantum level. Maudlin himself emphasizes that a violation of Bell's theorem implies neither a superluminal matter or energy transport nor the possibility of superluminal signalling (1994), in particular the summary in Chapter 9).

---

30 As regards this and other differences between Lewis' criterion and outcome dependence and an argument for the irrelevance of these differences, see Butterfield (1992b).

31 Redhead (1987), pp. 102–106; Redhead (1989a), pp. 438–440; Redhead (1989b), pp. 148–151; Redhead (1992). For a criticism of Redhead's argument, see Cartwright and Jones (1991); Healey (1992); Maudlin (1994), pp. 150–154.

The most prominent strategy for explaining the correlations between the measurement outcomes without coming into conflict with the principle of local action consists in the following idea: since the two systems do not have a separate spin state each, a measurement interaction with the one system directly changes the state of the whole. Qua change of the state of the whole, this local interaction is relevant to the probabilities for the outcome of a measurement on the other system. But this change of the state of the whole is not a local change in the other system. A local change in the other system in the sense of acquiring a definite numerical value of the local property in question occurs only in a local interaction of that system with a measurement device.

One proposal along these lines is put forward by Richard Healey. This proposal is in the framework of the so-called modal interpretation. It yields definite numerical values as measurement outcomes without working with the assumption that a reduction of the singlet state to a product state occurs in measurement. Healey, however, envisages the possibility that his proposal is a causal explanation of the Einstein–Podolsky–Rosen correlations (1989, pp. 129–136, 148–173; 1994, pp. 359–373). Another proposal is the one of Shimony: he acknowledges state reductions and conceives measurement as an actualization of potential properties of quantum systems. But he maintains that in a Bell experiment a measurement of the spin of the one system is not sufficient for an actualization of a definite numerical value of the spin in the same direction of the other system too, even if one such value of the other system has probability one.[32]

Even if we acknowledge that there is a state reduction in measurement, according to the idea under consideration, there is no question of a physical interaction between, say, the one system acquiring a state in which it has spin up and the other system acquiring a state in which it has spin down. For these are not changes in separate states of the two systems, but a change from a non-separable state of the whole to separate states of the parts. The principle of local action presupposes separability: local action is a constraint on the way in which changes in the separate states of physical systems are carried out [7.2.1]. Applied to this case, that is to say: the question of a superluminal physical interaction on the quantum level can arise only if we

32 See "Events and Processes in the Quantum World" (1986) in Shimony (1993), Chapter 11, pp. 151–154. See also Rohrlich (1987), p. 175, and furthermore Hawthrone and Silberstein (1995) who employ the notion of holistic connections between events (pp. 115–117).

presuppose, contra quantum theory, that the two systems have a separate spin state each prior to measurement.

Furthermore, if the measurements in the two wings of a Bell experiment are separated by a space-like interval, there is no fact of the matter as to which of these measurements occurs first. To accommodate this point within an account that acknowledges state reductions, there is an elaborate suggestion by Gordon Fleming according to which the local observables of spin of the photons in the Bell experiments are properties which are relative to a reference frame – in the sense that they are relative to a space-like hyperplane.[33] There is no space here for an examination of this suggestion and the other mentioned proposals. The only point that is needed here is this one: based on quantum non-separability there is a strategy available for explaining the correlations between the outcomes of a Bell experiment which does not have recourse to superluminal interaction on the quantum level. This strategy supports the view that there is peaceful coexistence between quantum theory and special relativity.

Coming back to Lewis' theory of causation, this strategy for an explanation of the Einstein–Podolsky–Rosen correlations shows the following: not all cases of causation à la Lewis are realized as some sort of a physical interaction. There are two main options to accommodate this result:

1. One can divorce Lewis' causation from physical interaction. Lewis' counterfactual criterion tells us when there is causation, but it leaves entirely open what the means of causation is. In this perspective, if one acknowledges definite measurement outcomes, quantum entanglement or non-separability simply is a means of causation that is distinct from physical interactions (forces). This is not to say that quantum entanglement is itself a causal relation. But if a measurement interaction occurs, quantum entanglement can act as a means of causation.[34] However, one can object to this strategy that by divorcing causation from a realization in terms of a physical interaction, one divorces this concept from widespread pre-philosophical intuitions about what causation is.

---

33 See, in particular, Fleming (1996). For a non-technical description and critical evaluation of this view, see Maudlin (1994) pp. 204–212, 233–234; Maudlin (1996), pp. 298–303. See furthermore the criticism of Dorato (1996), pp. 593–595.

34 Berkovitz (1998a), pp. 203–219, and Berkovitz (1998b), section 2, argues for a position along these lines.

2. One can take the correlations between the measurement outcomes in a Bell experiment as a case that calls for an amendment of a theory of causation such as the one of Lewis:
a) One possibility is to include further physical conditions for a relation of causal dependence that can be formulated within a theory such as the one of Lewis. This is the strategy of, for instance, Redhead, as mentioned above.
b) One can go beyond the framework of a theory such as the one of Lewis and demand outright that physical interaction is a necessary condition for causation over and above counterfactual dependence. However, if one introduces physical interaction into the definition of causation, then physical interaction is likely to be sufficient on its own to do the job of accounting for causation. We then get to a theory of causation according to which a causal process posseses or transmits a conserved quantity, and an interaction is an exchange of a conserved quantity.[35] This theory is a sophisticated regularity account of causation.

Whatever stance one takes on this issue, this example highlights Quine's thesis in "Two Dogmas of Empiricism" that science and metaphysics are interwoven [2.1.1]: which stance one takes on this issue depends on which way one regards as most plausible when it comes to outbalancing new scientific evidence with metaphysical considerations such as those ones that speak in favour of a counterfactual account of causation.

7.3.3 "Experimental Metaphysics"

Let us apply Quine's philosophy of science to the discussion on quantum holism. Einstein's view of physical reality accords with common sense. When introducing what has become known as the principle of separability in the long quotation at the beginning of [7.2.1], Einstein himself says that this principle "originates in everyday thought", and he considers it to be necessary for "physical thought in the sense familiar to us" to be possible. An analogous argument applies to the principle of local action. Recall that even Newton regards action at a distance as "so great an absurdity that I believe no man who has in philosophical matters any competent faculty of thinking can ever fall into it".[36] Separability and local action are

35 See, for example, "Causality without Counterfactuals" in Salmon (1998), essay 16; Kistler (1998) and (1999). See furthermore Dowe (1992), pp. 210–215, and Dowe (1995).

36 Letter 406 to Bentley, 25 Feb. 1692/3, in Newton (1961), p. 254.

metaphysical principles in the sense that they are a precise formulation of assumptions at the centre of our common sense view of nature. We take it for granted that physical systems have a state each which completely determines their local properties and that interactions propagate with some finite velocity.

The answer to the question whether quantum theory is complete therefore seems to be a purely metaphysical matter. It depends on whether we lay stress upon separability and local action as metaphysical principles which are at the very foundation of empirical science or whether we base ourselves on quantum physics and claim that separability or local action are to be abrogated consequent upon new experimental evidence. The predictions of quantum theory and their empirical confirmation are not in dispute.

However, as a result of Bell's theorem and the Bell experiments, the apparently clear distinction between physics and metaphysics in the interpretation of quantum theory has become blurred. Physics – the predictions of quantum theory – challenges a metaphysics that bases itself on separability and local action as principles for the very possibility of physical science. We cannot simply accept the predictions of quantum theory and claim when it comes to metaphysics that all physical systems nevertheless conform to separability and local action, taking quantum theory to be incomplete. Metaphysics – Einstein's realism – has been shown to have exactly computable, empirical consequences which can be put to the test in experiments.

To describe this situation, Shimony uses the term "*experimental metaphysics*".[37] This term has not been invented by Shimony. To my knowledge, it has been introduced by Michele Besso (1948), a friend of Einstein in his Swiss years, in the context of Ferdinand Gonseth's philosophy. Gonseth's position is similar to Quine's, at least insofar as Gonseth also rejects a separation between mathematics, science and philosophy and considers every part of our knowledge to be revisable.[38] But

---

37 See, for instance, Shimony (1989), p. 27. See also Jarrett (1989). Compare furthermore the title of Redhead (1995a): "From Physics to Metaphysics". Lecture 3 in that book is entitled "Experimental Metaphysics". That title is intended to be an allusion to Shimony's claim. See furthermore the conclusion on p. 87.

38 See, in particular, Gonseth (1948), pp. 123–124. On the analogy with Quine, compare Gochet (1977), p. 121, and Esfeld (forthcoming c). See also the reference to Gonseth in Specker (1960), p. 239.

the term "experimental metaphysics" is not yet applied to the interpretation of quantum theory there.

This historical background puts the talk of experimental metaphysics in the interpretation of quantum physics into perspective: it makes clear that speaking of experimental metaphysics does not at all imply that metaphysical issues are settled by experiment.[39] On the contrary, *speaking of experimental metaphysics highlights that the discussion on Einstein's misgivings about quantum theory can be comprehended along the lines of the Duhem–Quine thesis*: this discussion shows that there is no separation between physics and metaphysics in the interpretation of quantum theory. We can with reason take the Bell experiments to be a test of two hypotheses, i.e., the conjunction of parameter independence and outcome independence. But the point is: What the Bell experiments are taken to confirm or to refute depends on which background assumptions we base ourselves. Consequently, we have a number of metaphysical options for accommodating our system of knowledge to the results of these experiments, depending on which of these background assumptions we retain and which ones we are prepared to give up.[40] All these options account for the experimental evidence. Nonetheless, these options can be evaluated in the light of their consequences for our system of knowledge as a whole. In an interview, Bell himself describes the experiment of Aspect in terms of Quine's confirmation holism:

> I think that it is very difficult to say that any one experiment tells you about any isolated concept. I think that it's a whole world view which is tested by an experiment, and if the experiment does not verify that world view, it is not so easy to identify just which part is suspect and has to be revised. Certainly the experiment says that Einstein's world view is not tenable. (in Davies and Brown (1986), pp. 46–47)

According to the scholarly work of Howard (1990), Einstein did not know Quine's "Two Dogmas", but he knew Duhem's holism and endorsed it.[41]

If we base ourselves on quantum theory without introducing hidden variables, we have to give up separability; but we can avoid a conflict with local action, as argued in the preceding section. This position leads to the option for quantum holism. I will elaborate on quantum holism in the next

---

39 Jones and Clifton (1993) construe that term in that way and argue against it.

40 For a good overview of proposals for the philosophical impact of Bell's theorem, see the papers in Cushing and McMullin (1989).

41 See also Fine (1986), pp. 86–90; Hentschel (1987).

chapter. However, we can also keep separability and give up local action by assuming that there is a superluminal signal which connects the two wings of a Bell experiment.[42] The price for this option is that we have to admit a new kind of interaction which does not fit in with the known kinds of interaction, because it propagates with a superluminal velocity.

What is more, even if we accept that the experimental evidence shows a violation of Bell's theorem, we can endorse both separability and local action if we are prepared to make drastic enough changes to other background assumptions in our system of knowledge. As mentioned at the beginning of [7.3.1], Bell's theorem takes for granted a presupposition of experimental science, namely that the setting of the parameters which are to be measured on a system is independent of the state of the system that is prepared at the source of the experimental set-up. If one is willing to give up this presupposition, one can introduce hidden variables which satisfy separability as well as local action and retain the predictions of measurement outcomes that quantum theory yields. There are two candidates for a causal explanation of the correlations between the measurement outcomes in a Bell experiment which satisfy both separability and local action, but which infringe upon background assumptions in our conception of causality and experimental science: *common cause* and *backward causation.*

The violation of Bell's theorem implies that the preparation of the singlet state at the source in a Bell experiment cannot be a sufficient *common cause* which screens the one measurement outcome off from the other measurement outcome. Nonetheless, one can opt for a common cause if one is willing to go beyond the experimental arrangement: the hidden variables are a factor in the intersection of the past light-cones of the two measuring events which lies in the past of the preparation event and which makes it that the two measuring events are correlated as predicted by quantum theory. However, it is not sufficient for an account of these correlations to assume a factor in the common past of the two systems which influences their behaviour. The correlations between the measurement outcomes crucially depend on the parameters that are measured. In the experiment of Aspect, the parameters are adjusted only after the two systems have been emitted from the source. No property of the preparation of the state of the two systems has an influence upon which parameters will be measured.

---

42 See, for instance, Chang and Cartwright (1993), pp. 181–189. Popper (1982), pp. 22–27, also contemplates non-local interaction to explain the correlations in a Bell experiment.

Therefore, the factor in the past which is supposed to influence the behaviour of the two systems must furthermore be seen as having an influence upon the parameter that is measured on each of the systems.

Because of the last point, Bell himself and others object to the option for a common cause that it amounts to conspiracy:[43] the mentioned factor has to coordinate the behaviour of the two systems with the adjustment of the two parameters in such a way that the predictions of quantum theory are satisfied. This conspiracy can include the decisions of the experimenters, because we can imagine a Bell experiment in which not two automatic switches, but two experimenters arbitrarily adjust the parameters. Therefore, it is sometimes objected that this option contradicts the assumption that the experimenters have free will. Bell himself voices such an objection.[44] However, in metaphysics, it is in dispute whether determinism in the physical realm really clashes with free will.

Nonetheless, the common cause option infringes upon a presupposition of experimental science.[45] It is a presupposition of experimental science that there is no common cause which coordinates in a deterministic or probabilistic way both the behaviour of the system under examination and the decisions of the experimenter or a machine on which parameters to measure. It may be compatible with experimental science that everything, including the decisions of experimenters, is determined by physical factors. However, if there were a factor which coordinates the decisions on the parameters that are to be measured with the behaviour of the systems under examination, one could argue that the very possibility of experimental science would be undermined.

The common cause option does not fall victim to the consequences of Bell's theorem and the theorem of Kochen and Specker, because it introduces a common cause which has a salient influence on the behaviour of the examined systems as well as the measurements that are performed on

---

43 "Bertlmann's Socks and the Nature of Reality" in Bell (1987), p. 154. See furthermore Shimony, Horne and Clauser, "Comment on Bell's Theory", in Shimony (1993), p. 168; Kronz (1990), pp. 424–431. See against this objection Shanks (1993).

44 Interview in Davies and Brown (1986), p. 47. See furthermore "The Theory of Local Beables" in Bell (1987), p. 61; "Free Variables and Local Causality" in Bell (1987), pp. 100–102; "Bertlmann's Socks and the Nature of Reality" in Bell (1987), p. 154. See Kronz (1990), pp. 424–431, against invoking free will in this context.

45 Compare also Shimony, Horne and Clauser, "Comment on Bell's Theory", in Shimony (1993), p. 168.

them. This option does not contradict local action either: the influence of the common cause can be taken to propagate with a velocity that is not higher than the velocity of light. But a conspiracy between the measured systems and the experimental arrangement including the decisions of the experimenters due to a common cause outside the experimental arrangement is a high price to pay for this option. This option is entirely *ad hoc*: there is no concrete suggestion for a common cause, since it is completely unclear how a common cause could carry out the necessary determination.[46]

If one wishes to avoid taking recourse to a factor outside the experimental arrangement, but intends to retain both separability and local action, one can work with *backward causation*: if one assumes that events in the future light-cone such as measurements and the future choice of parameters that are to be measured contribute to determine the hidden state of the pair of photons at the source in a Bell experiment,[47] one can explain these experiments without coming into conflict with separability and local action.

Assuming a causal influence of future events prevents the option for backward causation from falling victim to the theorem of Kochen and Specker. Over and above that, backward causation does not contradict local action, because it can be maintained that the effects of future events, like the effects of past events, propagate with a velocity that is not higher than the velocity of light. However, the option for backward causation involves an alteration of the view of causation that is connoted with local action: local action is connoted with a distinction between the past and the future light-cone of an event in the sense that effects propagate from the past light-cone to the future light-cone. This option also infringes upon the mentioned presupposition of experimental science: future decisions on which parameters to measure causally influence the state that is prepared at the source of the experimental set-up. Furthermore, to show that the assumption of backward causation is not *ad hoc*, it would be preferable to have first firm evidence for backward causation independent of quantum theory and then examine whether this assumption can be applied to quantum physics.

Summing this chapter up, we should bear in mind two principles: separability and local action. Quantum theory violates separability. Unless

---

46 See furthermore the objection in Maudlin (1994), pp. 63–64, and the recent papers by Kowalski and Placek (1999) and Placek (forthcoming). But see also the suggestion by Hofer-Szábo, Redei and Szabo (1999).

47 Price (1996), Chapter 9. Compare also Dowe (1996) and already Costa de Beauregard (1987), Chapter 4.6.13.

we are willing to pay the price of backward causation or of a common cause that coordinates the behaviour of physical systems with the parameters that are measured on them, Bell's theorem implies that no theory with hidden variables which conform to separability and local action can match the predictions of quantum theory. Consequent upon Bell's theorem, the apparently clear distinction between physics and metaphysics in the interpretation of quantum theory has become blurred. The claim that there is holism in quantum physics is based on quantum entanglement and the violation of separability to which entanglement amounts.

CHAPTER 8
THE MEANING OF QUANTUM HOLISM

Summary

If one endorses quantum holism, one is committed to a minimal requirement for an ontological interpretation of quantum theory: a system has those properties at a given time of which its state is an eigenstate. If one accepts this minimal requirement, one has to acknowledge that entanglement extends as far as the whole of matter at the level of quantum systems. Entanglement is distinct from interactions [8.1]. My proposal for a characterization of quantum holism combines the suggestions which exist in the literature and which employ the notions of non-supervenience and non-separability [8.2]: a quantum system has some of the properties which belong to the family of properties that make something a quantum system in the last resort only taken together with all other quantum systems. Consequently, the whole of matter is one holistic system at the quantum level. This proposal can leave open whether or not quantum systems are individuals; the issue of quantum holism thus is distinct from the issue of individuals [8.3]. Finally, I contrast this holism with the holism that can be built upon general relativity and that fits into a Cartesian philosophy of physics [8.4].

8.1 Commitments in the Interpretation of Quantum Theory

8.1.1 An Ontological Interpretation without Hidden Variables

When we base ourselves on entanglement in order to make a case for holism in quantum physics, we do not mean that our knowledge of quantum systems is somewhat entangled, but that the states of quantum systems themselves are entangled. In speaking of quantum holism in contrast to a holism that concerns our knowledge we are thus committed to an ontological interpretation of quantum theory, i.e., a realistic interpretation. That is an interpretation which takes quantum theory to tell us something about nature independently of experimental arrangements and measurements. According to an ontological interpretation, the formalism of quantum theory is not a mere means to make predictions of experimental results; instead it refers to quantum systems themselves. In particular, in an ontological interpretation, quantum theory is regarded as referring to the states of single systems in distinction from ensembles.

One can cite Bohr's interpretation as an objection to the claim that one is committed to an ontological interpretation if one speaks of holism in quantum physics. Bohr does not approve of an ontological interpretation of quantum theory. Instead, he maintains that the state vector and its development in time are merely a means to calculate probabilities for the outcomes of measurements.[1] According to Bohr, we can attribute state-dependent properties to a quantum system only relative to a measurement including a description of the whole experimental arrangement.[2] Bohr thus is a realist in the sense that he acknowledges the existence of quantum systems independent of observers and their means of observation such as experimental arrangements and measurements.[3] But he is an anti-realist in the sense that he endorses propositions which attribute state-dependent properties to quantum systems only together with a description of the means for the verification of these propositions, i.e., a whole experimental arrangement including a measurement.[4]

Although Bohr does not advance an ontological interpretation, it may seem that he proposes a holism. Bohr regards an experimental arrangement including the measuring instruments and the measured quantum systems as a whole in which a separate treatment of the components is not possible.[5]

---

1 "Introductory Survey" (1929) in Bohr (1934), pp. 12, 18; "The Atomic Theory and the Fundamental Principles underlying the Description of Nature" (1929) in Bohr (1934), p. 111; Bohr (1948), p. 314; "Light and Life" (1932) in Bohr (1958), pp. 5, 7; "Discussion with Einstein on Epistemological Problems in Atomic Physics" (1949) in Bohr (1958), p. 64; "Atoms and Human Knowledge" (1955) in Bohr (1958), p. 90; "Physical Science and the Problem of Life" (1957) in Bohr (1958), p. 99.

2 Bohr (1937), pp. 291–293; Bohr (1948), pp. 313, 317; "Introduction" in Bohr (1958), pp. 1–2; "Light and Life" (1932) in Bohr (1958), p. 7; "Natural Philosophy and Human Cultures" (1938) in Bohr (1958), pp. 25–26; "Discussion with Einstein on Epistemological Problems in Atomic Physics" (1949) in Bohr (1958), pp. 39–40, 46–47, 52, 61, 64; "Unity of Knowledge" (1954) in Bohr (1958), pp. 71–73; "Physical Science and the Problem of Life" (1957) in Bohr (1958), p. 98; "The Unity of Human Knowledge" (1960) in Bohr (1963), p. 10.

3 Bohr's realism is emphasized by Honner (1987) and Favrholdt (1994). Folse (1985), Chapters 7 and 8, goes as far as ascribing to Bohr a distinction between a phenomenal object and an object behind the phenomenon. For a criticism of that distinction, see Faye (1991), pp. 203–211. Compare the reply of Folse (1994).

4 Bohr's anti-realism is emphasized by Faye (1991), Chapter 8; Faye (1994).

5 "Introductory Survey" (1929) in Bohr (1934), pp. 4–5, 12; Bohr (1937), p. 291; Bohr (1948), pp. 313, 316–317; "Introduction" in Bohr (1958), p. 1; "Discussion with Einstein on Epistemological Problems in Atomic Physics" (1949) in Bohr (1958), pp. 39–40, 44–47, 50–52, 57, 61; "Unity of Knowledge" (1955) in Bohr (1958), pp. 71–72; "Atoms and Human Knowledge" (1955) in Bohr (1958), pp. 85, 90; "Physical Science and the Problem of Life" (1957) in Bohr (1958), pp. 98–99; "Quantum Physics and Philosophy" (1958) in Bohr (1963), pp. 2, 4.

Bohr's refusal to ascribe state-dependent properties to quantum systems independently of measurements excludes a treatment of the measured system without referring to the measuring instruments. According to Bohr, a separate treatment of the different components of the experimental arrangement is not possible, because changing one component can amount to changing the whole set-up in the sense that another arrangement is installed by means of which another observable is measured. Subdividing an arrangement by making, for example, a position measurement behind a double slit means for Bohr setting up a new experimental arrangement.

Bohr considers incompatible observables to be complementary, because no experimental arrangement can be installed in which more than one of these observables is measured in such a way that it has a definite numerical value.[6] Furthermore, Bohr speaks of the complementarity of experimental arrangements and of the complementarity of the information which is obtained through such arrangements.[7] He introduces the term "phenomenon" as referring to observations that comprise a description of the whole experimental arrangement including the measured system.[8] Therefore, he speaks of the complementarity of phenomena.[9] The wholeness of an experimental arrangement is a consequence of complementarity. Bohr traces complementarity and this wholeness back to the indivisibility of the

6 Bohr (1935), p. 700.

7 Bohr (1937), pp. 291–294, 296; "Introduction" in Bohr (1958), p. 2; "Discussion with Einstein on Epistemological Problems in Atomic Physics" (1949) in Bohr (1958), pp. 39–40; "Quantum Physics and Philosophy" (1958) in Bohr (1963), p. 4; "The Unity of Human Knowledge" (1960) in Bohr (1963), p. 12.

8 Bohr (1948) p. 317; "Discussion with Einstein on Epistemological Problems in Atomic Physics" (1949) in Bohr (1958), p. 64; "Unity of Knowledge" (1954) in Bohr (1958), p. 73.

9 "The Quantum Postulate and the Recent Development of Atomic Theory" (1927) in Bohr (1934), pp. 56–57; Bohr (1935), pp. 699–700; Bohr (1948), pp. 314, 316–317; "Biology and Atomic Physics" (1937) in Bohr (1958), p. 19; "Discussion with Einstein on Epistemological Problems in Atomic Physics" (1949) in Bohr (1958), pp. 41, 44–47, 51; "Atoms and Human Knowledge" (1955) in Bohr (1958), p. 90; "Physical Science and the Problem of Life" (1957) in Bohr (1958), p. 99. As regards Bohr's view of complementarity in quantum theory see in particular Meyer-Abich (1965), pp. 140–159; Scheibe (1973), Chapter 1, especially p. 32; Folse (1985), Chapters 7 and 8; Murdoch (1987), Chapters 4 and 5; Faye (1991), Chapter 6; Held (1994) and (1998).

quantum of action.[10] Instead of the term "indivisibility", he also uses the term "individuality", whose literal sense is "indivisibility".[11]

Do Bohr's claims about the wholeness of an experimental arrangement amount to holism in a substantial sense? These claims imply that each of the parts of such an arrangement depends on all the other parts for fulfilling its function. An alteration of one of the parts can mean a change in the function of all the other parts, since such an alteration can mean that a complementary arrangement is set up. But the dependence on there being other things with which something is arranged in a suitable way in order to fulfill a certain function does not amount to holism on pain of trivializing holism [1.3]. All the parts of an experimental arrangement have properties independent of such an arrangement which make them fulfill a certain function in an experimental set-up, given a suitable arrangement with other things.

Surely there is more to Bohr's claims about the wholeness of an experimental arrangement than such functional dependence. But a relativistic claim such as the thesis that quantum systems require complementary descriptions relative to different experimental arrangements does not in itself amount to holism.[12] Holism is distinct from relativism. Whether or not this thesis amounts to holism depends on the reason that is given for quantum systems being such that they require complementary descriptions relative to different experimental arrangements. Why does a measurement in quantum physics – in distinction from a measurement in classical physics – bring in complementarity and the wholeness of an experimental arrangement in Bohr's sense? The reason why quantum theory calls for complementarity

---

10 "Introductory Survey" (1929) in Bohr (1934), pp. 4–5, 12; Bohr (1937), p. 291; Bohr (1948), pp. 313, 316–317; "Introduction" in Bohr (1958), p. 1; "Discussion with Einstein on Epistemological Problems in Atomic Physics" (1949) in Bohr (1958), pp. 39–40, 44–47, 50–52, 57, 61; "Unity of Knowledge" (1955) in Bohr (1958), pp. 71–72; "Atoms and Human Knowledge" (1955) in Bohr (1958), pp. 85, 90; "Physical Science and the Problem of Life" (1957) in Bohr (1958), pp. 98–99; "Quantum Physics and Philosophy" (1958) in Bohr (1963), pp. 2, 4.

11 "Introductory Survey" (1929) in Bohr (1934), pp. 7–8; "The Quantum Postulate and the Recent Development of Atomic Theory" (1927) in Bohr (1934), p. 53; Bohr (1935), p. 697; Bohr (1937), p 291; Bohr (1948), p. 313; "Natural Philosophy and Human Cultures" (1938) in Bohr (1958), p. 24; "Discussion with Einstein on Epistemological Problems in Atomic Physics" (1949) in Bohr (1958), pp. 34, 62; "Physical Science and the Problem of Life" (1957) in Bohr (1958), p. 101; "Light and Life Revisited" (1962) in Bohr (1963), p. 24.

12 Compare, by contrast, Meyer-Abich (1965), p. 186.

and the wholeness of an experimental arrangement cannot itself be an epistemological one. In particular, it cannot have anything to do with a limited accessibility of quantum systems, if an account in terms of hidden variables is to be excluded.

The case for holism in quantum physics thus rests on an argument from the ontology of quantum systems which explains the wholeness of experimental arrangements that Bohr claims for quantum physics in distinction from classical physics. Even if we accepted Bohr's thesis that complementarity and the wholeness of an experimental arrangement can be traced back to the indivisibility of the quantum of action, we would need an interpretation in which the indivisibility of the quantum of action is embedded in a suggestion on what quantum theory tells us about the microphysical realm. Otherwise the objection that the indivisibility of the quantum of action merely indicates a limit to the accessibility of quantum systems is not precluded. Hence, we have to go beyond Bohr's interpretation to some sort of an ontological interpretation if we are to make a case for holism in quantum physics.[13]

Not any ontological interpretation of quantum theory implies holism. Holism can be avoided if one changes quantum theory by introducing hidden variables, such as superluminal signals, backward causation, or a common cause which coordinates the behaviour of physical systems with the choice of the parameters that are measured on them [7.3.3]. However, the most elaborate and most known approach in terms of hidden variables, the theory of David Bohm, is associated with holism too. Bohm has worked on his theory for four decades from the early fifties until his death in 1992.[14] Although Bohm does not defend a deterministic world view,[15] his alternative to standard quantum theory is deterministic. Bohm maintains that a quantum system such as an electron is a particle that has at any time a definite numerical value of position. This value is the so-called hidden variable. Consequently, each quantum system has a definite trajectory; this

---

13 For a similar conclusion on the basis of a positivistic reading of the reply of Bohr (1935) to Einstein, Podolsky and Rosen (1935) see Beller and Fine (1994), in particular pp. 23–27.

14 The first publication is Bohm (1952), the last one Bohm and Hiley (1993). For recent expositions of Bohm's theory, see Holland (1993) and Cushing (1994), in particular Chapter 4. For a recent collection of papers discussing Bohm's theory, see Cushing, Fine and Goldstein (1996).

15 See Bohm and Hiley (1993), pp. 2–3, Chapter 9, pp. 323–324.

trajectory is causally determined.[16] The behaviour of a quantum system in measurement is causally determined as well.[17] Therefore, Bohm's theory is sometimes referred to as the causal interpretation of quantum mechanics.

Nonetheless, Bohm claims that his theory does not fall victim to the theorem of Kochen and Specker [7.3.1], because, apart from position, all other observables are contextual.[18] Independently of the context of a measurement, a particle has only potentially a value of these observables. In the context of an appropriate experimental arrangement, the one or the other of these properties are actualized in the sense that the system acquires a value of the observable that is measured.[19] For instance, in a Bell experiment, the two systems acquire a value of spin in a given direction only after an experimental arrangement including the parameters of the polarization analyzers has been set up.[20] Already in his textbook on traditional quantum theory, Bohm suggests considering properties of which the system does not have an eigenvalue in a given state as dispositional properties in that state.[21] This proposal is independent of the issue of hidden variables.

Bohm maintains that there is a quantum potential or quantum field. This potential is also referred to as pilot wave. It causally determines the trajectory of a quantum system. Since Bohm has set out his theory for the first time, he conceives the quantum potential as non-local.[22] He thereby departs from Einstein's position. Unlike a classical potential, the quantum potential does not have to fall off with the distance. The way in which a quantum system is guided by the quantum potential can immediately depend on factors which are far apart in space. These factors may include a whole experimental arrangement which stretches over a considerable distance in space such as the arrangement in a Bell experiment.[23] Bohm's theory

---

16 Bohm and Hiley (1993), Chapter 3, in particular p. 29, and furthermore pp. 105, 109–110, 130, 154–156.

17 Bohm and Hiley (1993), Chapter 6 and pp. 78, 107 in particular.

18 Bohm and Hiley (1993), Chapter 6.5.

19 Bohm and Hiley (1993), pp. 18, 107, 120, 149. See furthermore Bohm (1980), pp. 137–138.

20 Bohm and Hiley (1993), pp. 106–114, 149, 177.

21 Bohm (1951), pp. 132–133, 138–139, 175, 385, 609–610, 620–622, 625–627.

22 Bohm (1952), pp. 170, 186–187.

23 Bohm and Hiley (1993), Chapter 3, in particular pp. 29–38, 42, 49; pp. 57, 108; Chapter 7, in particular pp. 134, 150–151, 157–158; pp. 176–179, 202–203, 352.

violates parameter independence, whereas standard quantum theory violates outcome independence [7.3.1]. The fixing of the parameter that is to be measured in the one wing in a Bell experiment induces immediately a change in the situation in the other wing via the quantum potential.

The quantum potential is supposed to be a new type of interaction (force) which is peculiar to the quantum domain. Bohm refers to the quantum potential as quantum force.[24] He speaks of non-local interaction.[25] For example, referring to the Bell experiments, he says in his last book that "the behaviour of the second particle was explained by the nonlocal features of the quantum potential which could provide for a direct interaction between the two different particles that does not necessarily fall off with their separation and that can be effective even when there are no interaction terms in the Hamiltonian".[26] Hence, the quantum potential implies a violation of local action [7.2.1], because two such particles can be separated by a space-like interval.[27] In an earlier publication, however, Bohm says that the Einstein–Podolsky–Rosen correlations between the outcomes on both wings of a Bell experiment are not causal ones; but he thereby means simply that the outcomes are not causally connected by a signal that propagates with a finite velocity through space and time such that it admits of a detailed causal explanation following its trajectory from one region to the next region.[28]

Does Bohm's theory amount to a substantial holism? On the one hand, he regards the quantum system and the quantum potential including all the factors that determine it (which may include a whole experimental arrangement) as a single undivided whole.[29] According to him, the quantum potential cannot be defined in terms of preassigned interrelationships among the parts.[30] The quantum potential is not determined by the positions of the particles alone; it does not satisfy a philosophical locality claim such as Lewis' Humean supervenience which will be considered in [8.2.1] below. On the other hand, Bohm identifies non-locality with indivisible wholeness

---

24 Bohm and Hiley (1993), p. 30. Compare as well Bohm (1952), p. 170.

25 Bohm and Hiley (1993), pp. 49, 57, 95, 150, 179; compare p. 157.

26 Bohm and Hiley (1993), p. 202.

27 But see the differentiations in Berkovitz (1998a), pp. 205–209.

28 Bohm (1980), pp. 129, 133, 136, 175–176, 186–188, 211.

29 Bohm and Hiley (1993), pp. 38, 42, 108; Chapter 7, in particular pp. 134, 150–151, 157–158; pp. 176–177, 352.

30 Bohm and Hiley (1993), pp. 58–59, 92, 95.

and non-separability.[31] He does not take into account the distinction between non-separability and the failure of local action [7.2.1]. He introduces a new, non-local interaction in the shape of the quantum potential. But non-local interaction is not holism, however comprehensive that interaction may be and however odd the features of the potential which carries out that interaction may be compared to a classical potential and the requirement of local action in classical field theories.

It is hence ambiguous to say the least whether one can speak of holism in a substantial sense with respect to Bohm's quantum theory. In his broader philosophical considerations, Bohm undoubtedly envisages a substantial and comprehensive holism. But he lacks the conceptual tools to establish this claim: he nowhere clearly distinguishes holism from mere non-local interaction. In the following, I build the case for quantum holism on an ontological interpretation of quantum theory without hidden variables.

The point of an ontological interpretation of quantum theory is not to defend an outdated metaphysics. In particular, a realistic interpretation of quantum theory is not tied to Einstein's realism which works with the notion of a one to one correspondence between elements of physical reality and elements of a physical theory and according to which physical concepts are related to sense impressions [7.2.1]. The point of a realistic approach to quantum physics is to regard quantum theory as a whole as referring to the microphysical level of nature and to suggest ontological commitments on this basis. The experimental evidence does not confirm the predictions of quantum theory in isolation from the means in the formalism by which these predictions are reached; instead, it confirms quantum theory as a whole.[32]

8.1.2 A Minimal Requirement for an Ontological Interpretation

It is a minimal requirement for an ontological interpretation of quantum theory that we accept observables of which the state of a quantum system is an eigenstate at a given time as properties which the system has at this time. The reason for attributing these properties to the system is their definite numerical value. To repeat, such a value is known as an eigenvalue. Note that I consider the eigenstate–eigenvalue link only as a sufficient condition

---

31 Bohm and Hiley (1993), pp. 176–178; compare furthermore p. 157. In his exposition of Bohm's theory, Cushing (1994), Chapter 4.4, also identifies non-locality with non-separability.

32 On the debate about realism in quantum physics and different variants of realism, see Falkenburg (1994).

for attributing a property to a system at a time; I do not claim that this is also a necessary condition. This minimal requirement is all that we need to make a case for holism. We do not have to commit ourselves to one specific ontological interpretation of quantum theory. In particular, there is no need for a change in logic.

What does this minimal requirement for an ontological interpretation of quantum theory mean in the case of entangled states of two or more systems? Consider again the simplest example of entanglement: the singlet state that is prepared in the Bell experiments. Two systems of spin 1/2 or two photons are emitted together from a source. Their spin states are entangled. The singlet state is a superposition of the product state "first system spin up and second system spin down" with the product state "first system spin down and second system spin up". Neither system is in an eigenstate of spin in any direction. Nonetheless, the singlet state is an eigenstate of a global spin observable, namely the total spin:

(8.1) $$S^2 = (S_{1x} + S_{2x})^2 + (S_{1y} + S_{2y})^2 + (S_{1z} + S_{2z})^2$$

In this formula, the indices *1* and *2* refer to the first and the second system respectively, and the indices x, y and z refer to the spin in the three orthogonal spatial directions. The spin S^2 is a significant physical observable. It has the eigenvalue zero in the singlet state.

An analogous consideration applies to the thought experiment of Einstein, Podolsky and Rosen: they focus on two systems in one-dimensional motion whose states are entangled with respect to position and momentum. The state which they consider can be represented as a superposition of infinitely many product states of position as well as momentum of the two systems. Neither system has an eigenvalue of position or momentum. Nonetheless, the state which Einstein, Podolsky and Rosen conceive is an eigenstate of two physically significant global observables, namely the relative distance and the total momentum:

(8.2) $$\mathrm{Q}_1 - \mathrm{Q}_2$$

(8.3) $$\mathrm{P}_1 + \mathrm{P}_2$$

In these formulae, Q_1, Q_2 stand for the position and P_1, P_2 stand for the momentum of the two systems respectively. These two global observables are compatible.[33] They both have a definite numerical value at once. The value of the relative distance changes in time in a deterministic way. The value of the total momentum is constantly zero. As indicated in [7.1.2], I pretend that there are eigenstates and eigenvalues of position as well as momentum and skip the technical difficulties that are created by quantities which have a continuous spectrum of values. Furthermore, I ignore the qualifications which have to be introduced in a relativistic treatment of the case which Einstein, Podolsky and Rosen consider.[34] Entanglement with respect to spin does not imply entanglement with respect to position and momentum. Nonetheless, for the sake of simplicity, I treat the case of entanglement with respect to position and momentum together with the case of entanglement with respect to spin.

If we adopt an ontological interpretation of quantum theory without hidden variables, we have to acknowledge that the quantum state description in terms of entanglement describes the state in which these two systems are. Consequently, the two systems taken together have at least those properties of which this state is an eigenstate, i.e., properties such as relative distance, total momentum, or total spin. But neither of the two systems has a position, a momentum, or a spin in any direction in distinction from the other system.

Nonetheless, the state of the whole is more than just an eigenstate of global observables such as relative distance, total momentum, or total spin. These properties are represented as combinations of local observables of the parts in the Hilbert space formalism. Having these properties includes probability distributions for these local observables and correlations among the probability distributions of the local observables of both parts. If we accept an ontological interpretation of quantum theory, we have to regard these probability distributions as telling us something about the systems.

However, in order to make a case for holism, we are not committed to one particular way of spelling out what these probability distributions tell us about quantum systems. Over and above the mentioned global properties, we may admit dispositions or propensities (or potentialities or latencies) for each of the systems to acquire a position, a momentum, or a spin in a given direction in distinction from the other system; in this case, we take the

33 See Einstein, Podolsky and Rosen (1935). For an elaboration, see Redhead (1987), pp. 78–81. Healey (1991), pp. 418–419, disputes that in the state which Einstein, Podolsky and Rosen consider there is a precise value of the distance.

34 As regards these qualifications, see Smith and Weingard (1987).

probability distributions to indicate the strength of propensities.[35] The main philosophical problem for such an interpretation is that it may imply a commitment to what is known as bare dispositions, that is, dispositions which do not have a non-dispositional ground. If one accepts bare dispositions, it can be objected that one has to countenance a bare relation of similarity between possible worlds, i.e., a relation of similarity that does not supervene on qualitative features.[36]

Alternatively, we may go beyond the link between eigenstates and properties which have a value. One possibility of doing this is the so-called modal interpretation. According to this interpretation, observables can have a definite numerical value even if the state of the system is not an eigenstate of the observable in question.[37] Another possibility is to associate the probability distributions in quantum physics with values of properties that have some sort of a dispersion.[38] In this case, we have to make sense of the notion of one value that is not one number, but spread out so to speak to a considerable extent. Furthermore, we may go as far as maintaining that all

---

35 For an interpretation of the probabilities in quantum theory in terms of dispositions or propensities see Margenau (1950), in particular pp. 175–176, 335–337, 343, 354, 452–454 (latencies); Heisenberg (1989), Chapter 3, pp. 40–41 (potentialities); Popper (1967), pp. 30–34, on propensities in quantum theory, and Popper (1990), lecture 1, for his general theory of propensities; "Events and Processes in the Quantum World" (1986) in Shimony (1993), Chapter 11; Rohrlich (1987), in particular pp. 160–161, 204; Mainzer (1988), Chapter 4.23; Hughes (1989), Chapter 10.2 (latencies); Fleming (1992); Redhead (1992), p. 391; Stapp (1993), pp. 18–20, 38–40, 43, 80, 91–96, 117–126, 134–138, 148–150, 193–194, 199–200, 212–213, 220–221. For such an interpretation in the context of quantum field theory, see Teller (1995), in particular pp. 7–8, 32, 105–106, 109–111, 124. For a general theory of propensities as objective, non-relational properties see Mellor (1971), in particular Chapter 4.

36 See Smith and Stoljar (1998), pp. 90–99.

37 See in particular Kochen (1985); Healey (1989); van Fraassen (1991), Chapter 9; Dieks (1994). For an assessment of the modal interpretation, see the papers in Dieks and Vermaas (1998).

38 See in particular Bunge (1985), Chapter 2, sections 4 to 6. See already Schrödinger (1935b), pp. 811–812. See also Kanitscheider (1988), pp. 172–173, as well as Penrose (1989), pp. 251–256; Rohrlich (1986), in particular pp. 374–375; Rohrlich (1987), pp. 162–163, 172–174, 178–179, 204. Rohrlich (1987), however, does not distinguish between what he calls blurred observables and potential properties. On values that have some sort of a dispersion, see furthermore Fine (1971); Teller (1979); Teller (1984), pp. 387–393; Busch and Lahti (1985); Busch, Grabowski and Lahti (1995) (unsharp properties); Auyang (1995), pp. 63–80.

the terms of a superposition actually exist and conceive the probability distributions in such a way that they do not relate to definite numerical values that can be acquired in measurement at all [9.2.1], etc.

The point as regards holism is: in whatever way we construe the state descriptions in quantum theory, no interpretation without hidden variables can provide for properties which each of the two systems whose states are entangled has independently of the other system and which could form a basis on which global properties such as the mentioned ones could supervene. The so-called mixed state which one may contemplate attributing to each of the parts and which is the same for the two parts in the mentioned cases does not contradict this point. In the case which Einstein, Podolsky and Rosen consider, the description in terms of a mixed state can be taken to supply information about the relative distance: the probability distribution for position of each of the two systems is concentrated around two peaks with a distance between them. However, this description does not contain the information about the correlations between the local observables of position of the two systems.

Hence, in whatever way we interpret the description in terms of a mixed state, this description does not completely specify the local observables of the parts. It is therefore possible that two systems can be described in terms of the same mixed state as in the mentioned cases without being in the singlet state or the state which Einstein, Podolsky and Rosen consider.[39] Consequently, in the case of entanglement, only the pure state of the whole completely determines the local properties of the parts and their relations (to the extent that these properties and relations are determined at all in quantum physics). It is this point that we need to make a case for holism.

To sum up, in order to speak of holism in the domain of quantum physics, we have to endorse a minimal ontological interpretation of quantum theory. Such an interpretation recognizes those global observables of which the joint state of two or more systems whose states are entangled is an eigenstate as properties which these systems taken together have. These properties indicate an internal structure of the whole. That structure includes probability distributions for the local observables of the parts. An ontological interpretation regards the probabilities in quantum theory as objective probabilities. But in order to make a case for holism, we are not

---

39 Compare also the exposition in van Fraassen (1991), pp. 73–76, 193–194, 201, 290–294, 298.

committed to a particular version of an ontological interpretation of quantum theory or a particular conception of objective probabilities.

8.1.3 Entanglement at the Microphysical Level of Nature

If entanglement concerned merely pairs of quantum systems, one could not build a case for a wide-ranging holism upon entanglement. However, engaging in an ontological interpretation of quantum theory implies an ontological commitment which reaches beyond the treatment of the paradigmatic examples of entanglement: *we are committed to countenance an objective entanglement wherever quantum theory describes the states of two or more systems as being entangled within its domain of validity.* The domain of validity of quantum theory comprises at least the microphysical level of nature, i.e., the level of physical systems such as electrons and photons, protons and neutrons including their constituents (quarks) as well as whole atoms. I apply all the considerations in this chapter only to the microphysical level. I shall discuss the relation between the microphysical and the macrophysical in the next chapter.

Entanglement is not confined to states of quantum systems of the same kind. Whenever we consider a whole which is composed of two or more quantum systems of whatever kinds, quantum theory tells us that, apart from very exceptional cases, the states of these systems are entangled. This point applies also to the whole of matter at the microphysical level. It thus applies to any quantum system in nature which we take into consideration. The paradigmatic examples of Einstein, Podolsky and Rosen and the singlet state are cases of maximal entanglement. The outcomes of measurements of the same local observables of both systems are maximally correlated: given the one outcome, we can predict the other outcome with certainty. Entanglement does not have to be maximal. But it touches in principle the states of all quantum systems. Consequently, an ontological interpretation of quantum theory is committed to the conclusion that *entanglement is ubiquitous among the states of quantum systems; if anything, only the whole of all quantum systems taken together is in a pure state.*[40]

This conclusion is sometimes established by means of the following argument: the states of the two systems in the Bell experiments are

---

[40] As regards this conclusion, see, for instance, Primas (1983), Chapters 3.7 and 5.6; Mainzer (1988), pp. 433–434; d'Espagnat (1991), pp. 77–78, 116, 146, 155–162 and also d'Espagnat (1995), Chapters 14.12 and 15; Kanitscheider (1993), pp. 5, 112–115; Redhead (1995a), pp. 51, 61–62, 86–87.

entangled, because these systems have interacted in the past. Whenever two systems interact and their interaction dies down, their states are entangled after their interaction according to quantum theory. It is to be expected that all quantum systems in the world have directly or indirectly interacted with each other. Therefore, we have to conclude that entanglement touches the states of all quantum systems.[41]

Interaction gives rise to entanglement. But we do not need to have recourse to interaction in an argument which sets out to establish that entanglement is ubiquitous among the states of quantum systems.[42] When we regard quantum theory and the superposition principle in particular as referring to physical systems independently of measurements, we are committed to this consequence. Hence, we do not have to base ourselves on relations of causal dependence in order to make a case for quantum holism. We cannot reconstruct what quantum theory says about nature in the following way: "We begin with quantum systems each of which has a state that completely determines its state-dependent, local properties. These systems would continue to have such states if there were no interaction among them. However, since there is interaction, the states of these systems subsequently become entangled." There is no justification in quantum theory for starting with systems that are like classical atoms.

Entanglement is not a relation of causal dependence. The two systems in the Bell experiments are removed in space in such a way that there is no longer any considerable interaction between them. Nonetheless, their spin states are entangled. If entanglement were a causal relation, it would be an interaction at a distance by means of a new, hitherto unknown force. A causal relation between two or more systems presupposes that the related systems have a state each that completely determines their state-dependent, local properties. If this were not the case, a causal dependence between a change in state-dependent properties of the systems in question could not be formulated. In the case of entanglement, by contrast, there is no such separability. Since entanglement thus is not a causal relation, we can maintain: *Entanglement of the states of several systems is a relation which is at least as basic as causal relations.*

---

41 As regards this argument, see, for instance, Schrödinger (1935b), pp. 811–812, 827, 848–849; Teller (1986), p. 81; Howard (1989), p. 248; Lockwood (1989), pp. 213–214, 228–229; Lockwood (1996), p. 163.

42 See Scheibe (1991), p. 228.

8.2 The Failure of Supervenience and Separability

8.2.1 Non-Supervenient Relations

The most elaborate proposals for a philosophical characterization of quantum holism employ the notion of a failure of supervenience. In the paradigmatic examples of Einstein, Podolsky and Rosen as well as the singlet state, two systems taken together are in a state in which they have properties such as relative distance, total momentum, or total spin. The pure state in which these two systems taken together are is not specified by the description in terms of a mixed state which can be applied to each of these systems. There are no properties which each of these systems has independently of the other one and which could form a basis on which the mentioned global observables could supervene.

The first suggestion that characterizes quantum holism in terms of a failure of supervenience is the one of *Paul Teller*. Teller makes use of the notion of *non-supervenient relations*. For instance, according to him, being in the singlet state is a relation between two systems. This relation does not supervene on the non-relational properties of these systems. Teller coins the term "*relational holism*" in his first paper on that matter.[43] In this paper, he furthermore employs the term "inherent relations" (1986, p. 73). But this term could provoke the association of substrata in which these relations inhere and which have a being beyond instantiating such relations (and some non-relational properties). Holism, however, is not committed to a theory of individuals as bare substrata.

Teller sets out the position which he calls "relational holism" by contrasting it with particularism. Particularism, for him, is the view that the world is composed of individuals which are characterized only by non-relational properties. All relations supervene on non-relational properties of the related individuals (1989, p. 213). However, we apparently have to make an exception for spatial or spatio–temporal relations. Furthermore, we should exclude non-qualitative, individual relational properties whose instantiation depends on the existence of a particular thing; owning a particular thing or being the father of a particular person are examples of such properties.[44] Given such qualifications, Teller's particularism matches the position which was introduced as atomism in [1.1], making clear that the things in question

---

43 Teller (1986). See furthermore Teller (1989), pp. 213–216, and French (1989).

44 Compare Healey (1991), pp. 401–402, 409.

do not have to be atoms in the literal sense of very small, indivisible bodies. The term "particularism" is misleading: the concept "particulars" is normally used in contrast to the concept "universals". Thus, Peter Simons, for instance, considers particularism as the denial that there are universals (1994, p. 557). However, the parts of holistic systems and holistic systems themselves are particulars as well.[45] Holism is, like atomism, not committed to any sort of realism about universals,[46] although both doctrines are compatible with some sorts of that position.

Before Teller's paper was published and independently of quantum physics as well as the issue of holism, Carol Cleland formalized the concept of supervenient relations by pointing out two conditions:

> A dyadic relation R is ***supervenient*** upon a determinable nonrelational attribute P if and only if
> 1. $\Box\,(\forall\, x, y) \sim \Diamond\,[R\,(x, y)$ and there are no determinate attributes P_i and P_j of determinable kind P such that $P_i\,(x)$ and $P_j\,(y)]$;
> 2. $\Box\,(\forall\, x, y)\ \{R\,(x, y) \supset$ there are determinate attributes P_i and P_j of determinable kind P such that $P_i\,(x)$ and $P_j\,(y)$ and $\Box\,(\forall\, x, y)\ [(P_i\,(x)$ and $P_j\,(y)) \supset R\,(x, y)]\}$. (p. 25)

The first condition states: it is necessary for a supervenient relation that each of the related entities instantiates a determinate non-relational property of a certain determinable kind. The second condition adds that these instantiations of non-relational properties determine the relation in question. It is necessary that any two individuals which instantiate the same non-relational properties stand in the same relation to each other. Cleland gives the example of mass relations. An individual which has a mass of 8g is heavier than an individual which has a mass of 6g. Any two individuals which have the same masses respectively instantiate the same mass relation. Therefore, mass relations are supervenient on the non-relational property of mass (pp. 23–27).

Stating these two necessary and sufficient conditions for supervenient relations enables Cleland to distinguish between two kinds of non-supervenient relations. A relation is *weakly non-supervenient* if and only if it satisfies the first, but not the second of these conditions. That is to say: it is necessary that all the related entities instantiate non-relational properties of a certain kind. But these non-relational properties are not sufficient to

---

45 Healey (1991), p. 400, adopts the term "particularism" from Teller, although he voices a similar reservation.

46 French (1989), pp. 18–19, claims the contrary.

determine the relation in question (see also Cleland (1985), p. 291). Cleland gives the following example: if two individuals are to bear a relation of spatial distance to each other, each of them must have a size or shape. However, their size or shape does not determine the spatial distance between them. Therefore, Cleland claims that the relation of spatial distance is weakly non-supervenient: this relation satisfies the first condition for a supervenient relation, but it fails to conform to the second one (1984, pp. 27–28). Accordingly, a relation is *strongly non-supervenient* if and only if it does not satisfy the first condition for a supervenient relation either (1984, pp. 28–29). It is thus not necessary that two entities have non-relational properties of a certain kind, if they bear a strongly non-supervenient relation to each other.

Steven French convincingly argues that the non-supervenient relations in the quantum case are strongly non-supervenient in Cleland's sense (1989, pp. 8–18, in particular pp. 17–18). Consider the singlet state. Being in the singlet state can be regarded as a way in which two systems are related with each other. We can assume that the two systems have some non-relational properties each. But being in the singlet state is not tied to having non-relational properties of a certain kind. This point was made above by saying that there are no properties which each of these systems has independently of the other one and which could form a basis on which the mentioned global observables could supervene. Even if we take the description of each of the two systems in terms of a mixed state into consideration, this description is an incomplete specification of the local properties of each of the two systems and not a description of non-relational properties which each of the two systems has independently of the other one.

There may be some argument about exactly which properties should be admitted in a basis for supervenience so that, if there is quantum holism, it implies a failure of supervenience in a substantial sense.[47] But this is not a real problem for those proposals which construe quantum holism in terms of a failure of supervenience. Consider the thesis which David Lewis puts forward under the name of *Humean supervenience*:

> Humean supervenience is named in honor of the greater denier of necessary connections. It is the doctrine that all there is to the world is a vast mosaic of local matters of particular fact, just one little thing and then another. ... We

---

47 Compare Robinson (1992). Liu (1996) goes as far as including the correlations among the measurement outcomes in the supervenience basis and then argues for particularism. See especially p. 277.

> have geometry: a system of external relations of spatio-temporal distance between points. Maybe points of spacetime itself, maybe point-sized bits of matter or aether or fields, maybe both. And at those points we have local qualities: perfectly natural intrinsic properties which need nothing bigger than a point at which to be instantiated. For short: we have an arrangement of qualities. And that is all. There is no difference without difference in the arrangement of qualities. All else supervenes on that.[48]

There is thus a distribution of intrinsic and hence non-relational properties at space–time points. Between these points, there are relations of spatio-temporal distance. These relations do not supervene on the non-relational properties of the related entities. The properties of everything else supervene on the distribution of non-relational properties at space–time points. The world is compartmentalized into local matters of particular fact, and everything else supervenes on these local matters.

If we interpret quantum theory as referring to physical systems, it implies a failure of Humean supervenience. First of all, apart from very exceptional cases, quantum systems are not localized at a point or an arbitrarily small region in space–time. Consequently, their properties are not instantiated at space–time points. This consideration also applies to their state-independent properties such as rest mass and charge, which always have a definite numerical value. What is much more, there is ubiquitous entanglement of the states of quantum systems. There are no non-relational properties on which entanglement supervenes. Entanglement is furthermore independent of spatio–temporal relations. In the example of the singlet state the entanglement is unaffected by how far apart in space or space–time the two systems are removed. Consequent upon quantum entanglement, there is no foundation for the formulation of supervenience claims such as Humean supervenience according to which everything supervenes on local matters of particular fact.[49] What Lewis describes is a substantial basis for supervenience. Quantum holism therefore implies an important failure of supervenience.

There is a strategy for a change to Humean supervenience that acknowledges quantum holism and attempts to rescue Humean supervenience from the challenge which quantum holism poses: one can replace the concrete four-dimensional space–time continuum with abstract mathematical state space. Barry Loewer advocates this strategy (1996, p.

---

48 Lewis (1986a), Introduction, pp. IX–X. See furthermore Lewis (1986b), p. 14.

49 Compare Teller (1986), pp. 71–73; Healey (1991), in particular pp. 405–406.

104). The state of any quantum system can be represented as a point in state space. One can conceive the representation of the state as an ascription of non-relational properties to a point in state space. If one contemplates the whole world being represented in a state space, then one can defend Humean supervenience for that state space despite quantum holism.

However, state space is not real space in the same way as the concept "water" is not water. We use state space to represent things and events in real space in the same way as we employ the concept "water" to say something about water. If a realistic interpretation of the quantum state vector is called for, what is intended is not that the state vector itself is something real, but that it is a referring expression, i.e., that it refers to something in the spatio–temporal, physical world.[50] Hence, this strategy fails to rescue Humean supervenience as a thesis about the physical world.

In his thesis of Humean supervenience, Lewis admits only spatio–temporal relations as external and hence non-supervenient relations. In *On the Plurality of Worlds*, however, Lewis considers whether there can be external relations which are not strictly or analogically spatio–temporal and which unify a world. He imagines the following test case:

> We tend to think that positive and negative charge are natural intrinsic properties of particles; but suppose not. Suppose instead there are natural external relations of like-chargedness and opposite-chargedness. ... On this view, as opposed to the standard view, the relations of like- and opposite-chargedness do *not* supervene on the intrinsic natures of two particles taken separately; an electron and a positron may be perfect intrinsic duplicates. That is the point of calling the relations external. They are natural *ex hypothesi*. They are pervasive (at least, given the appropriate laws) in that whenever two particles are connected by a chain of such relations, they are connected directly. But they are very far from discriminating (again, given the appropriate laws): if there are as few as three particles, there must be two of them that are alike as far as these relations are concerned. If this story, or something like it, could be true, then here we have external relations that are not strictly or analogically spatio–temporal. (1986b, p. 77)

If we accept Teller's account of quantum holism, then we have non-supervenient relations that are not strictly or analogically spatio–temporal, namely relations of entanglement. In the quantum case as in Lewis' case, these relations are strongly non-supervenient in Cleland's sense: standing in

---

50 Compare Campbell (1994), p. 29.

these relations is not tied to instantiating non-relational properties of a certain kind.

Lewis argues that these imaginary charge relations are unable to unify a world, because there is no convincing reason against considering two things in different worlds as also standing in relations of like- or opposite-chargedness to each other (1986b, pp. 76–78). In the case of the quantum relations of entanglement, by contrast, there is, like in the case of spatio–temporal relations, no question of these relations holding also among things across different worlds. However, spatio–temporal relations (as well as Lewis' imaginary charge relations) are state-independent in the following sense: any two things in a world bear some spatio–temporal relation to each other whatever their state is. By contrast, it is state-dependent whether two things bear a relation of entanglement to each other. Nonetheless, since entanglement is ubiquitous at the microphysical level in all the worlds to which quantum theory applies, we can say: our world is unified not only by spatio–temporal relations, but also by quantum relations of entanglement. Both are non-supervenient relations that connect only things in one world.

However, because the relations of entanglement are state-dependent, it is questionable whether, strictly speaking, they qualify as external relations in Lewis' sense. According to Lewis, a relation is external if and only if (a) it does not supervene on the non-relational properties of the *relata*, but (b) it supervenes on the nature of the composite of the *relata* taken together (1986b, p. 62). Since the relations of entanglement are state-dependent, one can dispute that they satisfy (b): it belongs to the nature of quantum systems that their states admit of entanglement; but whether or not the states of two systems are entangled does not seem to be a question of their nature. I therefore refer to the quantum relations of entanglement simply as non-supervenient relations. However that may be, the point is that, consequent upon quantum entanglement, we have to admit non-supervenient relations over and above the spatio–temporal ones at the microphysical level, and these additional non-supervenient relations are ubiquitous at that level.

Lewis' imaginary charge relations come closer to quantum entanglement than spatio–temporal relations in the following respect: in the case of these charge relations as well as in the quantum case we expect the relations to be determined by properties which each of the related systems has independently of the other systems; but it then turns out that there are no non-relational properties which accomplish such a determination. Nonetheless, there is nothing like entanglement in Lewis' example. Lewis simply invites us to imagine a case in which the properties of being likecharged or being opposite-charged with respect to certain other things are irreducibly relational. Hence, Lewis' imaginary charge relations show the following: even if we make an exception for spatio–temporal relations, speaking of non-supervenient relations does not capture what is peculiar to quantum holism.

Non-supervenient relations over and above the spatio–temporal relations are a necessary condition for quantum entanglement; but that they fall short of being a sufficient condition.[51] Entanglement is a relation between two or more systems. Each of these systems has the relational property of being entangled with one or more other systems. This relation is strongly non-supervenient in the explained sense. However, we need a conceptual distinction within non-supervenient relations that picks out what is peculiar to quantum entanglement.

How can we reach such a distinction? Consider the following point: It is not specific enough to say of two systems that their states are entangled or that they stand in a relation of entanglement to each other. The question is: with respect to which properties are these systems entangled? The singlet state is an example of entanglement with respect to spin, and the state which Einstein, Podolsky and Rosen consider is an example of entanglement with respect to position and momentum. Let us therefore focus on the properties to which entanglement extends.

8.2.2 Non-Supervenient Properties of the Whole

Apart from Teller's account, there is another elaborate proposal for a characterization of quantum holism by non-supervenience. Whereas Teller analyses a quantum whole in terms of relations among its parts, *Richard Healey* does not conceive a state like the singlet state as a relation between two systems at all.[52] Basing himself on quantum entanglement, he makes the following suggestion for a conception of holism in physics: a physical whole in the sense of a holistic system has some significant, intrinsic properties which its parts do not have and which do not supervene on the qualitative, intrinsic properties of the parts and some of their relations such as their spatial relations. Healey applies this conception to the paradigmatic examples of quantum entanglement such as the singlet state and the state which Einstein, Podolsky and Rosen consider (1991, in particular pp. 400–403, 408–409, 415–421).

---

51 Teller agrees with this claim in conversation.

52 Confirmed in communication by e-mail.

Healey's proposal brings out an important feature of quantum holism: *if the states of two or more quantum systems are entangled, then the whole which consists of these systems has physically significant properties which do not supervene on the intrinsic properties of its parts.* In the case of the state which Einstein, Podolsky and Rosen consider as well as in the case of the singlet state, properties such as relative distance, total momentum, or total spin are non-relational properties of the whole. These properties do not supervene on the non-relational properties of the parts.

Healey considers these properties as properties of the whole which are different in kind from the properties of the parts. This position is disputable. Properties such as relative distance, total momentum, or total spin are global observables of the whole. Nonetheless, in the Hilbert space formalism of quantum theory, they are represented as a certain combination of local observables of the parts. One can therefore argue that relative distance, total momentum, or total spin are properties of the kind position, momentum, or spin. What is more, these properties of the whole are significant only insofar as they say something about the parts. Relative distance indicates the way in which the parts are related to each other with respect to position, although it is not the case that each of the parts has a position in distinction from the other one. Total momentum zero indicates the way in which the two systems are related with respect to momentum, although neither of them has a momentum in distinction from the other one. The same goes for total spin zero: this global observable indicates the way in which the local spin observables of the parts are related, although neither part has a value of spin in any direction in distinction from the other part.

Hence, in quantum entanglement *we have properties of the whole which do not supervene on the local properties of the parts, but which indicate the way in which the parts are related with respect to some of their local properties*. These properties of the whole contain *all* that can be said about the local properties of the parts, and *only* these properties of the whole contain all that can be said about the local properties of the parts. In which manner one elaborates on this notion of non-supervenient properties of the whole depends on which of the above mentioned options for an ontological interpretation of quantum theory without hidden variables [8.1.2] one takes to be most plausible. For instance, if one endorses the option for dispositions, one can say the following: these properties of the whole are the non-dispositional ground which determines the dispositions for a definite numerical value of local observables of each system and the correlations between these values that can be acquired in measurement. Thus, considering the properties of the whole enables one to argue that the option for quantum dispositions does not fall victim to the objection according to which a disposition requires a non-dispositional ground.

Does this account distinguish quantum entanglement from all cases of non-supervenient spatio–temporal relations? Consider Max Black's example of a possible world in which there are only two spheres that have the same non-relational properties (1952, p. 156). There is a definite distance between these two spheres. But neither sphere has a position in distinction from the other sphere (unless one presupposes absolute space). What is the difference between this example in the framework of classical physics and the case which Einstein, Podolsky and Rosen consider?

In Black's example, each sphere has the property of being at a distance of, say, ten metres from another sphere. In the quantum example, by contrast, even if the global observable of relative distance has the definite numerical value of, say, ten metres, it is not permissible to attribute to each of the two systems the property of being at a distance of ten metres from the other system. Imagine a simplified experimental realization of the example of Einstein, Podolsky and Rosen. If we attributed to the one system the property of being ten metres apart from the other system and attributed to that other system the property of being ten metres apart from the first system, we would imply that one system is on the one side of the experimental arrangement and the other system on the other side. However, we can ask about the local observable of position of each of the two systems, and we get a probability distribution which is identical for the two systems and which extends to both sides of the experimental arrangement. Hence, in the case of Einstein, Podolsky and Rosen, there is no warrant in the formalism of quantum theory for assuming that the one system is on one side of the arrangement and the other system on the other side of the arrangement.[53]

The point of the mentioned global observables in the case of quantum entanglement (such as relative distance, total momentum, or total spin) is that these observables indicate the way in which the systems under consideration are related with respect to some of their local observables; but nothing like one definite numerical value of a local observable can be attributed to each of these systems, not even relative to the other system. This crucial feature of quantum entanglement is not paralleled by Black's

---

53 See also the argument in Cartwright (1989), p. 263.

example. Let us therefore pursue this feature one step further in order to shape it into a characterization of quantum holism.

8.2.3 Non-Separability

There is a third elaborate proposal for a characterization of quantum holism. *Don Howard* does not talk in terms of a failure of supervenience. Starting from the principles of Einstein's view of physical reality, he employs the closely related notion of a failure of separability. To recall, according to Howard, separability means that (1) spatially separated systems possess their own, distinct physical state each and that (2) the joint state of two or more spatially separated systems is wholly determined by their separate states [7.2.1]. Consequently, Howard maintains that quantum entanglement exhibits non-separability. He regards non-separability as a sufficient condition for ontological holism.[54]

In [7.2.1], I argued that if the failure of separability is to be employed in a philosophical characterization of quantum entanglement, we should jettison the condition of spatial separation. There are cases of entanglement where there is no question of a spatial separation of the systems concerned. The following characterization of the principle of separability was suggested: Physical systems have a state each in the sense that (1) this state completely determines the state-dependent, local properties of the system and (2) the joint state of two or more systems supervenes on the states which each of these systems has. This proposal suggests the following characterization of non-separability:

Non-separability

> The states of two or more systems are non-separable if and only if it is only the joint state that completely determines the state-dependent, local properties of each system and the relations among these systems (to the extent that these are determined at all).

Howard's characterization of separability implies a supervenience claim: if the joint state of two or more systems is wholly determined by their separate states, then the joint state supervenes on the separate states. In the same vein, the suggestion made now for a characterization of non-separability implies a failure of supervenience. To be more precise, this proposal implies that there

---

54 Howard (1989), pp. 225–231. Healey, by contrast, uses the term "nonseparability" in such a way that it refers to processes in the first place. See Healey (1991); Healey (1994); Healey (1997), pp. 26–29.

are relations among the systems in question which do not supervene on the way in which the state-dependent, local properties of these systems are registered by the descriptions which can be attributed to each system, i.e., the descriptions in terms of a mixed state. Furthermore, since the description of the joint state of these systems taken together attributes properties to the whole, there are some properties of the whole that do not supervene on the way in which the properties of the parts are registered by the descriptions which can be attributed to each part. However, non-separability can say more than these suggestions in terms of a failure of supervenience. This will become clear in the next section.

Over and above considering non-separability to be sufficient for ontological holism, Howard takes separability to be the foremost principle of individuation for physical systems. Consequently, he suggests that in the case of entanglement, there are not two or more physical systems, but only one system.[55] The failure of separability certainly raises the question whether there are in the case of entanglement two or more *individuals*. However, we cannot content ourselves with speaking of one holistic system in the sense of one individual. Speaking of holism in a substantial sense requires the reference to an internal structure which makes it that the system under consideration is a holistic system. Even if we maintain that in the case of entanglement there is only one individual, i.e., the whole, this does not mean that we cannot refer to parts. Entanglement of the states is predicated of at least two parts of a whole. We can speak of entanglement if and only if we can apply the predicate "is entangled with" to at least two parts. In the formalism of quantum theory, there is a case of quantum holism or non-separability if and only if a state cannot be represented as a product state, but only as a superposition of product states; speaking of a product state or a superposition of product states requires that we refer to at least two parts. Furthermore, the properties that are attributed to the whole in the form of global observables are combinations of local observables of parts. Hence, if there is a case of quantum holism, there have to be at least two parts of a quantum whole.

Turning to the experimental evidence, in whichever way one explains the correlations of the measurement outcomes in a Bell experiment like the one of Aspect, Dalibard and Roger (1982), these correlations are no reason for abrogating the following suppositions: In this experiment, pairs of photons

---

55 Howard (1985), pp. 173, 191–196; Howard (1989), pp. 225–231; Howard (1997), in particular p. 122. See also Carrier (1992), p. 101; Folse (1996), pp. 133–134.

are generated. After the emission from the source, the two photons fly apart in opposite directions. Consequently, when the measurement takes place, there is no interaction between these photons any more that is worth considering.

I will continue to refer to the parts as physical systems; when I speak of physical systems in the following, I intend to leave open whether or not these systems are individuals. Recall that the term "system" is used here in a very broad sense according to which it is necessary and sufficient for something to be a system that properties can be predicated of it – be it properties such as "is entangled with", "is related with another system in such a way that only these systems together have a definite numerical value of an observable of the kind spin", etc. Quantum systems as treated in quantum mechanics are such that not only can properties be predicated of them, but they can also not be predicated as properties of anything. Thus, they satisfy the main criterion for a first substance which Aristotle sets out in the fifth chapter of the *Categories*. Let us first carry on with the enquiry into the internal structure of a quantum whole by pursuing the thread of non-separability and then consider whether or not the parts of a quantum whole are individuals.

8.3 Proposal for a Characterization of Quantum Holism

8.3.1 No Separate Properties

Quantum holism consists in the states of two or more quantum systems being entangled. States are attributed to physical systems. The state of a system at a given point of time is the way in which the system has properties whose value is time-dependent at this point of time [7.1.1]. These properties are the state-dependent properties. Thus, quantum holism concerns only some of the properties of physical systems, namely the state-dependent ones such as position, momentum, or spin in any direction.

Recall the notion of a family of properties that make something a system of a certain qualitative kind [1.2.2]. Let us apply this notion to quantum systems. What is the family of properties that are common to all kinds of quantum systems and that thus make something a quantum system? State-dependent properties such as position, momentum, and spin in any direction as represented in the quantum formalism belong to the family of properties that make something a quantum system. Furthermore, we have to include rest mass and charge. These latter, state-independent properties are sufficient to distinguish one kind of quantum systems from other kinds of quantum systems. But for something to be a quantum system more properties are needed than the state-independent ones. Having rest mass or charge is tied to being spatio–temporal. If there are things which are not spatio–temporal such as abstract objects, it does not make sense to attribute rest mass or charge to them. Nothing which does not have position, momentum, and spin in any direction in some sense therefore is a quantum physical system.

However, in case the state of a quantum system is entangled with the state of another quantum system such as in the paradigmatic examples of Einstein, Podolsky and Rosen as well as the singlet state, these systems fail to have a position, a momentum, or a spin in any direction separately. Nonetheless, these systems taken together have properties such as relative distance, total momentum, or total spin. If someone asks about position, momentum, or spin in a given direction of one of the two systems in such a state, we can therefore say the following: one system is related with another system in such a way that solely the two systems taken together have properties of the kind position, momentum, or spin – in the sense of a relative distance with a definite numerical value, total momentum zero, or total spin zero. The property of relative distance of the whole indicates the way in which the parts are related with respect to position, because it contains the correlations between the probability distributions of the local observables in question, in whatever way one may interpret these correlated probability distributions [8.1.2]. The property of total momentum of the whole indicates the way in which the parts are related with respect to momentum. Accordingly, the property of total spin of the whole indicates the way in which the parts are related with respect to spin in any direction. A similar consideration applies to any case of entanglement: there are global observables of the whole that have a definite numerical value in the state in question; these observables indicate correlations between probability distributions of local observables of the parts.

Recall the distinction between two types of holism within the proposed general conception [1.4]. The second type is this one: The parts instantiate some of the properties that belong to the family of properties which make something a part of a system of the kind in question only taken together. Nonetheless, the resultant properties of the whole indicate the way in which the parts are related with each other with respect to the properties in question. Quantum holism fits into this type. I thus suggest an account of quantum entanglement in three steps:

1) Quantum entanglement shows that there are non-supervenient relations over and above the spatio–temporal relations.

2) The non-supervenient relations of entanglement among the parts of a quantum whole amount to the following: the whole has non-relational properties which do not supervene on those properties of the parts which are taken into account in the description that can be attributed to each of the parts (i.e., the description in terms of a mixed state).
3) Quantum entanglement is baffling, because these properties of the whole come to the following: *the parts have some of the properties that belong to the family of properties which make something a quantum system not separately, but only in this way: there are properties of the whole which indicate the manner in which the parts are related with each other with respect to these properties.*

8.3.2 Matter as a Holistic System at the Quantum Level

Combining the account in the preceding section with the argument in [8.1.3] for entanglement being ubiquitous among the states of quantum systems, we have to conclude that in the last resort, if anything, only the whole of all quantum systems taken together is in a pure state. In the terms of the proposed account, this is to say that some of the properties which make something a quantum system are in the last resort completely determined only by the pure state of all quantum systems taken together (to the extent that these properties are determined at all). Consequently, all matter at the level of quantum systems is one holistic system. This is an *ontological* consequence of quantum physics, given reasonable commitments in the interpretation of quantum theory. It is not expected or demanded that anyone can or should give a concrete description of the state of all quantum systems taken together.

We can now set out a suggestion for a characterization of quantum holism:

Quantum holism

> State-dependent properties such as position, momentum, and spin in any direction belong to the family of properties which make something a quantum system. As far as these properties are concerned, any quantum system is ontologically dependent in a generic way on there actually being other quantum systems with which it constitutes the material realm at the microphysical level: since only the whole of all quantum systems is in a pure state, a quantum system has these properties only taken together with all other quantum systems – in the sense that there are only properties of the whole which indicate the

way in which the parts are related with respect to these properties. Therefore, the whole of matter is one holistic system at the level of quantum systems.

This description applies the proposed general conception of holism [1.2.3] to quantum holism. It not only combines the three proposals that exist in the literature. It also has a number of advantages over the suggestions for a characterization of quantum holism in terms of a failure of supervenience. To start with (1), entanglement is not a relation of causal dependence [8.1.3]. What sort of dependence holds among quantum systems whose states are entangled? Speaking of a failure of supervenience does not tell us positively in which way systems with entangled states depend on each other. By contrast, the proposal here makes clear the sort of dependence among systems with entangled states: it is ontological dependence with respect to those properties to which the entanglement relates. The sort of dependence is ontological, because quantum theory depicts every possible world to which it applies in such a way that the states of the quantum systems are entangled independently of whatever causal relations may hold among them.

When it comes to candidates for holism in physics, it is reasonable not to apply the notion of ontological dependence to all metaphysically possible worlds, but to limit it to all physically possible worlds. The laws of quantum theory including the superposition principle are not metaphysically necessary. A world that conforms to the principle of separability is *metaphysically* possible; but it is not *physically* possible, since it contradicts quantum theory [1.3]. If Einstein is wrong, he is wrong for physical reasons. The claim is thus that in all the possible worlds to which quantum theory applies, i.e., all the physically possible worlds, the material realm is one holistic system at the microphysical level.

Furthermore, (2) recall the opposition which Teller sets out between relational holism and what he identifies as particularism [8.2.1]. Because Teller explains quantum holism in terms of relations that do not supervene on the non-relational properties of the related systems, he is committed to conceiving the opposing view in terms of individuals that are characterized solely by non-relational properties. This conception accords with Lewis' Humean supervenience. But it is questionable whether it matches the targeted view in general: the point-like particles of classical physics are often taken to be characterized only by position, momentum, charge, and mass. Position obviously is a relational property. Momentum arguably is a relational property as well. A point-like charge is embedded in a whole field.

Taking considerations in connection with relativity physics into account, it is even disputable that mass is a non-relational property.[56] One can therefore argue that the point at issue between quantum holism and the philosophy of classical physics is not about irreducibly relational versus non-relational properties.

Since the proposal made above does not define quantum holism in terms of non-supervenient relations, it paves the way for an adequate characterization of the contrast between quantum holism and the mainstream of the philosophy of classical physics. The point at issue is the opposition between the following two conceptions:

a) Some of the properties which belong to the family of properties that make something a quantum system are instantiated only by these systems taken together in the explained sense.
b) All these properties are instantiated by each of these systems.

The point thus is separability and its failure in quantum theory.

Over and above that, (3) independently of whether or not there are any interesting connections between quantum holism and forms of holism in other areas, it is not advisable to characterize quantum holism in a way that obstructs an enquiry into such connections. Characterizing quantum holism in terms of a failure of supervenience may be an impediment to an enquiry whether quantum holism can be connected with other prominent forms of holism. Consider holism in the philosophy of mind: according to holism about beliefs, intentional properties such as meaning are relational properties, and there are no non-relational, mental properties on which these properties supervene. But holism in the philosophy of mind does not speak against a supervenience of intentional properties on physical properties. It is compatible with Lewis' Humean supervenience. In case all intentional properties supervened on a level of physical properties that are non-relational, holism at the intentional level would not at all be disproved. The proposed characterization of quantum holism, by contrast, opens up a way to connect quantum holism with other forms of holism, since it subsumes quantum holism under a general conception of holism.

8.3.3 Are Quantum Systems Individuals?

Let us now consider whether or not quantum systems are individuals. The objects of quantum mechanics such as electrons, protons, neutrons, etc. are physical systems, because properties are predicated of them. Moreover, they

---

56 Compare Teller himself in Teller (1991), sections VI to VII.

are not predicated as a property of anything. However, the Aristotelian condition of being an ultimate subject of the predication of properties may not be sufficient for being an individual, because further conditions can be regarded as necessary for something to be an individual – in particular intertemporal identity owing to (a) what is known as primitive thisness or (b) distinguishability by the ascription of properties.

The claim that quantum systems instantiate some of the properties that make something a quantum system only taken together highlights why it is an issue whether quantum systems are individuals. Quantum systems can be indistinguishable in the sense that quantum theory does not acknowledge any properties that distinguish one system from other systems of the same kind. Two points are to be noted: (1) Indistinguishability is not a lack of knowledge. If we accept that quantum theory gives a complete description of the states of microphysical systems,[57] indistinguishability is an ontological matter. (2) Quantum systems of the same kind have the same state-independent properties such as rest mass and charge. But this is not sufficient to establish indistinguishability. It has to be shown that there are no state-dependent properties either that distinguish one system from other systems of the same kind.

Steven French and Michael Redhead (1988) consider two quantum systems of the same kind. All state-dependent properties of these systems are captured by the description of their states in quantum theory. Their states are entangled. The argument of French and Redhead applies to an arbitrary number of quantum systems of the same kind that fulfill this condition. French and Redhead show that there are no properties which distinguish the one system from the other system. The description in terms of a mixed state is the same for the two systems. Furthermore, French and Redhead assume, for the sake of argument, that each of the two systems has in fact an identity across time in distinction from the other system; they show that under this assumption there are not even relational conditional probabilities that could distinguish the one system from the other system. It is of no importance for this result whether these systems are fermions (such as electrons or neutrons) or bosons (such as photons).[58] There is, however, some controversy about the latter point: van Fraassen claims that fermions can be regarded as distinguishable, but only, as he makes clear, at the price of introducing

---

57 Compare Hacking (1975) who maintains that indistinguishability is relative to a description.

58 French and Redhead (1988). See furthermore Butterfield (1993), pp. 463–464.

characteristics that are not included in the state description of quantum theory and that have no empirical effect (1991, Chapter 11, in particular pp. 422–423, 432–433). Thus, the problem whether or not quantum systems are individuals is a consequence of quantum holism. Recall that entanglement can encompass all state-dependent properties; and if we consider quantum systems of the same kind, state-independent properties fail to distinguish one such system from other such systems.

If there is entanglement, there are at least two systems whose states are entangled. There is a definite number of systems which is greater than one. But it is not possible to identify these systems by attaching number-labels to them. These systems can be given a cardinal number in the sense that there is a definite number of them, but not an ordinal number in the sense of number-labels that can be attached to each of them. For example, in a given situation, it may be determined that there are five electrons; but one cannot say this is electron number one, that electron number two, etc.

Imagine that the original thought experiment of Einstein, Podolsky and Rosen is realized. Two position measurements are performed. One system is found on the left side and the other system on the right side of the experimental arrangement. But in quantum theory there is no fact of the matter whether the first system is on the left and the second system on the right side or the first system on the right and the second system on the left side. Referring to the states in the Hilbert space formalism as the states of system 1, system 2, etc. can be misleading in that it can be taken to suggest that these systems are distinguished from one another and that each of them has an identity across time. Within quantum theory, however, there is no basis for an identity across time of the one system in distinction from the other system.

Einstein regards separability as a necessary condition for there being individuals [7.2.1]. He endorses separability at the price of considering quantum theory to be incomplete. His main concern is to ensure that physics refers to distinct individuals. Einstein's claim that separability is a necessary condition for there being individuals thus presupposes that distinguishability by the attribution of properties, including spatio–temporal location, is a necessary condition for there being individuals. This presupposition can be associated with Leibniz' principle of the identity of indiscernibles. According to this principle, if there are two individuals, they have to be distinguishable by at least one qualitative property, including position in space–time; otherwise, there is only one individual.[59] However, instead of regarding Leibniz' principle as an *a priori* metaphysical truth, we can take quantum theory to require us to abrogate this principle for empirical reasons. After all, in his last letter to Clarke, Leibniz himself concedes that it is metaphysically possible that there are two indistinguishable individuals; it is the wisdom of God which precludes that the world he created contains such individuals.[60]

If we acknowledge quantum holism and thus acknowledge that there are quantum systems for which distinguishability by the ascription of properties fails, we can nevertheless regard these systems as individuals that have an intertemporal identity. In this case, we have to commit ourseleves to what is known as primitive thisness (haecceity) (Adams (1979)). Speaking of primitive thisness implies that individuals are bare particulars: their identity does not depend on which properties they have or fail to have. However, if one is not prepared to rescue the intertemporal identity of quantum systems by considering them as bare particulars and if one maintains that being an individual is tied to intertemporal identity, then one has to conclude that quantum systems are not individuals. Thus, Steven French and Décio Krause propose in recent papers that quantum systems can be regarded as non-individuals in the sense of entities that lack intertemporal identity.[61] Jonathan Lowe argues in favour of considering quantum systems as quasi-objects, because they exhibit determinate countability, but lack determinate identity. Furthermore, if one considers separability as the principle of individuation for physical systems, as Howard does [8.2.3], then one links quantum holism with the position that quantum systems are not individuals.

Nonetheless, even if we acknowledge quantum entanglement, we can maintain that quantum systems are individuals without thereby being committed to regarding individuals as bare particulars. Instead, we can give up the condition that intertemporal identity is necessary for something to be an individual. As long as a quantum whole contains a definite number of quantum systems as its parts, one can regard their definite number as sufficient for these systems being individuals. One can thus maintain that countability in the sense of there being a definite number of physical systems

---

59 See, for instance, Leibniz, New Essays, book 2, Chapter 27, and letter 4 to Clarke, §§ 4–6 in the edition of Gerhardt (1875–1890), Volume 7.

60 Leibniz, letter 5 to Clarke, § 25. See as well § 26. Compare Clarke's reply to Leibniz in letter 5, §§ 21–25.

61 French and Krause (1995); Krause and French (1995); French and Krause (1999).

of which properties are predicated is sufficient for these systems to be individuals, although no number-labels can be attached to these systems. For instance, Paul Teller's proposal from the eighties to construe entanglement in terms of non-supervenient relations presupposes that the physical systems of which relations are predicated are individuals.[62]

The proposal made above to construe entanglement in terms of properties that a quantum system has only taken together with other quantum systems does not imply a commitment to count these systems as individuals. Because one system has some of the properties that belong to the family of properties which make something a quantum system only taken together with other systems, one can maintain that one system is an individual only taken together with the other systems with which it has these properties.[63] From this point of view all quantum systems of the same kind are just one individual in the last resort. It is a further advantage of the suggested characterization of quantum holism that it explains why it is debatable whether quantum systems are individuals without being committed to considering them either as individuals or not as individuals. We can speak of – non-supervenient – relations among the parts of a quantum whole in the sense of relations of entanglement, expressed by predicates such as "is entangled with", without thereby presupposing that the systems in question are individuals; for, to repeat, the point is that in the case of entanglement a quantum system has some of the properties that make something a quantum system not separately in the sense explained.

Hence, whether or not quantum systems are individuals is not settled by quantum holism alone. It depends on which conditions one takes to be necessary and sufficient for something to be an individual: if intertemporal identity is necessary, then quantum systems are not individuals, unless one subscribes to a metaphysics of bare particulars. If countability in the sense of there being a definite number of the systems in question is sufficient, then quantum systems are individuals. Consequently, *the issue of quantum holism does not determine the issue of individuals: independently of whether or not we consider quantum systems to be individuals, we can speak of holism in a substantial sense.* My own point of view is that the objects of quantum

---

62 Teller (1986), pp. 73, 80. See also French (1989).

63 French and Krause (1995) and Krause and French (1995), however, go farther in that they work only with the notion of non-individuals and refrain from admitting individuals at all in the context of quantum theory.

mechanics are physical systems, because properties can be predicated of them; but they are not individuals, because they lack intertemporal identity.

8.3.4 A Brief Glimpse at Quantum Field Theory

Characterizing quantum holism in such a way that one is not committed to regarding quantum systems as individuals may pave the way for paying tribute to quantum field theory. Teller (1995), for instance, does not mention quantum holism at all in his book on quantum field theory. The reason seems to be that his proposal for a characterization of quantum holism in terms of non-supervenient relations commits him to regard quantum systems such as electrons, photons and the like as individuals, whereas in quantum field theory, these entities can no longer be considered as individuals.

As mentioned in the preceding section, the number labels for quantum systems in Hilbert space quantum mechanics can be misleading. In a later paper, Redhead and Teller argue that we should prefer the Fock space description, which is used in quantum field theory, because it does not denote quantum systems by number-labels.[64] Furthermore, quantum mechanics takes the number of quantum systems to persist. However, quantum systems can be created and annihilated. Quantum field theory treats the creation and annihilation of quantum systems. It is therefore the more basic theory in comparison to quantum mechanics. However, the philosophy of quantum field theory is in its infancy.[65] Quantum field theory conceives for each kind of quantum systems one field. What is regarded as single quantum systems in quantum mechanics is considered as field quanta in quantum field theory.

The most striking feature of quantum field theory in this context is that there is no fixed number of field quanta. The number of field quanta in a given state of a quantum field is an observable which is represented by a field operator. This observable has all the features which are characteristic of an observable in quantum theory. In particular, there is no need for it to have one definite numerical value in every state. There are states of quantum fields that are a superposition of states with different numbers of field

---

64 Redhead and Teller (1992). See also Dieks (1990), pp. 139–141; Teller (1995), Chapter 2; Teller (1998).

65 Recent books on the philosophy of quantum field theory include Auyang (1995); Teller (1995); Cao (1997) and the papers in Cao (1999). See already Stöckler (1984) and Stöckler (1988b). Compare furthermore the reservations which Howard (1989), pp. 232–240, voices.

quanta. Furthermore, since the number of field quanta in quantum field theory is one observable among others, there is no need to represent a quantum field in that basis.

Field quanta, to which a number operator applies, amount to a particle aspect in quantum field theory. But they cannot be considered as individuals.[66] One can maintain that two or more entities are individuals, although they are not distinguished from each other by any properties. But if there is not even a definite number of these entities, there is no point any more in claiming that there are individuals.[67] The number of field quanta simply is the number of times a field state is occupied. Field quanta are excitations of quantum fields. It is therefore appropriate to regard field quanta as properties of quantum fields. Hence, what is considered as single physical systems in quantum mechanics is treated as properties of quantum fields in quantum field theory. Because field quanta are not individuals, at most a whole quantum field – and in the last resort only the whole of matter at the level of quantum fields – can be considered as an individual.

To the extent that entanglement is an issue for philosophy, that same issue arises in quantum field theory. The difference that quantum field theory makes concerns the entities of which entanglement is predicated: in quantum mechanics, we can predicate entanglement of single physical systems. In quantum field theory, we cannot conceive single physical systems as the subjects of the predication of entanglement. The quantum correlations of entanglement occur even in the vacuum state, as is highlighted by the Reeh–Schlieder theorem.[68] Furthermore, since field quanta are properties of quantum fields, we cannot regard them as parts of quantum fields. Nonetheless, a quantum field has an internal structure given that it is embedded in space–time. That structure consists at least in the conditional probabilities of the values of the field operators at space–time points (or pointlike space–time regions). In quantum field theory, the Einstein–Podolsky–Rosen correlations, which express entanglement, are conceived as correlations between the conditional probabilities of the values of field operators at different space–time points. Given the outcome of a

---

66 For a claim to the contrary, see van Fraassen (1991), Chapter 12. For a criticism of van Fraassen's view, see Butterfield (1993), in particular pp. 473–474; Teller (1998), section 6.

67 Compare Redhead (1983) with Redhead (1988). See furthermore Stöckler (1988a); Stöckler (1988b), pp. 150–182; Falkenburg (1994), Chapters 6.5 and 6.6.

68 See, in particular, Redhead (1995b) and (1995c).

measurement of an operator at one space–time point, the probability for the outcome of a measurement of the same operator at a space-like separated point can be changed.[69]

We can say that first of all each quantum field is a holistic system; in the last resort we have to say that the whole of matter at the level of quantum fields is one holistic system. To make a case for holism, again, we have to consider the quantum probabilities as objective probabilities. Note that even if we interpret these probabilities as describing propensities, we are not committed to bare dispositions. It is the quantum state of the whole field that determines these probabilities. That state itself does not have to be regarded as a disposition. Endorsing some such interpretation, we can regard the probability distributions of the values of field operators at space–time points as constituent parts of a quantum field. Following Auyang (1995), who does, however, not elaborate on quantum holism, we can introduce "event" as a technical term for these probability distributions at space–time points. (Or it may perhaps be more appropriate to consider them as tropes).[70] Given such events as constituent parts of a quantum field, we can apply the proposed conception of holism to them: with respect to basic physical properties that make something a part of a quantum field, each of those events at a space–time point is ontologically dependent on there being other such events that run through the whole quantum field.

The idea in order to pay tribute to quantum field theory hence is this one: We have to translate quantum holism as set out for quantum mechanics into ontological dependence among events at space–time points with respect to properties that are treated as field operators at space–time points in quantum field theory. What has to be given up is the assumption that a holistic system in the physical realm always has parts which are individuals.

8.4 The Revision of the Cartesian Tradition in Physics

Let us return to the sort of holism that can be incorporated in Cartesian philosophy of physics. In what sense does quantum physics and quantum holism in particular constitute a revision of the Cartesian tradition in the philosophy of physics? The Cartesian tradition encompasses the whole philosophy of classical physics including special and general relativity. Because Descartes can be read in such a way that he identifies matter with

---

69 As to the Einstein–Podolsky–Rosen correlations in quantum field theory, see also Wald (1986).

70 See Kuhlmann (1999).

space (or space–time), the Cartesian tradition admits of a holism: holism about space (or space–time).

What is the relationship between the holism in Cartesianism and quantum holism? Both cases are substantial sorts of holism, although not in the same way: the claim that a part of matter can only exist and any physical property only be instantiated if there is the whole of matter is a substantial holism. The thesis that the state of a quantum system is entangled with the states of many other such systems is a substantial holism as well, because it amounts to a quantum system having basic properties such as position, momentum, or spin in any direction only taken together with other such systems in the explained sense. *That sort of holism which Cartesianism includes shows that holism in philosophy of physics is not tied to non-separability.* It is merely in cases of holism along the lines of the second type of the proposed general conception of holism [1.4] that there can be a question of a failure of separability, because it is only in these cases that the properties to which holism extends are instantiated only by several parts taken together. Holism about space, by contrast, can be conceived in such a way that it falls within the first type of the proposed general conception of holism, although it is not mandatory to conceive holism about space in that way [6.3].

In holism about space (or space–time), there is no ontological dependence among the parts of matter insofar as they instantiate specific physical properties. Holism about space automatically concerns all physical systems – because they are identical with points or regions of space – and all physical properties, because they are identical with properties of points or regions of space. Quantum holism, by contrast, concerns specific physical properties, namely the state-dependent ones. It is about ontological dependence among physical systems with respect to some of the specific properties that make something a quantum system, such as position, momentum, or spin in any direction.

If we take into account that Cartesian philosophy of physics incorporates a holism and if we pay tribute to the unification of space and time in relativity physics, we get to holism about space–time. However, we cannot marry quantum holism to holism about space–time. The programme to integrate a theory of quantum systems into a theory of the physical properties of space–time, i.e., Wheeler's original programme of quantum geometrodynamics, failed [6.5.2]. Quantum systems, including quantum fields, can hence not be seen as properties of space–time. Holism about space–time is not a viable option as a result of the failure of geometrodynamics. Consequently, we cannot avoid a profound revision of

Cartesianism in contemporary philosophy of physics. This revision concerns mainly three points:

1) *Localization*: According to holism about space–time, all instantiations of basic physical properties are localized at points or regions of space–time. If there are any physical connections over and above spatio–temporal relations, these are interactions which propagate with a finite velocity. By contrast, quantum systems are usually not localized at a point or an arbitrarily small region in space. Consequent upon this failure of localization, even state-independent properties of quantum systems such as mass and charge are usually not localized at a point or an arbitrarily small region in space. Quantum holism shows that what there is in the world is more tightly interwoven than just by spatio–temporal relations and causal relations among property instantiations that are localized at space–time points. Relations of quantum entanglement are at least as fundamental as spatio–temporal relations in unifying a world. One can even go one step further. An ontology that recognizes both space–time and matter as different basic entities can with reason be taken to be prodigious. An ontology which shows that one of these is sufficient on the basic level would be more plausible.[71] Since the programme of geometrodynamics failed, it is attractive to envisage trying the opposite programme, namely to get to a theory of space–time on the basis of a quantum theory of matter.

2) *Individuality*: Even if we translate the talk of bodies and their motion in space and time in Cartesian physics into the talk of physical properties of regions of space and, ultimately, into the talk of physical properties of space–time points, we still conceive particulars with physical properties that are distinct individuals. Space–time points are the ultimate individuals according to this conception. However, quantum systems are, if they are individuals at all, at least not *distinct* individuals. Consequent upon quantum holism we have to give up the assumption that the ultimate physical systems are distinct individuals.

3) *Separability*: Holism about space–time satisfies separability. We can attribute a state to each point of space or space–time, and this state completely determines the local properties of this point. The joint state of two or more points supervenes on the states that are attributed to each of these points. Einstein maintains that classical field theories including general relativity accomplish the principle of separability (1948, p. 80) [7.2.1]. Holism about space or space–time therefore is compatible with Lewis'

[71] Compare Lewis (1986b), p. 76 note.

Humean supervenience. If we start from Humean supervenience, as the first quotation in [8.2.1] shows, one version of Humean supervenience consists in conceiving the physical properties on which everything else supervenes as properties of points of space–time. In fact, this is Lewis' preferred version (1986b, p. 76 note). In this case, holism about space or space–time ensues.

Quantum holism implies a reversal of the principle of separability: the joint state of two or more systems is not determined by the states that can be attributed to each of these systems. On the contrary, it is only the joint state of two or more systems which completely determines the local properties of each of these systems and their relations (to the extent that these are determined at all). Taking into account that entanglement is ubiquitous among quantum systems, quantum holism has the following impact on our view of nature: a position such as Humean supervenience has to be replaced with, ontologically speaking, a position that starts from features which are properties only of matter as a whole on the level of quantum systems. We have to include the state of the whole of all quantum systems as a global feature in the supervenience basis.

Let us review from the perspective of this ontological consequence what has been said in this chapter about quantum entanglement and individuals, starting from the standard examples of entanglement of the states of two quantum systems. It is reasonable to assume that the global quantum state of the world does not have a natural decomposition into single quantum systems. Consequently, the ontological assumption that the world is at the basic level composed of single quantum systems is not justified. (The number of single quantum systems could, of course, be infinite – like the number of space–time points, if one conceives space–time as being composed of space–time points). Acknowledging the ontological consequence that only the global quantum state of the world is a pure state, we can no longer accept as basic in our ontology the notion of single quantum systems of which properties are predicated and which may be regarded as individuals, albeit not distinct ones.

This consideration is strengthened if we go beyond quantum mechanics and take quantum field theory into account. In quantum field theory, only a whole quantum field is a physical system. Field quanta are properties of a quantum field, but not themselves systems of which properties are predicated. Owing to entanglement, all quantum fields are one holistic system in the last resort. Consequently, we have to admit the state of all quantum fields taken together to the supervenience basis. To take all interactions into account, this state has to be conceived as including gravitation, although as yet we do not have an elaborate quantum field theory of gravitation at our disposal. If we pay tribute to special relativity and accept the ontology of a block universe [6.5.2], we have to conceive this state in the last resort as not developing in time, but as being defined over the whole of space–time. Considering what was said in point 1) above about localization, one may go as far as envisaging to build a theory of space–time on the basis of such a quantum ontology.[72]

In whatever way one may elaborate on such an ontology, this global quantum state includes an internal structure: the properties of the whole introduce local properties as described by field operators and specify correlations between these local properties. The properties of the whole thereby introduce something which can be considered as parts of matter and properties which make something a part of matter. These parts are entirely determined through properties of the whole. David Mermin maintains that quantum theory describes a world of correlations without presupposing underlying particulars as correlata.[73] This idea makes good sense, if we do not assume free floating correlations, but a quantum state of the universe as that entity which gives rise to and determines these correlations. Recall what was said at the end of [6.3] on a holism about space construed along the lines of the second type of holism and how parts of space are introduced in this conception. In contrast to the case of holism about space, in the quantum case there is a good systematic reason for such a conception: whatever quantum properties there may be, they are completely specified only by the state of the whole.

Although the three mentioned points constitute a profound change in the view of nature that dominates modern philosophy and science, these points do not amount to a rejection of Cartesianism, but to its revision. If we adopt an attitude of hermeneutic good-will towards Cartesian philosophy of physics, we have to recognize that Descartes – and Spinoza in particular – envisage a substantial holism about the physical realm. Given the physics of their time, it is reasonable to regard holism about space as the concrete shape that a holism about the physical realm can take within the framework of Cartesian philosophy of physics. Today, the holistic intent which Cartesian philosophy of physics incorporates proves sound: all matter is one holistic system at the microphysical level; but our physics suggests that holism in the physical realm takes another concrete shape. Pursuing Cartesianism in the

72 For an overview of different suggestions in that direction, see Monk (1997).

73 Mermin (1998). See also Rovelli (1996).

direction of holism about space or space–time, what is at issue today under this perspective is a change within holism, namely replacing the holism in Cartesianism with another, arguably much more momentous form of holism and spelling out its philosophical impact.

CHAPTER 9
THE EXTENSION OF QUANTUM HOLISM AND THE PHILOSOPHY OF MIND

Summary

Is quantum holism limited to more or less the microphysical realm? Or does it touch upon all physical systems? The issue of the scope of quantum holism is linked with the measurement problem [9.1]. When it comes to this issue, we should take the philosophy of mind into account – and in particular the discussion on holism and a revision of Cartesianism in today's philosophy of mind. All those interpretations that regard quantum holism as universal in the physical realm are committed to epistemic self-sufficiency of intentional states and a representational realism. Consequently, they are incompatible with the revision of Cartesianism that goes with the proposed social holism and holism about beliefs [9.2]. By contrast, if one countenances a transition to a level of macroscopic systems which are not touched by entanglement, one can still regard quantum physics as a universal physical theory. But one can accommodate a macroscopic realm with definite properties. This is a prerequisite for direct realism and externalism (including social holism) in the philosophy of mind. The moral of this chapter therefore is a plea for caution: The arguments for holism and a revision of Cartesianism in the philosophy of mind sharpen up the task for the interpretation of quantum theory, and they constitute a weighty reason against the option for universal quantum holism. There is hence not one comprehensive, substantial holism that includes both the philosophy of physics and the philosophy of mind and that leads to a revision of the Cartesian tradition in modern thought [9.3].

9.1 The Measurement Problem

Having considered the meaning of quantum holism, let us go into its scope. It is largely uncontroversial that quantum physics reveals a holism. The question is: Is there holism *only* at the microphysical level of nature? Or is quantum holism universal in the sense that it extends to all physical systems? If not, how is the microphysical level related to higher levels? There is a great variety of answers to these questions; but these answers more or less boil down to two principal options: (a) Either one regards *quantum holism as universal* and has to reject, as will be shown, the ontological commitments of higher level theories as well as common sense. (b) Or one takes *quantum holism to be limited to more or less the microphysical level*. In this case, one has to concede that we do not have a fully convincing solution at our disposal to the problem how a transition comes about from a level of quantum entanglement to higher levels where there is no entanglement. Because there is as yet no fully convincing *physical* solution to this latter problem, it is still an issue of *philosophical* argument which one of these two principal options is more plausible.[1]

Common sense and classical physics acknowledge nothing like entanglement. If one characterizes a scientific theory as classical in this context, one does not mean that it is a theory which has been developed before quantum theory arose. *All and only those theories are classical that do not admit entangled states among the systems which they describe.* Consequently, all and only those systems are classical which do not have states that admit of entanglement with the states of other systems. Since systems admit of entanglement if and only if they have incompatible observables each [7.1.3], all and only those systems are classical which do not have incompatible observables. Consequently, *classical observables* are all and only those observables that are compatible with all the other observables of the system in question.

On the one hand, we can apparently not dispense with our common sense ontology or a refined version of it along the lines of a classical physical theory. In any experiment in which we put quantum theory to the test, we assume that the experimenter is not entangled with the experimental arrangement and that there is no entanglement between the measuring instruments and the measured systems prior to the measurement. What is more, we describe every experimental arrangement including the measurement outcomes as they are indicated by the instruments in refined common sense terminology.

On the other hand, there is no obvious limit to the domain of applicability of the formalism of quantum theory including the superposition principle and the Schrödinger dynamics. Consider *Schrödinger's cat.* In a famous thought experiment subsequent to the paper of Einstein, Podolsky and Rosen in 1935, Schrödinger imagines a cat being locked into a steel chamber with a tiny amount of a radioactive substance and a container of cyanide. If one radioactive atom decays, a mechanism is triggered that breaks the container of cyanide, and the cat is killed. There is a probability of 0.5 that one atom decays within an hour. Schrödinger maintains that according to the formalism of quantum theory, the states of all these systems are entangled

---

1 For a description of the problem in a broad context, see the recent paper by Maudlin (1997).

after one hour. Consequently, the atom is in a superposition of not having decayed and having decayed together with the cat being in a superposition of being alive and being dead, etc. (1935b, p. 812). Thus, starting from the formalism of quantum theory, we apparently do not get to macroscopic objects as they are described by our common sense ontology or a refined version of it.

In the paper in which Schrödinger invents the thought experiment of the cat, he contemplates an ontological interpretation of quantum theory which admits values of properties that are not definite numerical values, but have a considerable dispersion. However, he rejects such an interpretation. Schrödinger maintains that the quantum formalism extends to macroscopic systems whenever such systems enter into interaction with quantum systems. Consequently, according to Schrödinger, if we interpret quantum theory realistically, we have to assume that the states of common sense objects become entangled with the states of quantum systems. Schrödinger then introduces the example of the cat in order to highlight what he considers to be an absurd implication of an ontological interpretation of quantum theory (1935b, pp. 811–812). Because of this absurd implication, he concludes that quantum theory cannot be interpreted realistically and that it is incomplete: in fact, the states of no systems which are involved in the thought experiment of the cat are entangled; there is no entanglement in reality.[2] Notwithstanding Schrödinger's argument, it is questionable whether, if we acknowledge entanglement on the level of quantum systems, we have to countenance an extension of entanglement to common sense objects.

Somewhat analogous to Schrödinger's thought experiment is the so-called *von Neumann chain*. When von Neumann turns to measurement in the book in which he codifies the traditional formalism of quantum theory in 1932, he describes what amounts to the following chain: we start with a quantum system an observable of which is to be measured. However, basing ourselves on the formalism of quantum theory and the Schrödinger dynamics in particular, we have to say that as a result of the interaction between the quantum system and the measuring instrument, the state of the quantum system is entangled with the state of the instrument. Consequently, the quantum system is not in an eigenstate of the measured observable, and the instrument does not indicate a definite numerical value of this observable. Von Neumann contemplates the extension of this chain to an observer. If we

---

2 Compare the reading of the thought experiment which Lockwood (1989), pp. 196–198, proposes.

take an observer into consideration, we end up with a description according to which the state of the body of the observer including her brain is entangled with the states of the instrument and the object (1932, Chapter VI.1; English translation: von Neumann (1955)). Thus, the observer is in a state that is similar to the state of Schrödinger's cat with respect to the observable in question.

The *measurement problem* can thus be conceived as the question how to adapt the following two claims so that they fit together: (a) A measurement yields a definite outcome. (b) Applying the formalism of quantum theory including the Schrödinger dynamics to a measurement situation leads to a description according to which the states of all the systems involved are entangled with one another.

Von Neumann maintains that when a measurement takes place, the state vector of the system in question no longer evolves according to the Schrödinger dynamics. It undergoes, as von Neumann says, a non-causal change into an eigenvector of the measured observable (1932, Chapter VI.1). This position is known as the projection postulate: it is postulated that the state of the system is reduced to one of the eigenstates of the measured observable in measurement. The measurement problem thus arises in von Neumann's codification of quantum mechanics in the form of a gap between the description of the development of the states of quantum systems according to the Schrödinger dynamics and the description of the results of measurements. When no measurement takes place, the state of the system evolves according to the Schrödinger dynamics. When a measurement takes place, this evolution is interrupted, and the state of the system is reduced to one of the eigenstates of the measured observable. This latter change is often referred to as a collapse.

However, instead of a collapse, I prefer to speak of a *state reduction*: a superposition is reduced to one of the eigenstates of which it is a superposition, and an entanglement of the states of two or more systems is reduced to one of the product states of which it is a superposition. Whereas employing the term "collapse" ties us down to conceiving an instantaneous event, a state reduction can be thought of as a process that takes a short time.

Von Neumann suggests that we have to introduce a cut in the described chain. By means of this cut we decide what counts as observed object and what counts as observing instrument in this chain. This cut is necessary in order to take into account that, as a result of a measurement, the state of all the systems in such a chain is not a superposition of product states, but reduced to one of the product states. According to von Neumann, it is only by means of introducing such a cut in our description that we can refer to a measured object which has an eigenvalue of the observable in question. We are free to place the cut anywhere in the described chain. We can even put it back to the observer's brain.[3]

Von Neumann is not alone in proposing the introduction of a cut. According to Heisenberg, a cut between measured system and measuring instrument within the experimental arrangement has to be introduced by the observer. The observer stands outside the experimental arrangement, and he cannot but introduce such a cut when he describes the whole arrangement; however, he can arbitrarily choose within certain limits where to place the cut.[4] Similar statements can be found in the writings of Pauli too.[5] Thus, for Heisenberg, in contrast to von Neumann, the cut has to be placed between the measuring instrument in the laboratory and the measured system; it is not admissable to put the cut back to the observer.

The proposal of such a cut has the following consequence: two or more observers can give descriptions of one and the same experimental arrangement which differ in what is regarded as the observation tools and what is regarded as the observed system. There is nothing wrong with any of these descriptions. The operational consequences of all these descriptions are identical: they all agree as far as the description of the states of all the systems involved after the measurement is concerned. Von Neumann therefore maintains that there is no point in trying to determine where the state reduction takes place (1932, Chapter VI.1). Nonetheless, the status of the proposed cut is unclear: Is the cut a mere peculiarity of our epistemic access to quantum systems? Or does the cut refer to an event which takes place during the measurement process?

As Schrödinger's cat and the von Neumann chain show, if we start from the formalism of quantum theory without further qualifications, we end up with the conception of an entanglement between the states of the instrument and the object. In the case of entanglement, each of the systems in question can be described in terms of what is known as a mixed state [7.1.3]; but this description does not admit of an ignorance interpretation: the problem is not

3 von Neumann (1932), Chapter VI.1. On a cut between the means of observation and the observed object which can be placed arbitrarily, see also Jordan (1951), p. 62.

4 "Prinzipielle Fragen der modernen Physik" (1936) in Heisenberg (1984), p. 116.

5 Pauli (1948), p. 309; Pauli (1952), pp. 164–165; "Die philosophische Bedeutung der Idee der Komplementarität" in Pauli (1961), p. 16; "Phänomen und physikalische Realität" in Pauli (1961), p. 99.

that we do not know the precise value of the property in question, but that there is no such value. Take the measurement of the spin of an electron in a given direction in a state in which the outcomes spin up and spin down are equally probable. To solve the measurement problem, it has to be shown that after the measurement the joint state of the electron and the instrument is either "electron spin down and instrument indicating spin down" or "electron spin up and instrument indicating spin up" with probability 0.5 each; but the joint state of the two systems is not a superposition of these states. The problem is not that it has to be found out by observation which one of these two states actually is the outcome of the measurement and that the outcome cannot be predicted from the formalism of quantum theory. What has to be achieved is a description that allows for each system to have a definite numerical value of the property in question.[6]

The most significant progress as regards the measurement problem since the days of Schrödinger and von Neumann are *decoherence theories*.[7] Assume that we start by considering an isolated system which is in a pure state. Let this system interact with its environment, and let in particular a measuring instrument be part of the environment. The state of the system becomes entangled with the states of the systems of its environment and the state of the measuring instrument in particular. Decoherence theories show the following: although only the whole of system, environment, and measuring instrument is in a pure state, each of the parts of this whole can be described in terms of a mixed state. The point is that this whole rapidly develops in such a way that it cannot operationally be distinguished from a mixture of states, i.e., an ensemble of systems each of which is in a pure state, whereby the observer is ignorant of the pure state in which each system is. Decoherence theories do not assume that a state reduction takes place. They do not introduce a modification of the Schrödinger dynamics.

It can be argued that decoherence theories do not admit of an ontological interpretation[8] and that they cannot yield a satisfactory solution to the

---

6 For an up-to-date systematic treatment of the measurement problem and the various suggestions to its solution, see Busch, Lahti and Mittelstaedt (1996) and Mittelstaedt (1998a).

7 For an overview and introduction, see Giulini et al. (1996), in particular Chapter 2 (Zeh (1996)). See also Omnès (1994a), in particular Chapters 6 to 8.

8 d'Espagnat (1995), Chapters 10.6 and 12.3.

measurement problem.[9] Such an argument can be extended to all the approaches that draw attention to the role of the environment. These approaches have a point in emphasizing that, strictly speaking, there are no isolated systems which are in a pure state and which can thus be described by a state vector in the traditional formalism of quantum theory. However, making this point does not solve the measurement problem: it does not provide for a description of the right sort of mixture as measurement outcome. Because decoherence leads to a mixed state that cannot be distinguished operationally from an ensemble of systems each of which is in a pure state, but, ontologically speaking, is not a mixture of states, decoherence theories leave the decisive question open: Shall we endorse an ontology that regards quantum holism as universal and welcome decoherence theories as a means to explain how it comes about that the world *appears* classically to an observer?[10] Or shall we go beyond the scope of decoherence theories and assume a dynamics that is able to dissolve superposed states and entanglement in order to account for the *existence* of a classical level of the world?

9.2 Is Quantum Holism Universal?

9.2.1 The Many Minds Interpretation

One can bite the bullet of Schrödinger's cat and the von Neumann chain and maintain that quantum holism is universal: entanglement touches not only the states of microphysical systems, but the states of all physical systems. Consequently, Schrödinger's cat objectively is in a superposition of being alive and being dead together with the whole arrangement. The task then is to explain how it appears to observers that superposed states including entanglements are reduced, i.e., how it appears to observers that a cat always is either alive or dead.

Taking quantum holism to be universal recalls the PhD thesis of Hugh Everett. Everett (1957) argues that the Schrödinger dynamics is the only

9 See, for instance, Penrose (1989), pp. 293–294; Penrose (1994), Chapter 6.6; Butterfield (1995), pp. 119–121; Maudlin (1996), pp. 303–304. Zeh (1996), p. 16, concedes that decoherence alone does not solve the measurement problem.

10 Compare the title of Giulini, Joos, Kiefer, et al. (1996): "Decoherence and the Appearance of a Classical World in Quantum Theory". See in particular Zeh (1996), pp. 23–25, and Stamatescu (1996), pp. 252–253, in that book. See also Omnès (1994a), in particular pp. 353–366, 391–400, 506–531; Omnès (1994b), in particular pp. 310–316, 351–359.

dynamics of quantum systems and that every physical system including macroscopic ones is a quantum system. These two points are the core of Everett's work. Over and above that Everett is the person to whom the many worlds interpretation of quantum theory goes back.[11] However, I shall not go into this interpretation. If the branching of worlds in measurement is considered as an objective physical event which is a replacement for a state reduction, we are confronted with all the problems which the acknowledgement of state reductions has. In particular, it is unclear how a branching of worlds could come about. If the branching of worlds is not seen as a replacement for state reductions, a version of the option for universal quantum holism is endorsed; this option does not need a commitment to many worlds. One can therefore say: the ontological extravagance of assuming that many worlds actually exist and that these worlds interact with one another is not outweighed by conceptual gains in a solution to the measurement problem.

In the eigthies, an interpretation has been developed which takes up the core of Everett's ideas. This is the so-called *many minds interpretation* of quantum theory. This interpretation has been set out in a paper by David Albert and Barry Loewer in 1988.[12] Its elaboration owes much to the work of Michael Lockwood.[13] My exposition is based on Lockwood's version.

Assume that the spin of an electron in a given direction is measured, and that the outcome is either spin up or spin down with probability 0.5 each. Assume furthermore that spin up corresponds to a state of the measuring instrument which can be called "pointer up" and that spin down corresponds to "pointer down". Finally, there is an observer who perceives the state of the pointer by interacting with the pointer. According to Lockwood's many minds interpretation, the state of the electron becomes entangled with the state of the pointer and ultimately with the state of the observer's mind. The state of the three systems which is relevant in this context is a superposition

---

11 As to the many worlds interpretation, see the papers in DeWitt and Graham (1973), which include Everett's PhD thesis, and the recent book by Deutsch (1997), Chapters 2, 8, 10 to 13. As to the different readings of Everett's thesis, see Barrett (1997).

12 Albert and Loewer (1988). See also Albert and Loewer (1989), pp. 179–184; Albert (1992), pp. 126–133.

13 Lockwood (1989), Chapters 12 and 13; Lockwood (1996). The second issue of Volume 47 (1996) of the *British Journal for the Philosophy of Science* is devoted to a discussion of the many minds interpretation. For a comprehensive exposition of the many minds interpretation, see Barrett (1999), Chapter 7.

of the product state "electron spin up and pointer up and observer conscious of pointer up" with the product state "electron spin down and pointer down and observer conscious of pointer down".

Even if we do not acknowledge a state reduction, we can introduce the notion of a *relative state*. This notion goes back to Everett (1957). In the example of the last paragraph, none of the systems is in an eigenstate of the observable in question. However, if we arbitrarily choose one of the eigenstates for one of the systems, we gain an eigenstate of each of the other systems too relative to that choice. If we choose, for example, the state "conscious of pointer down" for the observer, relative to that state of the observer we gain the state "pointer down" for the instrument and the state "spin down" for the electron.

According to the many minds interpretation, the observer is conscious of such a relative state. Thus, relative to the observer being in a state in which she is conscious of pointer down, the pointer appears in a state in which it is down, and the electron appears in a state in which it has spin down. Assume now furthermore that in fact the state of the pointer and the state of the electron are entangled with the states of many more systems. Consequently, relative to the pointer appearing in the state of pointer down to an observer, not only does the electron appear in the state spin down, but all other systems also appear in a state in which they have a definite numerical value of the observed properties, if we assume maximal correlations.

The pointer appearing in the state pointer down, etc. is the content of a point of view, a perspective, or a mind according to the many minds interpretation. However, there also is a point of view, a perspective, or a mind in which the pointer appears in the state pointer up, etc. These two disjunct scenarios are both simultaneously the contents of a point of view, a perspective, or a mind of one and the same observer each. Moreover, there are states which are a superposition of not only two, but many, even infinitely many states. If we think of such states as being represented in the basis of eigenstates of consciousness or mind, we have to assume the existence of infinitely many such eigenstates. Consequently, according to Lockwood as well as Albert and Loewer, each observer does not only have two points of view, perspectives or minds, but infinitely many points of view, perspectives or minds simultaneously. The introduction of infinitely many minds of each observer enables the many minds interpretation to take probabilities into account, as Lockwood and Albert and Loewer claim.[14] Thus, this interpretation is known as the many minds interpretation, because it proposes that each observer has many minds simultaneously in the sense that she has many mutually incompatible experiences simultaneously. Since these experiences are disjunct, they do not interfere with each other.

According to the many minds interpretation, quantum holism concerns all that exists, including persons and the mind of each person as a whole. All that exists is a huge quantum system, which is the only thing that is in a pure state. Separability is universally violated. But there is no non-locality, since no state reductions take place. This point underlines again that it is important to distinguish between separability and locality. Consequently, according to the many minds interpretation, there is no indeterminism: the whole universe is one big entangled system whose state evolves deterministically. This interpretation faces a number of physical and philosophical problems:

- Why does each of the product states in question appear to one mind each? Why is it impossible to be conscious of a state that is a superposition of product states and literally see, for instance, a cat in a superposition of being alive and being dead?
- There are infinitely many bases in which the state of the universe can be represented as a superposition of infinitely many product states. In this interpretation, one basis has apparently to be preferred – the basis of mind eigenstates of each observer. Can the introduction of such a preferred basis be justified?
- Can this interpretation do justice to quantum probabilities?
- Can this interpretation escape solipsism? Can it explain the agreement among observers?
- Can this interpretation pay tribute to the intertemporal identity of a person?

There are sophisticated strategies to answer all these queries.[15] There is no space to consider these strategies here. Let us therefore grant, for the sake of argument, that a satisfactory answer to all these questions can be given within some version of the many minds interpretation. Let us evaluate the many minds interpretation in the light of its consequences for the philosophy

---

14 Donald (1997), by contrast, advances a many minds interpretation according to which an observer has only finitely many minds. This interpretation also takes probability statements into account.

15 See, for instance, Lockwood (1989), in particular Chapters 12 and 13, and the second issue of Volume 47 (1996) of the *British Journal for the Philosophy of Science*.

of mind. This interpretation proposes an astonishing consequence for the philosophy of mind to say the least: we are expected to admit many minds of each person solely on the basis of arguments from the formalism of quantum physics. However, I do not intend to focus on this thesis, because it is peculiar to the many minds interpretation. The presentation of the many minds interpretation is intended here only as a concrete example of an interpretation that takes quantum holism to be universal. I will therefore concentrate on those implications for the philosophy of mind that are common to all the interpretations which regard quantum holism as universal in the physical realm.

9.2.2 The Option for Universal Quantum Holism

The many minds interpretation is the most elaborate version of what I propose to call the *option for universal quantum holism*: quantum theory including the superposition principle and the Schrödinger dynamics (or a relativistic generalization) is universally valid in the physical realm. The physical world is a huge quantum system whose internal structure consists in ubiquitous entanglement. This claim has the following implication: Theories of higher level systems, such as theories of chemistry, biology, physiology, etc., as well as common sense do not describe things as they are objectively, since nothing like entanglement is acknowledged by these theories.

The point is not simply that, according to the option for universal quantum holism, macroscopic systems admit of much more states than those states which are taken into account by common sense or any theory of chemistry, biology, physiology, etc. The point is that, according to universal quantum holism, macroscopic systems almost always objectively are in states like Schrödinger's cat, which are nothing like the states that common sense and our scientific theories of macroscopic systems ascribe to these systems. These theories and common sense contain an ontological commitment which says that these systems always have definite values within a minimal margin for all their properties. By contrast, if one regards quantum holism as universal, one is committed to an ontology according to which, as far as the time-dependent properties are concerned, the systems in question cannot have a definite value for all their properties at once and usually do not even have a definite value for a significant sample of them; for the only state description which applies to each of them is a description in terms of a mixed state that does not admit of an ignorance interpretation. If one subscribes to universal quantum holism, one can take the higher level theories including common sense as correctly describing the way in which macroscopic systems appear to observers. But the point is: One cannot both endorse the option for universal quantum holism and accept the ontological commitments of higher level theories including common sense.

Given this consequence of the option for universal quantum holism, the task is to square the ensuing view of physical reality with our experience of a level of the world that can be described in terms of classical physics and common sense. That is why the option for universal quantum holism calls the philosophy of mind into play. Two main strategies to meet this task can be distinguished in the literature:

- *many minds of the observer*: As explained in the preceding section, each observer has many or even infinitely many minds in the sense of many incompatible, but non-interfering experiences at once. Each term in a superposition is correlated with at least one mind of the observer.
- *abstraction from entanglement by the observer*: The observer has only one mind, and quantum superpositions extend in no way to her mind. But in making perceptual experiences, the observer abstracts from entanglement that is objectively there. As a result of this abstraction, the world appears to the observer as described by common sense and classical physics. It is possible to connect this option with decoherence theories [9.1]. According to one version of this option, only a subset of properties (observables) of physical systems is accessible to the observer. This limited accessibility is the reason why physical systems appear to the observer as described by common sense and classical physics.[16] According to another version of this option, the observer cannot but introduce a cut in the *von Neumann chain* [9.1] by means of which she puts aside entanglement that is objectively there; as a result of this cut, physical objects appear to the observer as described by common sense and classical physics.[17]

Note that in all these approaches the observer has to be a literal observer, i.e., a being who has perceptual beliefs about her environment. As long as quantum holism is taken to be universal, the observer cannot be replaced with a machine that simply processes information. Such a machine cannot behave like a classical system, unless one introduces a level of systems which are not touched by entanglement. The state of such a machine is entangled with the states of many other systems. Whatever information processing such a machine does, this simply leads to further entanglement.

---

16 Compare Landsman (1995).

17 Compare in particular Primas (1993).

Therefore, a classical level of the world can only be the way in which the world appears to a literal observer.

Whether it is plausible to endorse the option for universal quantum holism or whether it is plausible to admit a means that dissolves entanglement when it comes to the transition to macroscopic systems cannot be settled by referring to the state of the art of physics; for we do not have a fully convincing account of a dynamics for a dissolution of entanglement at our disposal. When a question arises in the interpretation of a physical theory that is not decided by the state of the art of physics, it is reasonable to take broader philosophical considerations into account. In particular, as regards the option for universal quantum holism, if a physical theory is construed in such a way that it has direct implications for the philosophy of mind, it is not only appropriate from a methodological point of view, but even a methodological requirement to judge these implications in the light of arguments that originate in the philosophy of mind. The next section will therefore relate the discussion on the extension of quantum holism to the discussion on holism and a revision of Cartesianism in the philosophy of mind.

9.2.3 Universal Quantum Holism and Cartesian Epistemology

What is intended by Cartesianism in today's philosophy of mind can be characterized by the following two theses [4.1]:

- *representationalism*: the conceptual content of intentional states such as belief states consists in mental representations. Mental representations act as an epistemic intermediary between the belief states of a person and the world. One can go as far as maintaining that mental representations are the immediate object of our belief states. I described this latter claim as *strong representational realism* in [4.1]. Accordingly, I considered as *weak representational realism* the position that there are mental representations which act as an epistemic intermediary without being the object of our belief states.
- *epistemic self-sufficiency of intentional states*: the individuation of intentional states and their identity depends only on factors that are immanent to the person which has these states. Since intentional states are individuated by their content, intentional states are epistemically self-sufficient if and only if their content is ontologically independent of the social and physical environment. That is to say: the content of the intentional states of a person could be the same even if there were no

other persons and no physical world with which the person in question is in touch.

The revision of Cartesianism in today's philosophy of mind consists in replacing these two theses with direct realism and externalism. *Direct realism* is the following claim: Causal intermediaries between things in the world and our belief states notwithstanding, there are no epistemic intermediaries (such as mental representations). When it comes to our epistemic access to the world, our beliefs – and our perceptual beliefs in particular – directly relate to the world. This direct realism accords with common sense realism. *Externalism* is the following thesis: It is necessary not just in a causal, but in an ontological or metaphysical sense to be embedded in a social and a physical environment for having intentional states. The individuation of intentional states and their identity depends on the qualitative character of the social and the physical environment in which the subject of these states lives.

Recall the distinction between the following two types of externalism about the physical environment:

a) *singular beliefs*: Having singular beliefs about particular things in the world depends on the existence of these things. For instance, one can be in the state of believing that there is a fat man over there only if there actually is a man over there [4.3.2].
b) *natural kind concepts*: The meaning of beliefs in which natural kind concepts are employed (such as "water" or "tiger") depends on the physical constitution of the things referred to. Recall Putnam's famous thought experiment in "The Meaning of 'Meaning'" (1975, essay 12): Independently of whether or not someone knows that water is H_2O, it is part of the meaning of her beliefs about water that water is H_2O. On the imaginary twin earth, there is a stuff that is indistinguishable from water on the surface. However, that stuff is not H_2O, but *XYZ*. The meaning of all beliefs about that stuff is different from the meaning of all beliefs about water [4.3.3].

The Anti-Cartesian developments in contemporary philosophy of mind have as yet by and large not been taken into account when claims about the observer and her mind or consciousness are made in the interpretation of quantum theory. To my knowledge, the only exception is Michael Lockwood's book *Mind, Brain and the Quantum*: Lockwood argues against central features of the current Anti-Cartesian stream in epistemology (1989, in particular Chapters 9 and 16). Nonetheless, these developments in today's philosophy of mind are pertinent to the interpretation of quantum theory.

What is the consequence of the option for universal quantum holism as regards our epistemic access to the world? Albert and Loewer, in their many minds interpretation, concede the following consequence: almost all our beliefs about perceptible objects are false.[18] The option for universal quantum holism has this consequence, because it implies that things do not objectively have the definite properties that we ascribe to them in our perceptual beliefs; their objective states are nothing like the manner in which we conceive them in our perceptual beliefs. Despite this consequence, some philosophers who subscribe to universal quantum holism accommodate *externalism* as regards the physical environment. Thus Lockwood endorses externalism about singular perceptual beliefs as well as Putnam's externalism (1989, pp. 144–145, 310–311). Albert and Loewer can be seen as intending to make room for externalism: they leave open whether intentional states are correlated merely with states of the brain or with states of the brain plus the environment.[19]

However, one may wonder whether externalism really fits into a philosophy that takes quantum holism to be universal in the physical realm. What about externalism as regards singular beliefs? Is the man over there fat? My belief that the man over there is fat is false anyway, because, in fact, the man is in a superposition of, say, being fat and being thin. What is more, his position is not limited to the position indicated by "over there", say, a position in the doorway; rather it is a superposition of being over there in the doorway with, for instance, being in the cellar and many other positions that moreover involve the states of other things including other men. Accordingly, in the region indicated by "over there", there is not one man; instead, a superposition of many things that include, among others, several men extends to this region. Given that as a consequence of a view which considers quantum holism to be universal every object including common sense objects has a state which is entangled with the states of many other objects, what can it mean to say that having a singular belief is ontologically dependent on there being the object referred to?

One can of course say that for every singular belief of an observer there is a decomposition of the state description of the universe according to which the object of the belief of that observer exists. That is to say in terms of

---

18 Albert and Loewer (1988), p. 209. See also Albert and Loewer (1989), pp. 182, 184; Albert (1992), pp. 127, 132–133.

19 Albert and Loewer (1988), pp. 206, 211. As to the relation with externalism, see the hint in Butterfield (1995), p. 146.

Everett's notion of a relative state [9.2.1]: relative to the state of the observer believing that there is a fat man over there, there is a state of the environment in which there is a fat man over there. This consideration is crucial when it comes to explaining how it appears to the observer that there is a fat man over there. However, this consideration is not sufficient to accommodate externalism: The relative state is nothing but a component of a mixed state that does not admit of an ignorance interpretation. It establishes a correlation between the components of the state of mind of the observer (assuming that her state of mind is to be described in terms of a mixed state) and the components of the state of the physical environment. But this correlation is nothing like giving priority to the physical environment in the sense that the definiteness of the environment contributes to making belief states have a definite content by contributing to individuate the belief states of the observer. This conception therefore does not capture the point of externalism about singular beliefs.

An analogous objection applies to the attempt to combine externalism as regards beliefs that employ natural kind concepts with universal quantum holism. Coming back to Putnam's argument in "The Meaning of 'Meaning'", assume, as Putnam does when he introduces his thought experiment (1975, p. 223), that earth and twin-earth coexist in the same universe. In this case, we have to presume that the state of the earth is entangled with the state of the twin-earth. Consequently, water is entangled with twater. Thus, in fact, there is a superposition of H_2O and *XYZ* Taking this superposition into account, how can the meaning of beliefs about the liquid in question depend on being causally related to either H_2O or *XYZ*? Again, by means of the notion of a relative state, one can establish a correlation between components of the mixed state of mind of the observer and components of the mixed state of the environment. However, again, this correlation is nothing like the environment contributing to individuate the beliefs of the observer.

Considering these objections I submit that externalism about the physical environment is incompatible with the option for universal quantum holism: if one regards quantum holism as universal in the physical realm, one cannot take the individuation and the identity of beliefs to be ontologically dependent on the properties of the physical environment of the subject of these beliefs. *As a result of universal entanglement the physical environment of an observer does not have that definiteness which is a prerequisite for it being able to contribute to the individuation of the observer's intentional states* (since it has to be described in terms of a mixed state which does not admit of an ignorance interpretation). Hence, the point is not simply that universal quantum holism commits us to regarding more or less all our common sense beliefs as false. There are other metaphysical positions that can be taken to have a similar consequence. For instance, one can claim that there are only occurrents which have both spatial and temporal parts, but no ordinary things (continuants) [6.5.2]. However, in this case one sets out to reconstruct the definite properties which we attribute to things in common sense as objective, definite properties of events or of space–time points. The point is that in the case of universal quantum holism, there is no such reconstruction available.

The last point makes clear that the option for universal quantum holism also clashes with social externalism and the account of meaning that was proposed in Chapter 3 and elaborated on in Chapter 5: According to this account, the meaning of beliefs is determined by social practices that involve dealings with physical things in the sense that we in our social practices let physical things constrain the meaning of beliefs. However, according to the option for universal quantum holism, the properties of physical things do not objectively have a definiteness such that we could in our social practices let the meaning of our beliefs be constrained by them. But the notion of letting the properties of physical things constrain the meaning of our beliefs is one cornerstone of the argument that social holism and holism about beliefs concern our epistemic access to the world instead of being a spinning in the void [5.3.2]. If one subscribes to universal quantum holism, one is therefore committed to regarding the individuation and identity of our beliefs as depending only on factors that are internal to the person who has these beliefs. One is hence committed to epistemic self-sufficiency of belief states.

A similar argument applies to *direct realism*: the option for universal quantum holism cannot go with direct realism as regards the objects of our perceptual beliefs. According to this option, the features which we ascribe to physical things in our perceptual beliefs are relative to the way in which we observe these things by abstracting from entanglement that is objectively there. If one endorses universal quantum holism, one is committed to the position that the objects of our perceptual beliefs – such as cats being always either alive or dead, etc. – are dependent on the conditions of our observation of them, i.e., the abstraction from entanglement. Thus, starting from the option for universal quantum holism, Lehner, for instance, sets out an elaborate argument in favour of giving up an observer-independent notion of perceptible reality.[20] Lockwood goes as far as endorsing what I termed strong representational realism in [4.1]: mental representations are the immediate object of our perceptual beliefs; the assumption that there is a physical world is an inference to the best explanation of our representations (1989, Chapters 9 to 13 and 16).

The option for universal quantum holism implies a commitment to at least weak representational realism. There is an epistemic intermediary between our beliefs and the world, namely the appearance of a classical level of the world to observers in perception. Furthermore, the meaning or conceptual content of our perceptual beliefs consists in this appearance. It is only this representation of physical things relative to an observer that exhibits the features which we take to be characteristic of perceptible things in common sense. Therefore, the option for universal quantum holism is incompatible with a holism about beliefs that begins with inferential relations among beliefs and explains representation in terms of these relations. Such a holism about beliefs leaves no room for representations that act as epistemic intermediaries; instead, it opens up a direct realism.

In whatever way the relationship of representation is spelled out in detail depending on which option one takes to be most plausible to square universal quantum holism with our perceptual beliefs, it is, however, nothing like representing physical things as they are in themselves. The commitment to a representational realism arises in the case of the option for universal quantum holism just because of the gap between on the one hand the physical world as it is according to this option and on the other hand the properties which we ascribe to physical objects in our perceptual beliefs. Note that interpreting quantum theory in terms of universal quantum holism contradicts only a direct realism about perceptible objects including common sense realism. Nonetheless, this interpretation is a form of scientific realism.

The option for universal quantum holism hence implies a Cartesian epistemology. But no commitment to a Cartesian ontology ensues, i.e., an ontological dualism in the philosophy of mind. Albert and Loewer develop a many minds interpretation which has a dualist consequence in the sense of denying that all aspects of the mental supervene on the physical.[21] Lockwood, by contrast, combines a many minds interpretation with an

20 Lehner (1997), in particular pp. 208–213. See also Landsman (1995).

21 Albert and Loewer (1988), pp. 210–211. See also Albert and Loewer (1989), p. 184.

identity theory of the mental and the physical.[22] What about interpretations that regard the observer as abstracting from entanglement that is objectively there? If the abstraction consists in a limited accessibility of the properties of physical systems to the observer, this position is also neutral as regards the controversy between a dualist and a materialist ontology of the mind.

What does this comparison of universal quantum holism with the debate about Cartesianism and its revision in today's philosophy of mind show? If one assumes in the interpretation of quantum theory that all that has to be accounted for is the appearance of a classical level of the world to observers, then one tacitly presupposes a Cartesian epistemology. One presupposes what has been described here as epistemic self-sufficiency of intentional states and a representational semantics including at least weak representational realism. However, since Cartesianism is called into question in today's philosophy of mind, representationalism cannot be considered to go without saying. As one cannot take traditional assumptions about physics for granted in today's philosophy of mind, so one cannot take a traditional philosophy of mind for granted when it comes to the philosophical significance of quantum theory. *The discussion on Cartesianism and Anti-Cartesianism in current philosophy of mind thus is relevant to the interpretation of quantum physics: taking this discussion into account sharpens up the task for the interpretation of quantum theory.*

We hence face two philosophical packages, so to speak, both of which include the interpretation of quantum physics and the philosophy of mind:

1) On the one hand, one can opt for universal quantum holism in the physical realm. But in this case one cannot simply seek to square the ensuing view of physical reality with the appearance of a classical level of the world to observers. One has to counter the arguments for direct realism and externalism in connection with holism in today's philosophy of mind.
2) On the other hand, if one approves of the arguments for direct realism and externalism, one cannot commit oneself to universal quantum holism in the interpretation of quantum theory. These arguments tie the meaning of our beliefs to the interaction with a classical environment. Hence, if one endorses this theory of meaning, one has to regard the

---

22 Lockwood (1989), compare in particular Chapter 5 with Chapters 12 and 13. According to the version of the many minds interpretation which Donald (1997) develops, mental phenomena supervene on brain states.

holism that quantum physics exhibits as being limited to more or less the microphysical level.

These two packages are not exhaustive. For instance, one can conceive quantum holism as being limited to microphysics and favour a representational realism when it comes to the philosophy of mind. The point is: *One cannot approve of a substantial social holism and holism about beliefs in the philosophy of mind and construe quantum holism as concerning all physical systems; for the holism in the philosophy of mind takes us to a revision of Cartesianism along the lines of direct realism and externalism.*

One may be inclined to construe the clash between the option for universal quantum holism and the option for a substantial holism in the philosophy of mind as an instance of an opposition between scientific realism and common sense realism. But consider the following two points: (a) The proposed argument for holism in the philosophy of mind and a revision of the Cartesian along the lines of direct realism and externalism does not accord a privileged status to common sense realism [5.2.3]. That argument is built on considerations in the philosophy of mind which are as sophisticated as the most refined moves in the interpretation of quantum theory. (b) If we base ourselves on scientific realism, we have .to acknowledge that there is more to science than just quantum theory and physics. Considering quantum holism to be universal in the physical realm is incompatible with acknowledging the ontological commitments of higher level scientific theories such as biological theories [9.3.2]. There is hence no reason to regard the mentioned clash as an instance of an opposition between scientific realism and common sense realism.

9.2.4 Complementarity of Mind and Matter?

One may raise the following objection against the conclusion of the last section: If the position that quantum holism is universal in the physical realm is embedded in a philosophy which applies the notion of complementarity to the relationship between mind and matter,[23] then a comprehensive holism seems in prospect, namely a holism that includes the physical and the mental. Such a comprehensive holism is advertised as an attractive replacement for the Cartesian tradition in both the philosophy of nature and the philosophy of mind. Accordingly, universal quantum holism in the physical realm is only one aspect of a holism that concerns the whole of

---

23 Compare the hint in Bohr, "Introductory Survey" in Bohr (1934), p. 24.

being. In this case, complementarity has to be employed as an ontological notion: things themselves are complementary. For instance, Wolfgang Pauli says that it would be most satisfactory if *physis* and *psyche* could be considered as complementary aspects of the same reality.[24] In the same vein, Bernard d'Espagnat maintains on the basis of quantum holism that the physical and the mental are two complementary aspects of a holistic reality.[25]

If one regards the mental and the physical as two complementary aspects of being as a whole, one is committed to an ontology of *psycho-physical parallelism*. The thesis of such a complementarity excludes the positions that the mental is identical with the physical or that the mental supervenes on the physical (without the physical, in turn, supervening on the mental). This thesis furthermore excludes any account according to which there is interaction between the mental and the physical, because there can be no relation of causation between complementary items.[26] The replacement for interaction that this thesis contains is parallelism in the following sense: the mental and the physical are two aspects of being as a whole which are correlated in such a way that they correspond to one another without there being causal relations between them. The commitment to psycho-physical parallelism obtains also in case one merely maintains that the mental and the physical are two complementary aspects of each person.[27] Hans Jonas elaborates on this commitment.[28]

Psycho-physical parallelism, however, is a culmination of the Cartesian programme of epistemic self-sufficiency of intentional states: parallelism implies that it is in principle possible to describe and explain every mental state including belief states by referring only to other mental states. Since

---

24 Pauli (1952), p. 164. As to Pauli's philosophical views, see Laurikainen (1988); van Erkelens (1991); Atmanspacher and Primas (1996).

25 d'Espagnat (1991), Chapters 9 and 10, in particular pp. 155, 160, 191. As to an endorsement of the thesis that the mental and the physical as a whole are complementary aspects of being, see furthermore Globus (1976); Kafatos and Nadeau (1990), in particular Chapter 8; Primas (1993).

26 Compare Margenau (1984), p. 45.

27 As to this position, see, for instance, Bernays (1948); Brody and Oppenheim (1969); Smith (1976), p. 310; Stent (1993), pp. 74–78. Hyland (1985) and Kirsch and Hyland (1987) claim to suggest only a methodological in distinction from an ontological complementarity between body and mind; but their position includes the ontological thesis that there is no causal relation between body and mind.

28 Jonas (1980), pp. 123, 128–130; Jonas (1987), pp. 101, 104–105.

there is no interaction between physical and mental states, every mental state is solely caused by other mental states. When it comes to accounting for mental states, it is therefore in principle sufficient to refer only to other mental states. This implication is a central weakness of psycho-physical parallelism: there is no prospect of a self-sufficient psychology and epistemology which can do without referring to physical things and events. Physics, by contrast, is self-sufficient: explanations in physics do not need to have recourse to anything but physical states.

The programme which envisages a comprehensive holism by considering the physical and the mental as complementary aspects of being does hence not constitute an objection against the thesis that endorsing universal quantum holism in the physical realm clashes with holism and the revision of Cartesianism in the philosophy of mind. This programme can give a concrete shape to holism in the philosophy of physics by relying on quantum holism. But it fails to give a concrete shape to holism in the philosophy of mind. If we consider the holism in current philosophy of mind and the revision of the Cartesian tradition in epistemology to which a substantial version of this holism leads, it becomes clear why this programme cannot provide for a concrete elaboration of the envisaged holism as far as the philosophy of mind is concerned: by conceiving quantum holism in such a way that this holism touches the whole physical realm, this programme is committed to a Cartesian epistemology. Contrary to what is intended, this programme in fact implies that there is a gap between the mental and the physical, because it amounts to a version of psycho-physical parallelism. The clash between taking quantum holism to be universal and pursuing the project of a revision of Cartesianism in the philosophy of mind is a decisive blow to all those who claim in popular literature that quantum theory gives rise not only to a new physics, but also to a new philosophy of mind that amounts to holism and that breaks with the Cartesian tradition in modern thought. *We thus have to conclude that we cannot melt quantum holism together with a substantial holism in the philosophy of mind in order to reach one comprehensive holism.*

9.3 Limited Quantum Holism

9.3.1 Limited Quantum Holism and the Measurement Problem

When discussing the measurement problem in [9.1], it was pointed out that there are two principal options for a solution to this problem: (a) One can maintain that all physical systems are subject to the formalism of quantum theory and that the Schrödinger dynamics is the complete dynamics of quantum systems, thus considering quantum holism to be universal in the physical realm. (b) One can assume a dynamics which is able to dissolve superposed states and entanglement and thus regard quantum holism as being limited to more or less the microphysical level of nature. Let us now consider this latter option, the *option for limited quantum holism*. According to this option, there is a basic level of quantum systems, characterized by entanglement, and there are higher levels of complex systems, among which there is no entanglement, although nothing hinders that these latter systems are realized as compositions of quantum systems. There are of course no strict demarcations among these levels. Furthermore, a higher level is not only defined by the size of the systems that are considered, but also involves border conditions such as a temperature within certain limits. Drawing attention to quantum effects in macroscopic systems such as superfluidity and superconductivity is not an argument against limited quantum holism.

In the formalism of quantum theory, superselection rules are a means to introduce classical observables and thus properties that are not subject to entanglement. However, given the current state of the art, we do not have superselection rules at our disposal that are able to solve the measurement problem in such a way that they account for the existence of classical features at the macroscopic level.[29] The problem is to conceive a dynamics that leads to a dissolution of entanglement.

It would be most satisfactory if one were able to reconstruct the existence of classical features of the world by relying solely on the Schrödinger dynamics. However, given that decoherence theories do not account for a dissolution of entanglement [9.1], it is reasonable to envisage a modification of the Schrödinger dynamics. The aim of such a modification is to set out a general dynamics for quantum systems which includes the Schrödinger dynamics as well as a dynamics for state reductions. Note the difference between proposals for a modification of the Schrödinger dynamics in order to account for state reductions and approaches in terms of hidden variables. In these latter approaches, quantum theory is considered to give an incomplete specification of the states of single microphysical systems. These former proposals, by contrast, acknowledge that quantum theory gives a complete description of the states of single microphysical systems; they seek to clear up the dualism of two different dynamics in von Neumann's codification of quantum theory.

---

29 Compare Robinson (1994); Landsman (1995); Thalos (1998).

There are two main types of proposals for a change to the Schrödinger dynamics in order to account for events of state reduction. The first type of such proposals remains within the scope of current quantum theory. The most elaborate proposal of this type was introduced by Ghiradi, Rimini and Weber in 1986.[30] The suggestion of these physicists accounts for state reductions independently of measurements and observations. However, it can be objected that the change to the Schrödinger dynamics that these physicists introduce is *ad hoc*. Among the problems which this proposal faces are furthermore the questions whether it takes the conservation of energy into account, whether it admits of a relativistic generalization,[31] and whether it can pay tribute to the observed macroscopic quantum effects such as superfluidity.

The other main type of proposals for a change to the Schrödinger dynamics envisages going beyond the scope of current quantum theory in order to solve the measurement problem. Its starting-point is the idea that the measurement problem in quantum theory and the problem of the unification of quantum theory with general relativity are interconnected. It is an open question how quantum theory and general relativity can be unified. Any theory that attempts to be a universal physical theory has to take gravitation into account. The central idea is that when gravitation becomes relevant, a state reduction occurs due to gravitation. Therefore, higher level systems are not subject to superpositions, and superposed states of quantum systems are reduced as soon as gravitation comes into play. This approach brings not only quantum theory and general relativity together; it has the further advantage that no problem with energy conservation arises. Roger Penrose is the most prominent advocate of this approach.[32] However, this approach is still in its infancy.

Furthermore, recall the additional problems that a state reduction in a Bell experiment raises. In view of these problems it is questionable whether it is

---

30 Ghiradi, Rimini and Weber (1986). For an overview which outlines the philosophical implications, see Ghiradi, Grassi and Benatti (1995). For a critical evaluation of the programme of Ghiradi, Rimini and Weber (1986) see Stamatescu (1996). For a possible combination of this programme with a philosophical realism, see Stöltzner (1999), section 7.

31 But see Ghiradi, Grassi and Pearle (1991) and furthermore Pearle (1996).

32 Penrose (1989), Chapter 8; Penrose (1994), Chapter 6, §§ 10–12; Penrose (1997), pp. 83–92. Compare Kanitscheider (1993), pp. 118–120, and the discussion of Penrose's position in Penrose (1997), Chapters 4 to 7.

plausible to conceive a dissolution of entanglement in such a way that a suitable local interaction with a quantum system causes not only this system to acquire a definite numerical value of a certain local observable, but also dissolves entanglement in such a way that definite numerical values of local observables are acquired at space-like separated locations. There is a change of probabilities for such values (outcome dependence). However, as mentioned in [7.3.2], there are options which do not force us to countenance a change of a local property at a space-like separated location. Consequently, it is possible to conceive a dissolution of entanglement as a local affair.

Countenancing a way of dissolving entanglement such as state reduction does not commit us to indeterminism in quantum physics. As long as there is no fully satisfactory account of the dynamics of state reduction, it is an open question whether or not state reduction really is indeterministic. For instance, if the suggestion that state reduction is induced by gravitation proves sound, gravitation may turn state reduction into a deterministic process. Thus, Penrose contemplates the possibility that state reduction due to gravitation is deterministic, but not computable (1989, pp. 431–432).

9.3.2 Limited Quantum Holism and Reductionism

When higher level systems, which are not affected by entanglement, are characterized here as complex systems, I mean systems that are composed of quantum systems, but in such a way that they have significant properties of new kinds in comparison to the properties of atoms and their constituents. For instance, already chirality (right handedness or left handedness) is an example of a property of a new kind of a complex system, because only molecules, but not atoms and their parts exhibit chirality. Let us introduce a broad notion of a *higher order property* (or a higher level property): a higher order property is any property which cannot be instantiated at a point or an arbitrarily small region of space–time. In other words, it requires more than a point-like or arbitrarily small system to be instantiated. Thus, not only properties of new kinds such as chirality, but also position and momentum of macroscopic systems are higher order properties.

The option for limited quantum holism is committed to higher level systems that have developed from the microphysical level and that have properties of new kinds in comparison to the properties of quantum systems. One can of course say that quantum theory and higher level theories such as biological theories simply apply to different areas of nature. Nonetheless, however much the current state of science may resemble a patchwork rather than exhibit unity, the task of metaphysics is not to draw up lists of things that there are, but to put forward proposals as to how things hang together.[33] Therefore, it is reasonable to demand an account of how exactly quantum theory and higher level theories are related to each other.

Supervenience alone is not sufficient to give a satisfactory account of the relationship between different levels: saying that a level of properties *A* supervenes on a level of properties *B* alone is not sufficient to explain the level of properties *A*.[34] Therefore, in addition to assuming supervenience, it is reasonable to claim that, as far as ontology is concerned, higher order properties are realized by patterns of microphysical properties. For instance, right handedness or left handedness of molecules (chirality) is realized as a certain arrangement of atoms, etc. As far as epistemology is concerned, reduction of higher level theories to lower level theories still is an attractive candidate for spelling out the relationship between theories of different levels. Reduction may be conceived in the classical sense of Ernest Nagel (1961, Chapter 11) according to which the concepts of the reduced theory T_1 are to be translated into the concepts of the reducing theory T_2, and the laws of T_1 are to be deduced from the laws of T_2 by means of certain bridging principles. Or reduction may be conceived in the network sense of Ramsey and Lewis according to which the relations in which the terms of the reduced theory stand are to be captured by the relations in which the terms of the reducing theory stand, and no independent bridging principles are needed.[35]

However, given the current state of the art, no reduction of higher level theories to quantum theory has been carried out.[36] But neither are there conclusive arguments against such a reduction.[37] Whether or not a reduction is feasible depends on the way in which the physical process that leads to higher levels where there no longer is entanglement is spelled out. Suggestions for state reductions of the type of Ghiradi, Rimini and Weber or

---

33 Compare, for instance, the clear formulation of this position by Jackson in "Armchair Metaphysics", Chapter 11 in Jackson (1998b), pp. 156–157. For a contrary position, see Cartwright (1983) and "Fundamentalism versus the Patchwork of Laws" in Cartwright (1999), Chapter 1.

34 Compare Horgan (1993), in particular pp. 556–560, 565–566, 574–582.

35 See in particular "How to Define Theoretical Terms" in Lewis (1983a), Chapter 6. Smith (1994), pp. 44–54, gives an easily accessible outline of this conception.

36 As to the difficulties which quantum theory poses for reductionism, see also Hedrich (1990), Chapter 5.

37 As to the programme to reduce chemistry to quantum theory, compare Primas (1983). See also the elaborate conception of reduction in physics which Scheibe (1997/99) proposes (volume 1) and applies to quantum mechanics (volume 2, Chapter X).

the type of Penrose certainly do not exclude the possibility to reduce higher level theories to quantum theory. We need first a detailed and satisfactory theory of the physical process of a dissolution of superposed states and entanglement before a judgement on whether or not higher level theories can be reduced to quantum theory is sensible.

The option for limited quantum holism is opposed to considering quantum holism as being universal in the physical realm. Nonetheless, this option can admit quantum theory as a universal physical theory:[38] a quantum theory that shows how a transition comes about to higher order properties which are not subject to superposition and entanglement can be a universal theory (provided that it can be unified with general relativity and take gravitation into account). The option for limited quantum holism does not even prevent us from going for an ontology according to which all there is in the physical world is a universal quantum state [8.4], provided that this state makes room for a macroscopic level without entanglement. That is to say: macroscopic properties including common sense properties, to which no entanglement extends, are realized as patterns in a local assemblage of quantum systems or as a local pattern in a quantum field. All that we need to keep the possibility of such a conception open is a local dissolution of entanglement in a concentration of quantum systems or a region in a quantum field that enables these systems or this region to realize macroscopic properties. Such a conception is not even committed to regarding higher level theories as being reducible to quantum theory; for the ontological claim of realization does not imply the epistemological claim of theory reducibility.

If we endorse the option for limited quantum holism, we do not only accept that there is no entanglement among higher level systems; but we also maintain that state reductions and hence cases of disentanglement occur at least locally on the level of quantum systems. It could therefore seem that limiting quantum holism infringes upon the thesis of quantum holism itself. However, the crucial point is this one: In a reconstruction of the transition to higher level systems and state reductions on the level of quantum systems, we have to start with considering matter as one holistic system at the microphysical level. Only on the basis that first of all there is entanglement among quantum systems can subsequently develop more and more higher level, complex systems that are not subject to entanglement; and only on that basis can state reductions occur on the level of quantum systems. Accepting

---

38 Compare Mittelstaedt (1998a) who discusses the different options which we have as regards the measurement problem on the basis of quantum theory being universal.

quantum theory as the theory of the microphysical level, we cannot but reconstruct in our ontology cases of separate states on the basis of entanglement, however negligible the entanglement may be for practical purposes. From the point of view of quantum theory, what is to be explained is not entanglement, but the failure of entanglement.

9.3.3 A Plea for Caution

What is the moral of this discussion of the two principal options as regards the scope of quantum holism? The question of the scope of quantum holism cannot be settled by referring to the state of the art of physics. This question enables us to establish a link between the interpretation of quantum theory and the philosophy of mind. However, this link does not consist in an extension of the notions of quantum theory – and quantum holism in particular – to the philosophy of mind. This link consists rather in the following: taking considerations from today's philosophy of mind into account sharpens up the task for the interpretation of quantum theory and shows that we face at least two packages so to speak each of which includes both the interpretation of quantum theory and the philosophy of mind. To recall the end of [9.2.3], on the one hand, (1) if one opts for universal quantum holism in the physical realm, one has to counter the arguments for direct realism and externalism in today's philosophy of mind. On the other hand, (2) if one approves of these arguments, one has to regard the holism that quantum physics exhibits as being limited to more or less the microphysical level.

The option for universal quantum holism calls the philosophy of mind into play, because this option has direct implications for the philosophy of mind (namely at least weak representationalism including epistemic self-sufficiency of intentional states). The option for limited quantum holism, by contrast, does not have direct implications for the philosophy of mind, because it regards the features that are peculiar to quantum theory – and quantum holism in particular – as being limited to more or less the microphysical level. This option is not committed to a particular account of the relationship between quantum theory and higher level theories including the philosophy of mind.

Nonetheless, some scientists who endorse state reductions, notably Roger Penrose,[39] propose that quantum theory is important to neurophysiological

---

39 Penrose (1989), Chapters 9 and 10; Penrose (1994), Chapters 7 and 8; Penrose (1997), Chapter 3.

research. However, given the current state of neurophysiology, there is no firm evidence for this thesis. A judgement on whether or not quantum theory is relevant to brain research seems premature. What is much more, even if it were established that quantum theory is relevant to brain research, it is not clear what, if any, would be the consequences for the philosophy of mind. Furthermore, since the advent of quantum theory, it has been suggested that quantum physics has implications for the issue of freedom of will in the philosophy of mind. All the proposals in that respect base themselves on an interpretation of quantum theory according to which there are events of state reduction; these events are moreover considered to be indeterministic.[40]

Note the difference between the argument in [9.2.3] and the suggestions for a connection between quantum indeterminism and free will. In the case of the option for universal quantum holism, a physical theory is interpreted in such a way that it has direct implications for the philosophy of mind. It is in this case even a methodological requirement to evaluate these implications in the light of arguments that originate in the philosophy of mind. If these implications are considered to be objectionable on the basis of arguments from the philosophy of mind, then, by way of consequence, these objections speak also against the proposed interpretation of quantum theory. A connection between quantum theory and the philosophy of mind arises in this case because the implications of a physical theory are taken to extend to the philosophy of mind. By contrast, in the case of the suggestions for a link between quantum indeterminism and free will, the starting-point is a position in the philosophy of mind, namely the view that (a) there is freedom of the will and that (b) freedom of the will clashes with determinism in physics. The possibility of an interpretation of quantum theory that acknowledges indeterminism is then made use of in order to lend support to this view in the philosophy of mind.[41]

Are the option for universal and the option for limited quantum holism on an equal footing? Taking the discussion on holism and a revision of Cartesianism in today's philosophy of mind into account makes out a strong case in favour of the option for limited quantum holism: the mere fact that the option for universal quantum holism commits us to a position in the philosophy of mind which can no longer be considered to go without saying is a weighty argument against this option and supports the option for limited quantum holism. The moral of this chapter thus is a plea for caution: If we

---

40 The most concrete proposal in that respect is Eccles (1994), Chapter 9.

41 Compare Esfeld (2000).

wish to keep an open mind as regards the debate about holism and the revision of Cartesianism in today's philosophy of mind, we should not commit ourselves to an interpretation of quantum theory which takes quantum holism to reach beyond more or less the microphysical level of nature. To conclude this chapter, consider these three theses:

1) The arguments for holism and a revision of Cartesianism in today's philosophy of mind along the lines of direct realism and externalism are pertinent to the interpretation of quantum theory: they sharpen up the task for the interpretation of quantum theory, and they are a weighty reason against taking quantum holism to be universal in the physical realm.
2) Envisaging on the basis of quantum holism one comprehensive and substantial holism that includes both the physical and the mental is doomed to fail from the start: we cannot marry quantum holism, extended to the physical realm as a whole, with a substantial holism in the philosophy of mind that leads to a revision of Cartesianism in the philosophy of mind. By contrast, extending quantum holism to the physical realm as a whole commits us to a Cartesian epistemology in the philosophy of mind. If we regard quantum holism as being limited to more or less the microphysical realm, we avoid a clash between quantum holism on the one hand and holism and the revision of Cartesianism in the philosophy of mind on the other hand. However, the option for limited quantum holism precludes us from conceiving direct implications of quantum theory for the philosophy of mind on clean methodological grounds.
3) Nonetheless, we can regard quantum theory as a universal theory of the physical realm provided that we endorse an interpretation of quantum theory which contains a means to preclude entanglement from extending to higher order properties. In this case, quantum theory as a universal physical theory is compatible with the definiteness of the macroscopic that is a prerequisite for direct realism and externalism.

CHAPTER 10
THE MUTUAL ENRICHMENT OF QUANTUM HOLISM AND HOLISM IN PHILOSOPHY OF MIND

Summary

What is meant by holism in quantum physics and what is meant by holism in the philosophy of mind has a substantial common conceptual content. However, we cannot melt these two sorts of holism together in order to reach one comprehensive holism. So what is the significance of this common conceptual content? I propose a mutual enrichment of these two sorts of holism. In particular, the common conceptual content of quantum holism and holism in the philosophy of mind shows that holism in the philosophy of mind fits into a view of the world that bases itself on scientific realism [10.1]. Quantum holism can be used to cut off one aspect of the alleged link between holism, indeterminacy and eliminativism in the philosophy of mind [10.2]. Furthermore, quantum holism and holism in the philosophy of mind both support the same supervenience claim, namely global supervenience in contrast to local supervenience [10.3]. Finally, both quantum holism and holism in the philosophy of mind point at the same new sort of realism which overcomes the opposition between metaphysical realism and relativism or instrumentalism. In conclusion, we can therefore say that both these sorts of holism make a significant common contribution to a contemporary philosophical view of the world and ourselves [10.4].

10.1 A Common Conceptual Content

According to the characterization of quantum entanglement set out in [8.3], both quantum holism and the various forms of holism in the philosophy of mind come under the general conception of holism developed in [1.2]. What holism in the philosophy of mind says about beliefs and what quantum holism says about microphysical systems has a common conceptual content – although the systems which are claimed to be holistic are very different. As far as quantum systems are concerned, we can say the following: With respect to the instantiation of some of the properties that belong to the family of properties which make something a quantum system (such as position, momentum, and spin in any direction), any such system is ontologically dependent on there being other systems with which it finally constitutes the whole of matter at the level of quantum systems. As far as beliefs are concerned, we can say the following: With respect to the instantiation of some of the properties that belong to the family of properties which make something a belief (such as meaning, confirmation, justification), anything is ontologically dependent on there being other things with which it constitutes the system of beliefs of a person.

Recall the two types of holism within the proposed general conception [1.4]. Quantum holism is a case of the second type: some of the properties that make something a constituent of a holistic system of the kind in question are instantiated only by the whole; but these properties of the whole indicate the way in which the constituents are related with each other with respect to the properties in question. As a result of entanglement, some of the properties which make something a quantum system are in the last resort completely determined only by the state of the whole of all quantum systems. The holism about beliefs which Quine proposes also is a holism of this type: strictly speaking, it is only a whole system of beliefs that instantiates some of the properties which make something a belief such as confirmation and meaning; nonetheless, the properties of the whole indicate the way in which the parts are related with respect to confirmation and meaning. In [2.1.3], it was argued that conceiving holism about beliefs along these lines is plausible as far as confirmation is concerned.

However, in contrast to quantum holism, holism about beliefs can also be considered as belonging to the first type of holism that was distinguished within the general conception. Such a view was favoured in [2.3] and [2.4] as far as meaning is concerned: each belief instantiates the property of meaning; but each belief is, insofar as it instantiates this property, ontologically dependent on there being other beliefs with which it constitutes the system of beliefs of a person. Furthermore, social holism as motivated by the rule-following considerations also is a case of this type of holism. Nonetheless, this type is not limited to cases of holism in the philosophy of mind. Holism about space or space–time is a holism in the philosophy of physics that is most plausibly conceived as belonging to this type [6.3].

Consequently, the comparison between holism in philosophy of mind and holism in philosophy of physics can be spelled out in terms of the two types of the proposed general conception of holism: quantum holism in philosophy of physics is a case of the same type of holism as holism about confirmation in philosophy of mind. Holism about meaning, by contrast, and social holism in philosophy of mind as well as holism about space or space–time in philosophy of physics are cases of the other type of holism within the proposed general conception. The possibility to carry out the comparison between these forms of holism in such a detailed manner confirms the thesis that the conceptual content of "holism" is the same in both areas. We can classify these forms of holism by means of a substantial conception of holism independently of the area to which they belong.

Revealing a common conceptual content of quantum holism and holism in philosophy of mind is an important result in itself; for it is usually assumed that these cases do not have more in common than the mere label "holism". Thus, Davidson, for instance, claims in "Belief and the Basis of Meaning" (1974) that holism sets beliefs apart from the physical:

> Each interpretation and attribution of attitude is a move within a holistic theory, a theory necessarily governed by concern for consistency and general coherence with the truth, and it is this that sets these theories forever apart from those that describe mindless objects or describe objects as mindless. (in Davidson (1984), p. 154)

According to the argument of this book, holism in philosophy of mind indeed implies anti-reductionism. But holism is not peculiar to the mind. It covers also the physical at the microphysical level. What is particular to the mind according to Davidson as well as according to the position adopted in this work is the normative relations in which intentional states stand to each other.[1] These normative relations can be construed an argument for holism in the philosophy of mind; but it is not holism in itself that distinguishes intentional states from the physical.

As with any received opinion, there are exceptions. Patrick Suppes criticizes Davidson's thesis that holism sets psychology apart from physics by claiming the following:

> It is common in physics as well as in psychology to study systems that are not closed, that are not deterministic, and that are holistic in character. (1985, p. 184)

Suppes mentions only cosmology and field theories, but not quantum holism in particular. The argument of this book agrees with Suppes' claim. It goes beyond the mere claim by elaborating on a common conceptual content of holism in the philosophy of mind and the philosophy of physics. Furthermore, it thereby disentangles holism from the issues of open systems and indeterminism. One may regard the whole universe as a closed quantum system; in this case, quantum holism applies to this system. Moreover, quantum holism is neutral as far as the question of indeterminism in quantum physics is concerned [9.3.1].

---

1 In some places, Davidson bases his claim of irreducibility only on the normative character of the intentional without explicitly mentioning holism. See, for instance, Davidson (1987), pp. 46–47, and Davidson (1991b), pp. 162–165.

However important the result of a common conceptual content may be in itself, we can and should go further and ask: What can we do with this result? Can we melt these two sorts of holism together so that we reach one comprehensive holism? Starting from the problem in the philosophy of mind of conceiving a rational constraint of the world on our beliefs, we can on the basis of social holism and holism about beliefs propose a comprehensive holism that includes the physical realm by considering the conceptual to be unbounded. However, this holism can take only that aspect of the physical realm into account which makes it that, according to this position, the physical realm is included in the conceptual. It cannot take a holism into account that concerns the physical realm insofar as it is the object of scientific research [5.3.1]. Holism in the philosophy of mind therefore is one thing, and a holism within the scope of the scientific view of nature, such as quantum holism, is another thing. Starting from holism in the philosophy of mind, we cannot reach a comprehensive holism that includes quantum holism.

We come to an analogous conclusion if we start from quantum holism. Considering the measurement problem, we can endorse an option according to which quantum holism touches the whole physical realm. However, if we endorse such an option, we are committed to representationalism and epistemic self-sufficiency of belief states in the philosophy of mind. We have in this case to oppose the consequences to which the outlined social holism and holism about beliefs lead, namely direct realism and external individuation of belief states [9.2.3]. Starting from quantum holism, we cannot reach a comprehensive holism that includes holism in the philosophy of mind.

Hence, both holism in philosophy of mind and quantum holism open up the perspective of a comprehensive holism each. But, paradoxically enough, we can pursue this perspective in both cases only at the price of excluding the other holism. Consequently, *the significance of the common conceptual content of quantum holism and holism in philosophy of mind is not that we can melt these two sorts of holism together to reach one comprehensive holism.*

Given this result, the following question is the more pressing: Why is it interesting that quantum holism and holism in philosophy of mind are both instances of the same substantial conceptual content? Even if we cannot melt these two sorts of holism together to gain one comprehensive holism, there can be a *mutual enrichment between these two sorts of holism*: when it comes to considering the one sort of holism, it is important to take the other sort into account. As regards quantum holism, the preceding chapter elaborated on one such interconnection: when it comes to the scope of quantum holism and thus the measurement problem, taking holism in philosophy of mind into consideration (a) sharpens up the task for the interpretation of quantum theory and (b) makes out a strong case in favour of the option to regard quantum holism as being limited to more or less the microphysical level of nature.

In this chapter, I therefore start from holism in philosophy of mind and ask: In what respect is quantum holism pertinent to the discussion on holism in philosophy of mind? The thesis is: *Since "holism" means the same in the case of quantum physics and the case of the philosophy of mind, social holism and holism about beliefs fit into a view of the world that bases itself on scientific realism.* By scientific realism, I mean the attitude that – if anything – it is our best scientific theories which tell us what there is in the world. As Sellars puts it, adapting the famous saying of Protagoras, "in the dimension of describing and explaining the world, science is the measure of all things, of what is that it is, and of what is not that it is not".[2]

The argument for this thesis proceeds in three steps: It is first shown that quantum holism can be used to cut off one aspect of the alleged link between holism, indeterminacy and eliminativism in the philosophy of mind. I then establish that quantum holism and holism in philosophy of mind both support the same supervenience claim. Finally, it is argued that both quantum holism and holism in philosophy of mind point at the same new sort of realism. In elaborating on these claims, further aspects in which holism in philosophy of mind enriches quantum holism will become apparent.

10.2 Holism, Indeterminacy and Eliminativism

Holism in philosophy of mind is sometimes linked with eliminativism.[3] In contrast to reductionism, eliminativism is compatible with holism about beliefs and social holism. One can maintain that the description of intentional states cannot be reduced to a description of physical states and adopt the attitude that this is the worse for intentional states. If one takes this attitude, one can employ holism about beliefs as an argument for eliminativism.

---

2 "Empiricism and the Philosophy of Mind", § 41, p. 173 in Sellars (1963).

3 Compare Fodor and Lepore (1992), in particular pp. 131, 187, 205, 213 note 20, and furthermore pp. 13–17.

An argument for a link between holism and eliminativism cannot be based on the meaning of "holism". If the meaning of "holism" were an argument for eliminating something that manifests holism, then one would have to eliminate not only the intentional, but also the microphysical. For "holism" means the same in quantum physics and the philosophy of mind. The fact that holism in quantum physics and holism in philosophy of mind have a substantial common conceptual content shows that holism fits into scientific rationality. The general conception of holism developed in the first chapter makes clear that holism can be treated by the conceptual tools of analytic philosophy and scientific rationality. Hence, taking quantum holism into consideration enriches holism in philosophy of mind by underlining that there is nothing which is contrary to scientific rationality in the meaning of "holism".

The argument from holism to eliminativism in philosophy of mind is based on consequences that holism is alleged to have for the philosophy of mind. Quine, in particular, can be taken to maintain the following: semantic holism implies both indeterminacy of translation (1960, Chapter 2) and anti-reductionism; these two implications, in turn, speak in favour of eliminativism.[4] If Quine is read along these lines, the problem that his indeterminacy thesis poses for his naturalism disappears: on the one hand, according to the indeterminacy thesis, there are translation manuals for the interpretation of any speech acts which are incompatible with each other, but which are empirically equivalent in the sense that there is no physical difference at all between them.[5] On the other hand, according to Quine's naturalism, if these manuals do not make any physical difference, then there is no difference between them.[6] However, if the indeterminacy thesis is seen as an argument in favour of eliminativism, the problem that incompatible translation manuals pose for Quine's naturalism passes simply because there are no intentional states to which these manuals could apply.

The holism about beliefs and social holism advocated in Chapters 2.3-4 and 3 start from indeterminacy considerations in the sense that behavioural evidence does not determine meaning. However, the claim then is that an inferential role semantics based on a normative, social pragmatics determines the meaning of our beliefs. Such a semantics is an argument how

---

4 See, in particular, Quine (1960), §§ 45, 54; "Epistemology Naturalized" in Quine (1969), essay 3.

5 See, in particular, "Things and Their Places in Theories" in Quine (1981), p. 23.

6 See Gemes (1991), pp. 105–107.

meaning can be determined for a linguistic community contrary to Quine's claims (compare [2.3.1]). Furthermore, as argued in [4.2], the proposed holism about beliefs opens up a direct realism that repudiates the dualism of conceptual scheme and given content. It thereby shows that there is no *a priori*, conceptual limit for getting into the practices of another community. The proposed holism consequently fits into the suggestion of Lance and O'Leary-Hawthrone who criticize Quine's view of translation on the basis that translation is not a descriptive business in the first place, but a practical, normative effort to merge two or more communities into one (1997, Chapter 1; see the end of [3.3]). Hence, as in the cases of indeterminacies of meaning, so in the case of indeterminacies of translation, the idea is to overcome these indeterminacies by focussing on social practices, described in normative vocabulary. Instead of implying a principal indeterminacy of translation, the holism advocated here is an argument against any objectionable indeterminacy.

Apart from the indeterminacy of translation, Quine's holism – and the Duhem–Quine thesis in particular – is linked with indeterminacy in the sense that if holism is right, there are no rational criteria for the adoption of beliefs. We can use the debate about holism and other options in the philosophy of quantum theory as an argument to cut off the alleged link between holism and this indeterminacy.[7] This debate is one example from the philosophy of physics which illustrates Quine's thesis that there is no separation between science and metaphysics: owing to Bell's theorem and the Bell experiments, scientific and philosophical questions dovetail in such a way that some philosophers of physics go as far as speaking of experimental metaphysics [7.3.3].

When it comes to changes to our system of knowledge subsequent to new experience, Quine proposes a pragmatic attitude: according to "Two Dogmas", it is rational to opt for those changes which imply the slightest disturbance within our system of knowledge as a whole in order to accommodate that system to new experience (in Quine (1980), pp. 43–44, 46). This pragmatism of Quine is not a logical part of confirmation holism; but it is a plausible strategy to cope with the fact that, according to this holism, there always are several options for adapting a system of knowledge to new experience. My thesis is: *As a result of the dovetailing between science and metaphysics consequent upon Bell's theorem, Quine's criterion*

---

7 For a general argument against this link in the philosophy of science see Papineau (1979), pp. 117–124.

for changes to our system of knowledge makes possible a precise, rational evaluation of the options in the interpretation of quantum physics. There is indeterminacy in the sense that mathematical calculations and experimental results do not determine the interpretation of quantum theory. We have several options for integrating quantum physics into our system of knowledge. All the options imply a revision of beliefs which are regarded as metaphysical. Nonetheless, Quine's criterion enables us to evaluate the ontological price of each of these options and thus to distinguish some options as being more plausible than others. Therefore, Quine's confirmation holism does not lead to indeterminacy in the sense that there are no rational criteria for belief change.

Quine's criterion for changes to our system of knowledge supports the positions in the interpretation of quantum theory which were adopted in the preceding chapters. The option for quantum holism is tied to an ontological or realistic interpretation of quantum theory: quantum theory refers to the states of physical systems, and it gives a complete description of these states. If we base ourselves on Quine's confirmation holism, suppositions such as localization, separability and the identity of indiscernibles are not a metaphysical sanctuary, but belong to our assumptions about the way the world is. They can hence be changed consequent upon experience. My first claim is: *It is more reasonable in Quine's sense to abrogate such suppositions than to refuse to endorse ontological commitments of quantum theory at all on the basis of such suppositions.*

Bohr's interpretation as well as an instrumentalistic interpretation of quantum theory endorse a commitment to a classically describable world including experimental arrangements; but they resist an ontological interpretation of quantum theory. If statements that refer to quantum systems are countenanced at all, they are accepted only as descriptions of measurement outcomes.[8] The formalism of quantum theory is not regarded as referring to something in nature, but as an instrument to calculate probabilities for measurement outcomes.[9]

From a Quinean point of view one can object that quantum theory thus interpreted does not fit into our system of knowledge. Bohr's interpretation and an instrumentalistic interpretation remove that sort of description of the microphysical level of nature that higher level theories presuppose. Taking

---

8 Compare what Scheibe (1964), pp. 21–22, calls an epistemic interpretation. See [8.1].

9 Cushing (1997) claims that there is underdetermination in the sense of Quine between the Copenhagen interpretation and Bohm's theory.

the statements of the theory of the quantum level to require the reference to experimental arrangements which are set up by humans is incompatible with endorsing a commitment to higher level theories that intend to explain the evolution of live and human beings: these theories presuppose the quantum level; quantum events have been important in the evolution. This incompatibility shows up in physics itself when it comes to applying quantum theory to cosmology. Hence, although Bohr's interpretation and an instrumentalistic interpretation start from endorsing the ontological commitments of classical theories, these interpretations do not succeed in making quantum theory cohere with these theories.

One can of course adopt an instrumentalistic attitude to science as a whole for philosophical reasons, and one may attribute such an attitude to Quine's "Two Dogmas" (see Quine (1980), pp. 44–45), although Quine cannot be tied down to instrumentalism; he arguably is a naturalist in the first place. But in this case one regards the ontological commitments of the whole of science including the claim that there are common sense objects and experimental arrangements as a mere instrument to predict connections among sense data or the like. It is incompatible with this attitude to single out one physical theory and to adopt an instrumentalistic attitude only to this theory for physical reasons. This is contrary to the simplification that the philosophical instrumentalist has in mind.

It is possible to endorse an ontological or realistic interpretation of quantum theory without being committed to quantum holism: one can claim that quantum theory does not give a complete description of the states of microphysical systems. There are further, hidden variables which are not taken into account by quantum theory as it stands. For instance, we can propose that there are hidden variables which establish an interaction between the two quantum systems in a Bell experiment. However, as a result of Bell's theorem and experiments such as the one of Aspect, we would have to conceive a superluminal interaction. We would thus have to admit a new kind of interaction which does not fit in with the known kinds of interaction.

Furthermore, we can even continue to uphold both separability and local action. We can suppose that there is a common cause of the correlations between the measurement outcomes in a Bell experiment. However, as Bell's theorem makes clear, such a common cause would have to coordinate the decision by the experimenter or a random generator on the parameters that are to be measured with the behaviour of the systems under examination. We can avoid the assumption of such a conspiracy. But then we would have to change our conception of causality by admitting hidden variables of backward causation without having any independent evidence for backward causation. Both the option for a common cause and the option for backwards causation infringe upon a presupposition of experimental science: according to these options, the decision which parameter is to be measured on the system is connected with the state of the system that is prepared at the source of the experimental set-up [7.3.3]. One can therefore maintain that all those options which assume hidden variables imply a greater disturbance of our system of knowledge as a whole than giving up separability and thereby endorsing quantum holism. My second claim thus is: *On the basis of Quine's criterion for changes to our system of knowledge it is reasonable to give up separability instead of endorsing hidden variables.*

Quantum holism is not one option, but a type of options. The main question which we face if we acknowledge quantum holism concerns its extension: Is quantum holism limited to more or less the microphysical level? Or is it universal in the sense that it extends to all physical systems? My third claim is: *Quine's criterion for changes to our system of knowledge supports the option for limited quantum holism in distinction to the option for universal quantum holism.* From the point of view of Quine's holism, quantum theory and higher level theories (such as biological theories) have the same ontological status in the following sense: to the extent that these theories are confirmed by experience, they are all justified in claiming to describe physical systems as they exist independently of our knowledge of them. The task of the philosopher is to give an interpretation of our best scientific theories which brings them together in a description of the world which is as coherent as achievable, given the state of our knowledge.

This task brings the position that quantum holism is universal in the physical realm into conflict with Quine's confirmation holism. It may be possible to build a coherent system of knowledge on the basis of the option for universal quantum holism. However, the point is that this option fails to bring quantum theory together with our other scientific theories. Instead of integrating the new experience in quantum physics into our system of knowledge, this option obliges us to reject the ontological commitments of our higher level scientific theories as well as common sense: there is nothing like the familiar definiteness of the macroscopic [9.2.2]. Quine's pragmatism therefore supports the option for limited quantum holism.

Hence, as regards the discussion on quantum holism, the pragmatic criterion for changes to our system of knowledge in Quine's holism can serve as a guideline for the interpretation of quantum physics. Quine's holism sets the scene for a judgement of the significance of Bell's theorem and the Bell experiments: it puts proposals such as the claim about experimental metaphysics in an appropriate perspective, and it provides for a criterion to evaluate the options in the interpretation of quantum theory. The option for limited quantum holism can be justified on the basis of Quine's view of our system of knowledge. As regards holism in the philosophy of mind, the discussion on quantum holism can serve as a paradigmatic example of the Duhem–Quine thesis. This example shows the following: Even the form of holism in the philosophy of mind which Quine proposes is not tied to indeterminacy in the sense that there are no rational criteria for changes to a system of knowledge consequent upon new experience.

One may object that one example is not sufficient to disprove the thesis of a connection between holism and indeterminacy in the sense that if holism is right, there are no rational criteria for the adoption of beliefs. One may always maintain that there can be two or more comprehensive systems of knowledge which both fit all known empirical evidence and which are incompatible with each other. However, on such a general level, one can also always say that this is not a matter of indeterminacy, but a matter of ignorance on our side: even if there were two or more systems which both account for all known empirical evidence, there could be rational criteria which single out one of them. If we are unable to make such a decision on the basis of rational criteria, we may simply not yet have developed the appropriate criteria.[10] Hence, on such a general level, the argument about indeterminacy ends in stalemate. Taking quantum holism into consideration supports the following thesis: As soon as we have a concrete case in which scientific and metaphysical questions dovetail, this dovetailing enables a rational evaluation of the options which we have to accommodate the evidence in question.

10.3 Holism and Supervenience

The aim of the last sub-chapter was to stress that holism fits into scientific rationality and to reject the argument which links holism in the philosophy of mind with indeterminacy claims that undermine scientific realism. Let us now consider in positive terms how holism in philosophy of mind can go with scientific realism. Let us turn to supervenience first. Burge and Putnam, in particular, have argued that the social and the physical environment contribute to the individuation of the belief states of a person [4.3]. As a

10 Compare David Lewis, "Radical Interpretation" (1974), in Lewis (1983a), p. 118.

result of their arguments, a materialism that takes belief states to be identical with or at least to supervene on brain states has lost much of its appeal. Most of today's materialists seek to accommodate external factors (e.g. Papineau (1993)). It is widely recognized that we have to broaden the supervenience basis for belief states in such a way that the social and the physical environment are included.

The holism in the philosophy of mind advocated in this book is part of this trend. Social holism is incompatible with the claim that belief states supervene on brain states. To the extent that holism about beliefs is linked with some sort of externalism (be it social externalism or social holism, be it externalism about the physical environment), the same goes for holism about beliefs. Holism in philosophy of mind contradicts only local supervenience claims [5.4.1]. Although holism is incompatible with reductionism in the philosophy of mind, (a) holism does not clash with claims of global supervenience of the mental as a whole on the physical as a whole, and (b) holism can go with the thesis that intentional states are realized as physical states of the person in question in relation to her environment.

Owing to quantum holism, a holism in the philosophy of mind that supports global in contrast to local supervenience is not opposed to, but accords with scientific realism. A theory of the basic level such as quantum theory neither implies nor contradicts claims to the effect that higher order properties supervene on or – what is more – are realized as properties at the basic level. Turning to epistemology, such a theory neither implies nor excludes that theories which treat higher order properties can be reduced to it [9.3.2]. However, quantum holism is incompatible with all those claims of supervenience according to which at the microphysical level, the world is compartimentalized into local matters of particular fact, and everything else supervenes on these local matters. In particular, quantum holism contradicts Lewis' thesis of Humean supervenience [8.2.1]. If we take quantum physics into account, we can go only for a thesis of global supervenience that admits the state of all quantum systems taken together to the supervenience basis [8.4].

Hence, when it comes to supervenience claims, both quantum holism and holism in philosophy of mind have a common consequence: they clash with local supervenience claims, but fit into global supervenience. Thus, in the last resort, i.e., at the microphysical level, science is only compatible with global supervenience. This common consequence of quantum holism and holism in philosophy of mind shows that holism in philosophy of mind fits into a view of the world that bases itself on scientific realism.

10.4 Holism and a New Sort of Realism

Some philosophers who promote the current renaissance of the later Wittgenstein prefer common sense realism to scientific realism: common sense realism is based on arguments from the theory of meaning such as those ones elaborated on in Chapters 3 and 4. But scientific realism meets with reservations. Recall in particular the position of McDowell, who goes as far as proposing a philosophical conception of nature beyond scientific realism [5.3.1].

However, as it is inappropriate to argue against the proposed holism in the philosophy of mind on the basis of scientific realism, so it is inappropriate to link this holism only with common sense realism and have reservations about scientific realism. If one considers science, one realizes that the science which digs most deeply into nature reveals itself a holism. This holism means the same for quantum systems as holism in the philosophy of mind means for beliefs, and it has the same consequence as regards supervenience. Quantum holism thus invalidates arguments from scientific realism against holism in the philosophy of mind. There is hence no need to rely only on common sense realism in distinction from scientific realism when it comes to holism and a revision of the Cartesian tradition in the philosophy of mind. As the argument for limited quantum holism intends to show [9.3], it is reasonable to interpret quantum physics in such a way that the ontological commitments of this science are compatible with a macroscopic level of the familiar definite properties.

We can go one step further: Holism has an effect on our conception of realism itself. This effect is the same in both quantum holism and holism in philosophy of mind. Thus, not only is holism in philosophy of mind not opposed to scientific realism; but that sort of realism to which this holism gives rise can also be employed in science, the interpretation of quantum theory being a paradigmatic example.

Holism in philosophy of mind opens up a new realism which is beyond the traditional antagonism between metaphysical realism and relativism [5.2.3]. This new realism is a direct realism, because it repudiates any epistemic intermediary between belief states and the world; and it is a pragmatic realism, because it is based on an account of meaning in terms of social practices. Social practices determine the meaning of one type of beliefs by determining an inferential context for beliefs of this type, i.e., by determining meaning for a whole network of beliefs. This position accounts for the representational feature of beliefs in terms of inferences. Beliefs represent items in the world only by being integrated into a whole system of beliefs. What goes for meaning also goes for representation: with respect to meaning as well as with respect to representation, a belief is ontologically dependent on there being other beliefs with which it constitutes a whole system of beliefs. Consequently, although this position is a realism in that it takes the truth value of our beliefs to be dependent on the way the world is, it does not invite us to contemplate a relation of correspondence between single beliefs and items in the world. Instead, it invites us to conceive all the beliefs which constitute a system of beliefs taken together as representing or being about the world. We attune ourselves to an independent reality by carrying out changes within a belief system as a whole in our dealings with physical things. These considerations do not have the force of a logical implication of the proposed holism; but they are a plausible way in which the realism that can be combined with that holism can be spelled out.

Approving of quantum holism presupposes a realistic interpretation of quantum theory [8.1]. But the realism presupposed is beyond the traditional opposition of realism and relativism or instrumentalism in the philosophy of science. It is remarkably different from Einstein's realism which seeks to establish a correspondence between single physical quantities and single elements of reality [7.2.1]. Looking for a correspondence between single physical quantities and single elements of reality fails when it comes to developing an interpretation of quantum theory (unless one introduces hidden variables). Therefore, the interpretation of quantum theory was dominated for decades by an attitude which refused to draw ontological consequences from quantum theory. That situation changed when Bell's theorem made possible a mathematical precision of what is now known as quantum holism and when the subsequent experiments confirmed the predictions of quantum theory. The non-classical correlations that violate Bell's theorem and that are confirmed by experiments call for an account which shows what nature is like if it can give rise to such correlations.

The realism which is involved in the acknowledgement of quantum holism takes quantum theory as a whole to tell us something about nature. It asks what the world is like at the microphysical level if quantum theory is right. But, considering quantum holism, there is no point in trying to establish a correspondence between single elements of the theory and single items in the world. The elements of quantum theory have a representational feature and are about the world by being elements of a theory which is as a whole a representation of and about the microphysical realm. Consequently, the experimental evidence does not confirm the predictions of quantum theory in isolation from the means in the formalism by which these predictions are reached; instead, it confirms quantum theory as a whole.

Quantum holism on the ontological level shows up in the mathematical structure of quantum theory. This structure in turn has the following consequence: If one goes for a realistic interpretation of quantum theory as it stands, one can only conceive of quantum theory as a whole as telling us something about nature. The realistic interpretation of quantum theory that is presupposed in speaking of quantum holism therefore fits into the realism which the outlined holism in the philosophy of mind suggests.

However, what goes for the comparison between the proposed holism in philosophy of mind and holism in quantum physics also goes for the comparison between these two cases of realism: the proposed social holism and holism about beliefs are cases of the first type of holism within the general conception – the parts of a system of beliefs have meaning in the first place. By contrast, quantum holism is a case of the second type – some of the properties that make something a quantum system are instantiated by the whole in the first place. The same difference arises when it comes to representation and aboutness: the realism to which the proposed account of meaning is tied allows of single beliefs that they have a representational feature and that they are about something in the world, but only in virtue of being part of a whole system of beliefs. According to the realism to which quantum holism is tied, on the contrary, it is the whole theory which has a representational feature and which is about the microphysical realm in the first place; nonetheless, the way in which the theory is structured accounts for this representational feature, although it is difficult to see how a representational feature could be attributed to each of its elements.

Despite this difference, as regards the philosophy of quantum theory, taking holism in philosophy of mind into account enriches the realistic interpretation to which one is committed by endorsing quantum holism: such a realistic attitude is supported by the realism that holism in the philosophy of mind opens up, because this attitude can be construed as an instance of that realism. As regards the philosophy of mind, taking quantum holism into consideration shows the following: The realism that can be combined with the proposed holism in the philosophy of mind includes not only a direct realism about the objects of perceptual beliefs. Moreover, this realism can be employed in science as well; it is thus also a scientific realism. An interpretation of quantum theory in terms of quantum holism is one concrete example of how the new realism that holism in the philosophy of mind opens up can be made use of in science.

One may envisage going even another step further. If we go beyond quantum mechanics and take quantum field theory into account, we realize that an ontology of single quantum systems, however entangled their states may be, is not the truth of the matter [8.4]. We may have to contemplate an ontology of only one global quantum state which is the state of a comprehensive quantum field. The structure of this global state is such that it introduces a differentiation into parts and properties that make something a part of matter. Thus, in the last resort, there may be only one thing, the global state of a comprehensive quantum field. If we spell out the position that the truth of our beliefs supervenes on the world in terms of truth-makers, there may hence in the last resort be only one truth-maker, namely the global quantum state. It is conceivable that this truth-maker would have an internal differentiation which is specific enough to make some beliefs, in distinction to others, true.

If such an ontology turns out to be defensible, it may lend support to the proposed holism about beliefs in the following way: This holism starts with relations among beliefs, namely inferential relations. It seeks to reconstruct relations between beliefs and the world – such as representing, being true of, referring to – on the basis of inferential relations among beliefs. In case there are good reasons for the sketched ontology, there would in the last resort be only one thing which our beliefs represent, of which our beliefs are true and to which our beliefs refer – under a description that is the content of the beliefs in question. Belief-world relations then would not seem to be a profitable way to approach semantics. Instead, a semantics in terms of inferences would seem to be the appropriate manner to represent the internal structure of that one thing. Thus, if such an ontology turned out to be plausible, this would be another aspect in which the proposed holism in the philosophy of mind fits into a world-view that is shaped by scientific realism.

However, in any case, the argument for an inferential semantics in contrast to a representational semantics has to be made in the philosophy of mind and should not change the subject by invoking such an ontology. Having established such a semantics, one may then set out to show in a further step that the advocated positions in the philosophy of mind fit into a world-view based on scientific realism. This project may then finally lead one to consider the matters touched upon in the last paragraphs.

At first glance, the analysis of positions that are known as holism in contemporary philosophy of mind as well as philosophy of physics does not seem to have much in common with what is known as holism in the history of philosophy – the positions of Parmenides, Spinoza, Hegel, Bradley, to name just the most prominent historical philosophers. But this impression may not be the whole story as the speculations at the end of [8.4] and in the preceding paragraphs show. There is a certain irony here: Analytic philosophy starts in Russell and Moore at the beginning of the twentieth century with rejecting the claims of Bradley in particular. Nonetheless, at the beginning of the twentyfirst century, analytic philosophy may make available an ontology which is somewhat similar to the one of Bradley (see, in particular, Chapter 13 in Bradley (1920)) – namely an ontology of a holism that acknowledges in the last resort only one being, the global quantum state. But note (a) the internal, physical structure of this state and (b) the arguments for this position which come from scientific realism and not from logic and pure metaphysics. This ontological holism matches a social holism and a holism about beliefs in the philosophy of mind – without deviating from the standards of rational, analytical argument, and without amounting to an idealism. However, the purpose of this book was not to embark on such a holism, but the much more limited task of developing a conception of holism and comparing two cases.

To conclude this book by using the Cartesian tradition that dominates modern thought as a background, holism is at the centre of the challenge to this tradition in today's philosophy. I have argued in favour of conceiving this challenge as a revision of the Cartesian tradition instead of a break with that tradition. Holism in philosophy of mind can be reconstructed as sharing a particularly modern starting-point with the Cartesian tradition, i.e., the reflection on what it means to have beliefs. On the basis of this common starting-point, holism in philosophy of mind comes to conclusions that are opposed to central tenets of Cartesian epistemology. As to the philosophy of physics, Cartesian physics already provides for a holism, although one that is not supported by today's physics. Nevertheless, this Cartesian holism is able to set the scene for a judgement on the philosophical impact of quantum holism.

There is not one single area that provides for a revision of the whole Cartesian tradition. Trying to extract a whole philosophy from quantum theory paradoxically implies a Cartesian epistemology. And the revision of Cartesianism in the philosophy of mind yields a direct realism, but is on its own unable to bring about a revision of Cartesian assumptions in our scientific view of nature. Instead of expecting one area to yield a revision of the whole Cartesian tradition, it is reasonable to look for such a revision arising from considerations which are immanent to each area and then to try combine these considerations. The point is not to replace or to complement modern science and rationality with a new paradigm that can be described as holism. The point is an immanent revision of the tradition that dominates modern thought. We should receive the holism both in the philosophy of physics and the philosophy of mind in this vein.

The points which have been considered in this chapter show that the discussions on these two paradigmatic examples of holism are not only relevant to each other, but that these two examples also make a significant common contribution to a contemporary philosophical view of the world and ourselves: they both come to global supervenience, and they both give rise to a new sort of realism. They pave the way for a view of ourselves in nature that is beyond traditional dilemmata such as reductionism versus ontological dualism and metaphysical realism versus relativism.

REFERENCES

Adams, Robert M. (1979): "Primitive Thisness and Primitive Identity". *Journal of Philosophy* **76**, pp. 5–26.

Alanen, Lilli (1992): "Thought-Talk: Descartes and Sellars on Intentionality". *American Philosophical Quarterly* **29**, pp. 19–34.

Albert, David Z. (1992): *Quantum Mechanics and Experience*. Cambridge (Massachusetts): Harvard University Press.

Albert, David Z. and Loewer, Barry (1988): "Interpreting the Many Worlds Interpretation". *Synthese* **77**, pp. 195–213.

Albert, David Z. and Loewer, Barry (1989): "Two No-Collapse Interpretations of Quantum Theory". *Noûs* **23**, pp. 169–186.

Antony, Michael V. (1993): "Social Relations and the Individuation of Thought". *Mind* **102**, pp. 247–261.

Armstrong, David M. (1997): *A World of States of Affairs*. Cambridge: Cambridge University Press.

Arrington, Robert L. and Glock, Hans-Johann (eds.) (1996): *Wittgenstein and Quine*. London: Routledge.

Aspect, Alain, Dalibard, Jean and Roger, Gérard (1982): "Experimental Test of Bell's Inequalities Using Time-Varying Analyzers". *Physical Review Letters* **49**, pp. 1804–1807.

Aspect, Alain and Grangier, Philippe (1985): "Tests of Bell's Inequalities with Pairs of Low Energy Correlated Photons: An Experimental Realization of Einstein-Podolsky-Rosen-Type-Correlations". In: P. J. Lahti and P. Mittelstaedt (eds.): *Symposium on the Foundations of Modern Physics. 50 Years of the Einstein-Podolsky-Rosen Gedankenexperiment*. Singapore: World Scientific. pp. 51–71.

Atmanspacher, Harald and Primas, Hans (1996): "The Hidden Side of Wolfgang Pauli. An Eminent Physicist's Extraordinary Encounter with Depth Psychology". *Journal of Consciousness Studies* **3**, pp. 112–126.

Audretsch, Jürgen (1988): "Ist die Raum-Zeit gekrümmt? Der Aufbau der modernen Gravitationstheorie". In: J. Audretsch und K. Mainzer (eds.): *Philosophie und Physik der Raum-Zeit*. Mannheim: Bibliographisches Institut. pp. 52–82.

Audretsch, Jürgen (1990): "Eine andere Wirklichkeit. Zur Struktur der Quantenmechanik und ihrer Interpretation". In: J. Audretsch und K. Mainzer (eds.): *Wieviele Leben hat Schrödingers Katze?* Mannheim: Bibliographisches Institut. pp. 15–61.

Audretsch, Jürgen und Mainzer, Klaus (eds.) (1988): *Philosophie und Physik der Raum-Zeit*. Mannheim: Bibliographisches Institut.

Audretsch, Jürgen und Mainzer, Klaus (eds.) (1990): *Wieviele Leben hat Schrödingers Katze?* Mannheim: Bibliographisches Institut.

Auyang, Sunny Y. (1995): *How is Quantum Field Theory Possible?* Oxford: Oxford University Press.

Baker, Gordon P. and Hacker, Peter M. S. (1984): *Scepticism, Rules and Language*. Oxford: Blackwell.

Baker, Gordon P. and Hacker, Peter M. S. (1985): *An Analytical Commentary on Wittgenstein's Philosophical Investigations. Volume 2: §§ 185–242. Wittgenstein: Rules, Grammar and Necessity*. Oxford: Blackwell.

Barcan Marcus, Ruth (1993): *Modalities: Philosophical Essays*. Oxford: Oxford University Press.

Barrett, Jeffrey A. (1997): "On Everett's Formulation of Quantum Mechanics". *The Monist* **80**, pp. 70–96.

Barrett, Jeffrey A. (1999): *The Quantum Mechanics of Minds and Worlds*. Oxford: Oxford University Press.

Bell, John S. (1964): "On the Einstein–Podolsky–Rosen-Paradox". *Physics* **1**, pp. 195–200.

Bell, John S. (1987): *Speakable and Unspeakable in Quantum Mechanics*. Cambridge: Cambridge University Press.

Beller, Mara and Fine, Arthur (1994): "Bohr's Response to EPR". In: J. Faye and H. J. Folse (eds.): *Niels Bohr and Contemporary Philosophy*. Dordrecht: Kluwer. pp. 1–31.

Belousek, Darrin W. (1999): "Bell's Theorem, Nonseparability, and Spacetime Individuation in Quantum Mechanics". *Philosophy of Science* **66**, pp. S28–46. Supplementary Volume Proceedings of the Philosophy of Science Association 1998. Part I: Contributed Papers. Edited by Don Howard.

Bennett, Jonathan (1988): *A Study of Spinoza's "Ethics"*. Indianapolis: Hackett. Second printing of the first edition 1984.

Bennett, Jonathan (1991): "Spinoza's Monism: A Reply to Curley". In: Y. Yovel (ed.): *God and Nature: Spinoza's Metaphysics*. Leiden: Brill. pp. 53–59.

Bennett, Jonathan (1994): "Eight Questions about Spinoza". In: Y. Yovel (ed.): *Spinoza on Knowledge and the Human Mind*. Leiden: Brill. pp. 11–25.

Berkovitz, Joseph (1998a): "Aspects of Quantum Non-Locality I: Superluminal Signalling, Action-at-a-Distance, Non-Separability and Holism". *Studies in History and Philosophy of Modern Physics* **29B**, pp. 183-222.

Berkovitz, Joseph (1998b): "Aspects of Quantum Non-Locality II: Superluminal Causation and Relativity". *Studies in History and Philosophy of Modern Physics* **29B**, pp. 509–545.

Bernays, Paul (1948): "Über die Ausdehnung des Begriffes der Komplementarität auf die Philosophie". *Synthese* **7**, pp. 66–70.

Besso, Michele (1948): "Métaphysique expérimentale". *Dialectica* **2**, pp. 45–46.

Bieri, Peter (1982): "Nominalismus und innere Erfahrung". *Zeitschrift für philosophische Forschung* **36**, pp. 3–24.

Bilgrami, Akeel (1992): *Belief and Meaning. The Unity and Locality of Mental Content*. Oxford: Blackwell.

Black, Max (1952): "The Identity of Indiscernibles". *Mind* **61**, pp. 153–164.

Blackburn, Simon (1993): *Essays in Quasi-Realism*. Oxford: Oxford University Press.

Block, Ned (1986): "Advertisement for a Semantics for Psychology". In: P. A. French, T. E. J. Uehling and H. K. Wettstein (eds.): *Studies in the Philosophy of Mind. Midwest Studies in Philosophy. Volume 10*. Minneapolis: University of Minnesota Press. pp. 615–678.

Block, Ned (1995): "An Argument for Holism". *Proceedings of the Aristotelian Society* **95**, pp. 151–169.

Bloor, David (1997): *Wittgenstein, Rules and Institutions*. London: Routledge.

Blumenberg, Hans (1988): *Die Legitimität der Neuzeit*. Frankfurt (M.): Suhrkamp.

Blumenberg, Hans (1991): *The Legitimacy of the Modern Age. Translated by Robert M. Wallace*. Cambridge (Massachusetts): MIT Press. First edition 1983.

Boghassian, Paul A. (1989): "The Rule-Following Considerations". *Mind* **98**, pp. 507–549.

Boghossian, Paul A. (1994): "Sense, Reference and Rule-Following". *Philosophy and Phenomenological Research* **54**, pp. 139–144.

Bohm, David (1951): *Quantum Theory*. Englewood Cliffs: Prentice-Hall.

Bohm, David (1952): "A Suggested Interpretation of the Quantum Theory in Terms of 'Hidden' Variables". *Physical Review* **85**, pp. 166–193.

Bohm, David (1980): *Wholeness and the Implicate Order*. London: Routledge.

Bohm, David and Hiley, Basil (1993): *The Undivided Universe. An Ontological Interpretation of Quantum Theory*. London: Routledge.

Bohr, Niels (1934): *Atomic Theory and the Description of Nature*. Cambridge: Cambridge University Press.

Bohr, Niels (1935): "Can Quantum-Mechanical Description of Physical Reality be Considered Complete?". *Physical Review* **48**, pp. 696–702.

Bohr, Niels (1937): "Causality and Complementarity". *Philosophy of Science* **4**, pp. 289–298.

Bohr, Niels (1948): "On the Notions of Causality and Complementarity". *Dialectica* **2**, pp. 312–319.

Bohr, Niels (1958): *Atomic Physics and Human Knowledge*. New York: Wiley.

Bohr, Niels (1963): *Essays 1958–1962 on Atomic Physics and Human Knowledge*. New York: Wiley.

Bonjour, Laurence (1985): *The Structure of Empirical Knowledge*. Cambridge (Massachusetts): Harvard University Press.

Born, Max (1969): *Albert Einstein, Hedwig und Max Born, Briefwechsel 1916–1955. Kommentiert von Max Born*. München: Nymphenburger Verlagshandlung.

Braddon-Mitchell, David and Jackson, Frank (1997): "The Teleological Theory of Content". *Australasian Journal of Philosophy* **75**, pp. 474–489.

Bradley, Francis Herbert (1920): *Appearance and Reality*. London: Allen and Unwin. Seventh impression. First edition 1893.

Brandl, Johannes (1993): "Semantic Holism is Here to Stay". In: J. Fodor and E. Lepore (eds.): *Holism: A Consumer Update. Grazer Philosophische Studien Volume 46.* Amsterdam: Rodopi. pp. 1–16.

Brandom, Robert B. (1983): "Heidegger's Categories in Being and Time". *Monist* **66**, pp. 387–409.

Brandom, Robert B. (1994): *Making It Explicit. Reasoning, Representing, and Discursive Commitment.* Cambridge (Massachusetts): Harvard University Press.

Brandom, Robert B. (1997): "Study Guide". In: R. B. Brandom and R. Rorty (eds.): *Wilfrid Sellars: Empiricism and the Philosophy of Mind. With an Introduction by Richard Rorty and a Study Guide by Robert Brandom.* Cambridge (Massachusetts): Harvard University Press. pp. 118–181.

Brandom, Robert B. (1998a): "Perception and Rational Constraint". *Philosophy and Phenomenological Research* **58**, pp. 369–374.

Brandom, Robert B. (1998b): "Action, Norms, and Practical Reasoning". In: J. E. Tomberlin (ed.): *Philosophical Perspectives 12: Language, Mind, and Ontology.* Oxford: Blackwell. pp. 127–139.

Brandom, Robert B. (2000): *Articulating Reasons. An Introduction to Inferentialism.* Cambridge (Massachusetts): Harvard University Press.

Brandom, Robert B., McDowell, John, Rosen, Gideon, Rorty, Richard and Rosenberg, Jay F. (1997): "Book Symposium on Robert B. Brandom 'Making It Explicit'". *Philosophy and Phenomenological Research* **57**, pp. 153–204.

Bremer, Józef (1997): *Rekategorisierung statt Reduktion: Zu Wilfrid Sellars' Philosophie des Geistes.* Göttingen: Vandenhoeck and Ruprecht.

Brewer, Bill (1999): *Perception and Reason.* Oxford: Oxford University Press.

Broad, C. D. (1933): *An Examination of McTaggart's Philosophy.* Vol. **1**. Cambridge: Cambridge University Press.

Brody, Nathan and Oppenheim, Paul (1969): "Application of Bohr's Principle of Complementarity to the Mind–Body Problem". *Journal of Philosophy* **56**, pp. 97–113.

Brown, Gregory (1989): "Mathematics, Physics, and Corporeal Substance in Descartes". *Pacific Philosophical Quarterly* **70**, pp. 281–302.

Brown, Harvey R. and Redhead, Michael L. G. (1981): "A Critique of the Disturbance Theory of Indeterminacy in Quantum Mechanics". *Foundations of Physics* **11**, pp. 1–20.

Bunge, Mario (1979): *Treatise on Basic Philosophy. Volume 4. A World of Systems.* Dordrecht: Reidel.

Bunge, Mario (1985): *Treatise on Basic Philosophy. Volume 7. Epistemology and Methodology III: Philosophy of Science and Technology. Part I: Formal and Physical Sciences.* Dordrecht: Reidel.

Burge, Tyler (1979): "Individualism and the Mental". In: P. A. French, T. E. Uehling and H. K. Wettstein (eds.): *Studies in Metaphysics. Midwest Studies in Philosophy. Volume 4.* Minneapolis: University of Minnesota Press. pp. 73–121.

Burge, Tyler (1982): "Other Bodies". In: A. Woodfield (ed.): *Thought and Object*. Oxford: Oxford University Press. pp. 97–120.

Burge, Tyler (1986): "Cartesian Error and the Objectivity of Perception". In: P. Pettit and J. McDowell (eds.): *Subject, Thought, and Context*. Oxford: Oxford University Press. pp. 117–136.

Busch, Paul, Grabowski, Marian and Lahti, Pekka J. (1995): *Operational Quantum Physics*. Berlin: Springer.

Busch, Paul and Lahti, Pekka J. (1985): "A Note on Quantum Theory, Complementarity, and Uncertainty". *Philosophy of Science* **52**, pp. 64–77.

Busch, Paul, Lahti, Pekka J. and Mittelstaedt, Peter (1996): *The Quantum Theory of Measurement*. Berlin: Springer. Second edition. First edition 1991.

Butterfield, Jeremy N. (1985): "Spatial and Temporal Parts". *Philosophical Quarterly* **35**, pp. 32–44.

Butterfield, Jeremy N. (1989a): "The Hole Truth". *British Journal for the Philosophy of Science* **40**, pp. 1–28.

Butterfield, Jeremy N. (1989b): "A Space–Time Approach to the Bell Inequality". In: J. T. Cushing and E. McMullin (eds.): *Philosophical Consequences of Quantum Theory. Reflections on Bell's Theorem*. Notre Dame: University of Notre Dame Press. pp. 114–144.

Butterfield, Jeremy N. (1990): "Causal Independence in EPR Arguments". In: A. Fine, M. Forbes and L. Wessels (eds.): *Proceedings of the 1990 Biennial Meeting of the Philosophy of Science Association*. East Lansing: Philosophy of Science Association. pp. 213–225.

Butterfield, Jeremy N. (1992a): "Bell's Theorem: What It Takes". *British Journal for the Philosophy of Science* **43**, pp. 41–83.

Butterfield, Jeremy N. (1992b): "David Lewis meets John Bell". *Philosophy of Science* **59**, pp. 26–43.

Butterfield, Jeremy N. (1993): "Interpretation and Identity in Quantum Theory". *Studies in History and Philosophy of Science* **24**, pp. 443–476.

Butterfield, Jeremy N. (1995): "Quantum Theory and the Mind. Worlds, Minds and Quanta". *Proceedings of the Aristotelian Society. Supplementary Volume* **59**, pp. 113–158.

Campbell, Keith (1976): *Metaphysics. An Introduction*. Encino: Dickensen.

Campbell, Keith (1994): "Selective Realism in the Philosophy of Physics". *Monist* **77**, pp. 27–46.

Cao, Tian Yu (1997): *Conceptual Developments of 20th Century Field Theories*. Cambridge: Cambridge University Press.

Cao, Tian Yu (ed.) (1999): *Conceptual Foundations of Quantum Field Theory*. Cambridge: Cambridge University Press.

Carrier, Martin (1992): "Aspekte und Probleme kausaler Beschreibungen in der gegenwärtigen Physik". *Neue Hefte für Philosophie* **32/33**, pp. 82–104.

Cartwright, Nancy (1983): *How the Laws of Physics Lie*. Oxford: Oxford University Press.

Cartwright, Nancy (1989): *Nature's Capacities and their Measurement.* Oxford: Oxford University Press.

Cartwright, Nancy (1999): *The Dappled World. A Study of the Boundaries of Science.* Cambridge: Cambridge University Press.

Cartwright, Nancy and Jones, Martin (1991): "How to hunt Quantum Causes". *Erkenntnis* **35**, pp. 205–231.

Chang, Hasok and Cartwright, Nancy (1993): "Causality and Realism in the EPR Experiment". *Erkenntnis* **38**, pp. 169–190.

Child, William (1994a): "On the Dualism of Scheme and Content". *Proceedings of the Aristotelian Society* **94**, pp. 53–71.

Child, William (1994b): *Causality, Interpretation and the Mind.* Oxford: Oxford University Press.

Clauser, John F., Horne, Michael A., Shimony, Abner and Holt, Richard A. (1969): "Proposed Experiment to Test Local Hidden-Variable Theories". *Physical Review Letters* **23**, pp. 880–884.

Cleland, Carol E. (1984): "Space: An Abstract System of Non-Supervenient Relations". *Philosophical Studies* **46**, pp. 19–40.

Cleland, Carol E. (1985): "Causality, Chance and Weak Non-Supervenience". *American Philosophical Quarterly* **22**, pp. 287–298.

Clifford, William K. (1876): "On the Space-Theory of Matter". *Proceedings of the Cambridge Philosophical Society* **2**, pp. 157–158. Reprinted in James R. Newman (ed.): *The World of Mathematics. Volume 1.* New York: Simon and Schuster 1956. pp. 568–569.

Clifton, Robert K., Redhead, Michael L. G. and Butterfield, Jeremy N. (1991): "Generalization of the Greenberger–Horne–Zeilinger Algebraic Proof of Nonlocality". *Foundations of Physics* **21**, pp. 149–184.

Coates, Paul (1997): "Meaning, Mistake and Miscalculation". *Minds and Machines* **7**, pp. 171–197.

Cohen, Jonathan (1999a): "Holism: Some Reasons for Buyer's Remorse". *Analysis* **59**, pp. 63–71.

Cohen, Jonathan (1999b): "Holism, Thought, and the Fate of Metaphysics: Counter-reply to Heal". *Analysis* **59**, pp. 79–85.

Collin, Finn (1998): "Semantic Holism in Social Science". *Philosophical Explorations* **1**, pp. 201–214.

Collins, Arthur W. (1992): "On the Paradox Kripke Finds in Wittgenstein". In: P. A. French, T. E. J. Uehling and H. K. Wettstein (eds.): *The Wittgenstein Legacy. Midwest Studies in Philosophy. Volume 17.* Notre Dame: University of Notre Dame Press. pp. 74–88.

Costa de Beauregard, Olivier (1987): *Time, the Physical Magnitude.* Dordrecht: Reidel.

Cottingham, John (1986): *Descartes.* Oxford: Oxford University Press.

Cottingham, John, Stoothoff, Robert and Murdoch, Dugald (1984–1991): *The Philosophical Writings of Descartes. 3 Volumes.* Cambridge: Cambridge University Press.

Curley, Edwin M. (1969): *Spinoza's Metaphysics: An Essay in Interpretation.* Cambridge (Massachusetts): Harvard University Press.

Curley, Edwin M. (1985): *The Collected Works of Spinoza. Volume 1.* Princeton: Princeton University Press.

Curley, Edwin M. (1988): *Behind the Geometrical Method. A Reading of Spinoza's "Ethics".* Princeton: Princeton University Press.

Curley, Edwin M. (1991): "On Bennett's Interpretation of Spinoza's Monism". In: Y. Yovel (ed.): *God and Nature: Spinoza's Metaphysics.* Leiden: Brill. pp. 35–51.

Cushing, James T. (1994): *Quantum Mechanics: Historical Contingency and the Copenhagen Hegemony.* Chicago: University of Chicago Press.

Cushing, James T. (1997): "It is the Theory Which Decides What We Can Observe". In: R. S. Cohen, M. A. Horne and J. Stachel (eds.): *Experimental Metaphysics. Quantum Mechanical Studies for Abner Shimony. Volume 1.* Dordrecht: Kluwer. pp. 13–23.

Cushing, James T., Fine, Arthur and Goldstein, Sheldon (eds.) (1996): *Bohmian Mechanics and Quantum Theory: an Appraisal.* Dordrecht: Kluwer.

Cushing, James T. and McMullin, Ernan (eds.) (1989): *Philosophical Consequences of Quantum Theory. Reflections on Bell's Theorem.* Notre Dame: University of Notre Dame Press.

d'Espagnat, Bernard (1971): *Conceptual Foundations of Quantum Mechanics.* Menlo Park: Benjamin.

d'Espagnat, Bernard (1991): *A la recherche du réel.* Paris: Gauthier-Villars. Third edition.

d'Espagnat, Bernard (1995): *Veiled Reality.* Reading (Massachusetts): Addison-Wesley. (French edition: *Réel voilé.* Paris: Fayard 1994).

Davidson, Donald (1980): *Essays on Actions and Events.* Oxford: Oxford University Press.

Davidson, Donald (1983): "A Coherence Theory of Truth and Knowledge". In: D. Henrich (ed.): *Kant oder Hegel?* Stuttgart: Klett-Cotta. pp. 423–438. (Reprinted in E. LePore (ed.): *Truth and Interpretation. Perspectives on the Philosophy of Donald Davidson.* Oxford: Blackwell 1986. pp. 307–319).

Davidson, Donald (1984): *Inquiries into Truth and Interpretation.* Oxford: Oxford University Press.

Davidson, Donald (1987): "Problems in the Explanation of Action". In: P. Pettit, R. Sylvan & J. Norman (eds.): *Metaphysics and Morality. Essays in Honour of J.J.C. Smart.* Oxford: Blackwell. pp. 35–49.

Davidson, Donald (1990): "The Structure and Content of Truth". *Journal of Philosophy* **87**, pp. 279–328.

Davidson, Donald (1991a): "Epistemology Externalized". *Dialectica* **45**, pp. 191–202.

Davidson, Donald (1991b): "Three Varieties of Knowledge". In: A. P. Griffiths (ed.): *A. J. Ayer Memorial Essays.* Cambridge: Cambridge University Press. pp. 153–166. Supplement to Philosophy **30** (1991).

Davidson, Donald (1995): "The Problem of Objectivity". *Tijdschrift voor Filosofie* **57**, pp. 203–220.

Davies, P. C. W. and Brown, J. R. (1986): *The Ghost in the Atom*. Cambridge: Cambridge University Press.

Descartes, René (1965–1973): *Œuvres. 11 volumes. Eds. Charles Adam and Paul Tannery.* Paris: Vrin.

Deutsch, David (1997): *Fabric of Reality*. Harmondsworth: Penguin.

Devitt, Michael (1985): "Critical Note of Evans, 'The Varieties of Reference'". *Australasian Journal of Philosophy* **63**, pp. 216–232.

Devitt, Michael (1995): *Coming to Our Senses. A Naturalistic Program for Semantic Localism*. Cambridge: Cambridge University Press.

DeWitt, Bryce S. and Graham, Neill (eds.) (1973): *The Many-Worlds Interpretation of Quantum Mechanics*. Princeton: Princeton University Press.

Dieks, Dennis (1990): "Quantum Statistics, Identical Particles and Correlations". *Synthese* **82**, pp. 127–155.

Dieks, Dennis (1994): "Modal Interpretation of Quantum Mechanics, Measurements, and Macroscopic Behaviour". *Physical Review* **A49**, pp. 2290–2300.

Dieks, Dennis and Vermaas, Pieter E. (1998): *The Modal Interpretation of Quantum Mechanics*. Dordrecht: Kluwer.

Diels, Hermann and Kranz, Walter (eds.) (1960/61): *Die Fragmente der Vorsokratiker. Band 1 und 2*. Berlin: Weidmannsche Verlagsbuchhandlung. 10. Auflage.

Dodd, Julian (1995): "McDowell and Identity Theories of Truth". *Analysis* **55**, pp. 160–165.

Dodd, Julian (1999): "Farewell to States of Affairs". *Australasian Journal of Philosophy* **77**, pp. 146–160.

Donald, Matthew J. (1997): "On Many-Minds Interpretations of Quantum Theory". *Preprint quant-ph/9703008 Nov. 97*.

Dorato, Mauro (1996): "On Becoming, Relativity, and Nonseparability". *Philosophy of Science* **63**, pp. 585–604.

Dowe, Phil (1992): "Wesley Salmon's Process Theory of Causality and the Conserved Quantity Theory". *Philosophy of Science* **59**, pp. 195–216.

Dowe, Phil (1995): "Causality and Conserved Quantities: A Reply to Salmon". *Philosophy of Science* **62**, pp. 321–333.

Dowe, Phil (1996): "Backwards Causation and the Direction of Causal Processes". *Mind* **105**, pp. 227–248.

Dretske, Fred I. (1995): *Naturalizing the Mind*. Cambridge (Massachusetts): MIT Press.

Duhem, Pierre (1954): *The Aim and Structure of Physical Theory. Translated by Philip P. Wiener*. Princeton: Princeton University Press.

Duhem, Pierre (1981): *La théorie physique. Son object – sa structure*. Paris: Vrin. Reprint of the second edition 1914.

Dummett, Michael (1974): "The Significance of Quine's Indeterminacy Thesis". *Synthese* **27**, pp. 351–397.

Dummett, Michael (1981): *Frege. Philosophy of Language*. London: Duckworth. Second edition. First edition 1973.

Dummett, Michael (1991): *The Logical Basis of Metaphysics*. Cambridge (Massachusetts): Harvard University Press.

Dummett, Michael (1993): *Origins of Analytical Philosophy*. London: Duckworth.

Earman, John (1972): "Some Aspects of General Relativity and Geometrodynamics". *Journal of Philosophy* **69**, pp. 634–647.

Earman, John (1989): *World Enough and Space-Time. Absolute versus Relational Theories of Space and Time*. Cambridge (Massachusetts); MIT Press.

Earman, John and Norton, John (1987): "What Price Spacetime Substantivalism? The Hole Story". *British Journal for the Philosophy of Science* **38**, pp. 515–525.

Ebbs, Gary (1997): *Rule-Following and Realism*. Cambridge (Massachusetts): Harvard University Press.

Eccles, John C. (1994): *How the Self Controls Its Brain*. Berlin: Springer.

Einstein, Albert (1948): "Quanten–Mechanik und Wirklichkeit". *Dialectica* **2**, pp. 320–324.

Einstein, Albert (1951): "Autobiographisches". In: P. A. Schilpp (ed.): *Albert Einstein: Philosopher–Scientist*. New York: Tudor. pp. 2–95. Second edition. First edition 1949 by the Library of Living Philosophers, Evanston (Illinois).

Einstein, Albert (1953): "Einleitende Bemerkungen über Grundbegriffe". In: *Louis de Broglie. Physicien et penseur*. Paris: Albin Michel. pp. 4–15.

Einstein, Albert, Podolsky, Boris and Rosen, Nathan (1935): "Can Quantum-Mechanical Description of Physical Reality be Considered Complete?". *Physical Review* **47**, pp. 777–780.

Engel, Pascal (1994): *Donald Davidson et la philosophie du langage*. Paris: Presses Universitaires de France.

Engel, Pascal (forthcoming): "Wherein lies the Normative Dimension in Meaning and Conceptual Content?". Forthcoming in *Philosophical Studies* **97** (2000).

Esfeld, Michael (2000): "Is Quantum Indeterminism Relevant to Free Will?", *Philosophia Naturalis* **37**, pp. 177–187.

Esfeld, Michael (forthcoming a): "Aristotle's Direct Realism in *de Anima*", forthcoming in *The Review of Metaphysics* **54** (2000).

Esfeld, Michael (forthcoming b): "What can Heidegger's *Being and Time* Tell Analytic Philosophy?", forthcoming in *Philosophical Explorations* **4** (2001), first issue.

Esfeld, Michael (forthcoming c): "Gonseth and Quine", forthcoming in *Dialectica* **55** (2001).

Evans, Gareth (1982): *The Varieties of Reference. Ed. John McDowell*. Oxford: Oxford University Press.

Everett, Hugh (1957): "'Relative State' Formulation of Quantum Mechanics". *Reviews of Modern Physics* **29**, pp. 454–462. Reprinted in B. S. DeWitt and N. Graham (eds.): *The Many-Worlds Interpretation of Quantum Mechanics*, Princeton: Princeton University Press 1973, pp. 141–149.

Ewing, A. C. (1934): *Idealism: A Critical Survey*. London: Methuen.

Falkenburg, Brigitte (1994): *Teilchenmetaphysik. Zur Realitätsauffassung in Wissenschaftsphilosophie und Mikrophysik*. Mannheim: Bibliographisches Institut.

Favrholdt, David (1994): "Niels Bohr and Realism". In: J. Faye and H. J. Folse (eds.): *Niels Bohr and Contemporary Philosophy*. Dordrecht: Kluwer. pp. 77–96.

Faye, Jan (1991): *Niels Bohr: His Heritage and Legacy. An Anti-Realist View of Quantum Mechanics*. Dordrecht: Kluwer.

Faye, Jan (1994): "Non-Locality or Non-Separability? A Defense of Bohr's Anti-Realist Approach to Quantum Mechanics". In: J. Faye and H. J. Folse (eds.): *Niels Bohr and Contemporary Philosophy*. Dordrecht: Kluwer. pp. 97–118.

Fine, Arthur (1971): "Probability in Quantum Mechanics and in Other Statistical Theories". In: M. Bunge (ed.): *Problems in the Foundations of Physics*. Berlin: Springer. pp. 79–92.

Fine, Arthur (1982): "Some Local Models for Correlation Experiments". *Synthese* **50**, pp. 279–294.

Fine, Arthur (1986): *The Shaky Game. Einstein, Realism and Quantum Theory*. Chicago: University of Chicago Press.

Fine, Arthur (1989): "Do Correlations need to be explained?". In: J. T. Cushing and E. McMullin (eds.): *Philosophical Consequences of Quantum Theory*. Notre Dame: University of Notre Dame Press. pp. 175–194.

Fine, Kit (1994): "Essence and Modality". In: J. E. Tomberlin (ed.): *Logic and Language. Philosophical Perspectives, Volume 8*. Atascadero: Ridgeview. pp. 1–16.

Fine, Kit (1995): "Ontological Dependence". *Proceedings of the Aristotelian Society* **95**, pp. 269–290.

Fleming, Gordon N. (1992): "The Actualization of Potentialities in Contemporary Quantum Theory". *Journal of Speculative Philosophy* **6**, pp. 259–276.

Fleming, Gordon N. (1996): "Just How Radical is Hyperplane Dependence?". In: R. K. Clifton (ed.): *Perspectives on Quantum Reality*. Dordrecht: Kluwer. pp. 11–28.

Flügel, Martin (1996): "Duhems Holismus". *Philosophia Naturalis* **33**, pp. 143–167.

Fodor, Jerry A. (1990): *A Theory of Content*. Cambridge (Massachusetts): MIT Press.

Fodor, Jerry A. (1998): *Concepts. Where Cognitive Science Went Wrong*. Oxford: Oxford University Press.

Fodor, Jerry A. and Lepore, Ernest (1992): *Holism. A Shopper's Guide*. Oxford: Blackwell.

Fodor, Jerry A. and Lepore, Ernest (eds.) (1993): *Holism: A Consumer Update. Grazer Philosophische Studien Volume 46*. Amsterdam: Rodopi.

Folse, Henry J. (1985): *The Philosophy of Niels Bohr*. Amsterdam: North-Holland.

Folse, Henry J. (1994): "Bohr's Framework of Complementarity and the Realism Debate". In: J. Faye and H. J. Folse (eds.): *Niels Bohr and Contemporary Philosophy*. Dordrecht: Kluwer. pp. 119–139.

Folse, Henry J. (1996): "Ontological Constraints and Understanding Quantum Phenomena". *Dialectica* **50**, pp. 121–136.

Frege, Gottlob (1984): *Collected Papers on Mathematics, Logic, and Philosophy. Edited by Brian McGuinness*. Oxford: Blackwell.

Frege, Gottlob (1986a): *Funktion, Begriff, Bedeutung. Ed. Günther Patzig*. Göttingen: Vandenhoeck und Ruprecht. Sixth edition. First edition 1962.

Frege, Gottlob (1986b): *Logische Untersuchungen. Ed. Günther Patzig*. Göttingen: Vandenhoeck und Ruprecht. Third edition. First edition 1966.

French, Peter A., Uehling, Theodore E. Jr. and Wettstein, Howard K. (1986): *Studies in Essentialism. Midwest Studies in Philosophy. Volume 11*. Minneapolis: University of Minnesota Press.

French, Steven (1989): "Individuality, Supervenience and Bell's Theorem". *Philosophical Studies* **55**, pp. 1–22.

French, Steven and Krause, Décio (1995): "Vague Identity and Quantum Non-Individuality". *Analysis* **55**, pp. 20–26.

French, Steven and Krause, Décio (1999): "The Logic of Quanta". In: T. Y. Cao (ed.): *Conceptual Foundations of Quantum Field Theory*. Cambridge: Cambridge University Press. pp. 324–342.

French, Steven and Redhead, Michael L. G. (1988): "Quantum Physics and the Identity of Indiscernibles". *British Journal for the Philosophy of Science* **39**, pp. 233–246.

Garber, Daniel (1992): *Descartes' Metaphysical Physics*. Chicago: University of Chicago Press.

Geach, P. T. (1972): *Logic Matters*. Oxford: Blackwell.

Gemes, Kenneth (1987): "The World in itself: neither Uniform nor Physical". *Synthese* **73**, pp. 301–318.

Gemes, Kenneth (1991): "The Indeterminacy Thesis Reformulated". *Journal of Philosophy* **88**, pp. 91–108.

Gerhardt, C. J. (ed.) (1875–1890): *Die philosophischen Schriften von G. W. Leibniz. 7 Volumes*. Berlin: Weidmannsche Verlagsbuchhandlung.

Gerrans, Philip (1998): "How to be a Conformist, Part II. Simulation and Rule Following". *Australasian Journal of Philosophy* **76**, pp. 566–586.

Ghiradi, Giancarlo, Grassi, Renata and Pearle, Philip (1991): "Relativistic Dynamical Reduction Models and Nonlocality". In: P. J. Lahti and P. Mittelstaedt (eds.): *Symposium on the Foundations of Modern Physics 1990. Quantum Theory of Measurement and Related Philosophical Problems*. Singapore: World Scientific. pp. 109–123.

Ghiradi, Giancarlo and Grassi, Renata (1994): "Outcome Predictions and Property Attribution: The EPR Argument Reconsidered". *Studies in History and Philosophy of Science* **25**, pp. 397–423.

Ghiradi, Gincarlo, Grassi, Renata and Benatti, F. (1995): "Describing the Macroscopic World: Closing the Circle within the Dynamical Reduction Program". *Foundations of Physics* **25**, pp. 3–38.

Ghiradi, Giancarlo, Rimini, Alberto and Weber, Tullio (1986): "Unified Dynamics for Microscopic and Macroscopic Systems". *Physical Review* **D34**, pp. 470–491.

Ghiradi, Giancarlo and Weber, Tullio (1997): "An Interpretation which is Appropriate for Dynamical Reduction Theories". In: R. S. Cohen, M. A. Horne and J. Stachel (eds.): *Potentiality, Entanglement and Passion-at-a-Distance. Quantum Mechanical Studies for Abner Shimony. Volume 2*. Dordrecht: Kluwer. pp. 89–104.

Gibbard, Allan (1994): "Meaning and Normativity". In: E. Villanueva (ed.): *Philosophical Issues 5: Truth and Rationality*. Atascadero: Ridgeview. pp. 95–115.

Gibbard, Allan (1996): "Thought, Norms, and Discursive Practice: Commentary on Robert Brandom, 'Making It Explicit'". *Philosophy and Phenomenological Research* **56**, pp. 699–717.

Ginet, Carl (1992): "The Dispositionalist Solution to Wittgenstein's Problem about Understanding a Rule: Answering Kripke's Objections". In: P. A. French, T. E. J. Uehling and H. K. Wettstein (eds.): *The Wittgenstein Legacy. Midwest Studies in Philosophy. Volume 17*. Notre Dame: University of Notre Dame Press. pp. 53–73.

Gisin, Nicholas (1991): "Bell's Inequality holds for all Non-product States". *Physics Letters* **A154**, pp. 201–202.

Giulini, Domenico, Joos, Erich, Kiefer, Claus, Kupsch, Joachim, Stamatescu, Ion-Olimpiu and Zeh, H. Dieter (1996): *Decoherence and the Appearance of a Classical World in Quantum Theory*. Berlin: Springer.

Glauser, Richard (1994): "Substance et attribut chez Spinoza". *Studia Philosophica* **53**, pp. 225–248.

Gleason, Andrew M. (1957): "Measures on the Closed Subspaces of a Hilbert Space". *Journal of Mathematics and Mechanics* **6**, pp. 885–893. Reprinted in Clifford A. Hooker (ed.): *The Logico-Algrebraic Approach to Quantum Mechanics*. Volume 1. Dordrecht: Reidel 1975, pp. 123–133.

Globus, Gordon G. (1976): "Mind, Structure, and Contradiction". In: G. G. Globus, G. Maxwell and I. Savodnik (eds.): *Consciousness and the Brain*. New York: Plenum Press. pp. 271–293.

Gochet, Paul (1977): "Being, Truth and Meaning in Quine's Philosophy". *Philosophica* **19**, pp. 119–135.

Goldfarb, Warren (1985): "Kripke on Wittgenstein on Rules". *Journal of Philosophy* **82**, pp. 471–488.

Goodman, Nelson (1983): *Fact, Fiction, and Forecast*. Cambridge (Massachusetts): Harvard University Press. Fourth edition. First edition 1954.

Gonseth, Ferdinand (1948): "A propos des exposés de MM. Ph. Devaux et E. W. Berth". *Dialectica* **2**, pp. 120–125.

Graham, Peter J. (1999): "Brandom on Singular Terms". *Philosophical Studies* **93**, pp. 247–264.

Graves, John C. (1971): *The Conceptual Foundations of Contemporary Relativity Theory*. Cambridge (Massachusetts): MIT Press.

Greenberger, Daniel M., Horn, Michael A., Shimony, Abner and Zeilinger, Anton (1990): "Bell's Theorem without Inequalities". *American Journal of Physics* **58**, pp. 1131–1143.

Haag, Rudolf (1992): *Local Quantum Physics*. Berlin: Springer.

Habermas, Jürgen (1999): *Wahrheit und Rechtfertigung. Philosophische Aufsätze*. Frankfurt (Main): Suhrkamp.

Hacker, Peter M. S. (1982): "Events and Objects in Space and Time". *Mind* **91**, pp. 1–19.

Hacking, Ian (1975): "The Identity of Indiscernibles". *Journal of Philosophy* **72**, pp. 249–256.

Hagley, E., Maître, X., Nogues, G., Wunderlich, C., Brune, M., Raimond, J. M. and Haroche, S. (1997): "Generation of Einstein-Podolsky-Rosen Pairs of Atoms". *Physical Review Letters* **79**, pp. 1–5.

Hanson, N. R. (1958): *Patterns of Discovery*. Cambridge: Cambridge University Press.

Harman, Gilbert (1982): "Conceptual Role Semantics". *Notre Dame Journal of Formal Logic* **23**, pp. 242–256.

Harris, Errol E. (1973): *Salvation from Despair: A Reappraisal of Spinoza's Philosophy*. The Hague: Nijhoff.

Hartz, Glenn A. (1989): "Leibniz on why Descartes' Metaphysics of Body is Necessarily False". In: N. Rescher (ed.): *Leibnizian Inquiries: a Group of Essays*. Lanham: University of America Press. pp. 23–36.

Haugeland, John (1982): "Heidegger on Being a Person". *Noûs* **16**, pp. 15–28.

Haugeland, John (1998): *Having Thought. Essays in the Metaphysics of Mind*. Cambridge (Massachusetts): Harvard University Press.

Harman, Gilbert (1982): "Conceptual Role Semantics". *Notre Dame Journal of Formal Logic* **23**, pp. 242–256.

Haukioja, Jussi (1999): *Rule-Following, Response-Dependence and Realism*. Ph.D. Thesis, University of Turku, Finland.

Hawthrone, James and Silberstein, Michael (1995): "For Whom the Bell Arguments Toll". *Synthese* **102**, pp. 99–138.

Heal, Jane (1989): *Fact and Meaning. Quine and Wittgenstein on Philosophy of Language*. Oxford: Blackwell.

Heal, Jane (1994): "Semantic Holism: Still a Good Buy". *Proceedings of the Aristotelian Society* **94**, pp. 325–339.

Heal, Jane (1999): "Thoughts and Holism: Reply to Cohen". *Analysis* **59**, pp. 72–78.

Healey, Richard A. (1989): *The Philosophy of Quantum Mechanics. An Interactive Interpretation*. Cambridge: Cambridge University Press.

Healey, Richard A. (1991): "Holism and Nonseparability". *Journal of Philosophy* **88**, pp. 393–421.

Healey, Richard A. (1992): "Discussion: Causation, Robustness, and EPR". *Philosophy of Science* **59**, pp. 282–292.

Healey, Richard A. (1994): "Nonseparable Processes and Causal Explanation". *Studies in History and Philosophy of Science* **25**, pp. 337–374.

Healey, Richard A. (1997): "Nonlocality and the Aharonov-Bohm Effect". *Philosophy of Science* **64**, pp. 18–41.

Hedrich, Reiner (1990): *Komplexe und fundamentale Strukturen: Grenzen des Reduktionismus*. Mannheim: Bibliographisches Institut.

Heil, John (1999): "Multiple Realizability". *American Philosophical Quarterly* **36**, pp. 189–208.

Heisenberg, Werner (1984): *Gesammelte Werke / Collected Works. Abteilung / Series C: Allgemeinverständliche Schriften / Philosophical and Popular Writings. Band I: Physik und Erkenntnis 1927–1955. Eds. Walter Blum, Hans-Peter Dürr, Helmut Rechenberg.* München: Piper.

Heisenberg, Werner (1989): *Physics and Philosophy*. Harmondsworth: Penguin. First edition New York: Harper and Row 1958.

Held, Carsten (1994): "The Meaning of Complementarity". *Studies in History and Philosophy of Science* **25**, pp. 871–893.

Held, Carsten (1998): *Die Bohr-Einstein-Debatte. Quantenmechanik und physikalische Wirklichkeit*. Parderborn: Schöningh.

Heller, Mark (1990): *The Ontology of Physical Objects: Four-Dimensional Hunks of Matter*. Cambridge: Cambridge University Press.

Hellman, Geoffrey (1987): "EPR, Bell and Collapse: a Route Around 'Stochastic' Hidden Variables". *Philosophy of Science* **54**, pp. 558–576.

Hentschel, Klaus (1987): "Einstein, Neokantianismus und Theorienholismus". *Kant-Studien* **78**, pp. 459–470.

Hesse, Mary B. (1961): *Forces and Fields. The Concept of Action at a Distance in the History of Physics*. London: Nelson.

Heywood, Peter and Redhead, Michael L. G. (1983): "Nonlocality and the Kochen-Specker Paradox". *Foundations of Physics* **13**, pp. 481–499.

Hösle, Vittorio (1990): *Die Krise der Gegenwart und die Verantwortung der Philosophie*. München: Beck.

Hofer-Szabo, Gábor, Redei, Miklós & Szabo, László E. (1999): "On Reichenbach's Common Cause Principle and Reichenbach's Notion of Common Cause". *British Journal for the Philosophy of Science* **50**, pp. 377–399.

Hoffman, Joshua and Rosenkrantz, Gary S. (1994): *Substance Among Other Categories*. Cambridge: Cambridge University Press.

Hoffman, Paul (1985): "Kripke on Private Languages". *Philosophical Studies* **47**, pp. 23–28.

Holland, Peter R. (1993): *The Quantum Theory of Motion. An Account of the de Broglie—Bohm Causal Interpretation of Quantum Mechanics*. Cambridge: Cambridge University Press.

Holtzman, Steven H. and Leich, Christopher M. (1979): "Introductory Essay: Communal Agreement and Objectivity". In: S. H. Holtzman and C. M. Leich (eds.): *Wittgenstein: to Follow a Rule*. London: Routledge. pp. 1–27.

Honner, J. (1987): *The Description of Nature. Niels Bohr and the Philosophy of Quantum Physics*. Oxford: Oxford University Press.

Hookway, Christopher (1988): *Quine. Language, Experience, and Reality*. Cambridge: Polity Press.

Horgan, Terence (1993): "From Supervenience to Superdupervenience: Meeting the Demands of a Material World". *Mind* **102**, pp. 555–586.

Horstmann, Rolf-Peter. (1984): *Ontologie und Relationen. Hegel, Bradley, Russell und die Kontroverse über interne und externe Beziehungen*. Königstein (Taunus): Athenäum.

Horwich, Paul (1998): *Meaning*. Oxford: Oxford University Press.

Howard, Don (1985): "Einstein on Locality and Separability". *Studies in History and Philosophy of Science* **16**, pp. 171–201.

Howard, Don (1989): "Holism, Separability, and the Metaphysical Implications of the Bell Experiments". In: J. T. Cushing and E. McMullin (eds.): *Philosophical Consequences of Quantum Theory. Reflections on Bell's Theorem*. Notre Dame: University of Notre Dame Press. pp. 224–253.

Howard, Don (1990): "Einstein and Duhem". *Synthese* **83**, pp. 363–384.

Howard, Don (1992): "Locality, Separability, and the Physical Implications of the Bell Experiments". In: A. v. d. Merwe, F. Selleri and G. Tarozzi (eds.): *Bell's Theorem and the Foundations of Modern Physics*. Singapore: World Scientific. pp. 306–314.

Howard, Don (1997): "Space-time and Separability: Problems of Identity and Individuation in Fundamental Physics". In: R. S. Cohen, M. A. Horne and J. Stachel (eds.): *Potentiality, Entanglement and Passion-at-a-Distance. Quantum Mechanical Studies for Abner Shimony. Volume 2*. Dordrecht: Kluwer. pp. 113–141.

Hughes, Richard I. G. (1989): *The Structure and Interpretation of Quantum Mechanics*. Cambridge (Massachusetts): Harvard University Press.

Humberstone, I. Lloyd (1996): "Intrinsic / Extrinsic". *Synthese* **108**, pp. 205–267.

Hurley, S. L. (1992): "Intelligibility, Imperialism, and Conceptual Scheme". In: P. A. French, T. E. J. Uehling and H. K. Wettstein (eds.): *The Wittgenstein Legacy. Midwest Studies in Philosophy. Volume 17*. Notre Dame: University of Notre Dame Press. pp. 89–108.

Husserl, Edmund (1970): *Logical Investigations. Translated by J. N. Findlay*. London: Routledge.

Husserl, Edmund (1984): *Logische Untersuchungen. Zweiter Band. Erster Teil. Husserliana Volume 19.1. Ed. Ursula Panzer*. The Hague: Nijhoff.

Hyland, M. E. (1985): "Do Person Variables exist in Different Ways?". *American Psychologist* **40**, pp. 1003–1010.

Jackson, Frank (1998a): *From Metaphysics to Ethics. A Defence of Conceptual Analysis*. Oxford: Oxford University Press.

Jackson, Frank (1998b): *Mind, Method and Conditionals. Selected Essays*. London: Routledge.

Jackson, Frank (1998c): "Reference and Description Revisited". In: J. E. Tomberlin (ed.): *Philosophical Perspectives 12: Language, Mind, and Ontology*. Oxford: Blackwell. pp. 201–218.

Jacob, Pierre (1997): *What Minds can Do. Intentionality in a Non-Intentional World*. Cambridge: Cambridge University Press.

Jammer, Max (1974): *The Philosophy of Quantum Mechanics*. New York: Wiley.

Jarrett, Jon P. (1984): "On the Physical Significance of the Locality Conditions in the Bell Arguments". *Noûs* **18**, pp. 569–589.

Jarrett, Jon P. (1989): "Bell's Theorem: A Guide to the Implications". In: J. T. Cushing and E. McMullin (eds.): *Philosophical Consequences of Quantum Theory. Reflections on Bell's Theorem*. Notre Dame: University of Notre Dame Press. pp. 60–79.

Johnston, Mark (1989): "Dispositional Theories of Value". *Proceedings of the Aristotelian Society. Supplementary Volume* **63**, pp. 139–174.

Johnston, Mark (1993): "Objectivity Refigured: Pragmatism Without Verificationism". In: J. Haldane and C. Wright (eds.): *Reality, Representation, and Projection*. Oxford: Oxford University Press. pp. 85–130.

Johnston, Mark (1998): "Are Manifest Qualities Response-Dependent?". *Monist* **81**, pp. 3–43.

Jonas, Hans (1980): "Parallelism and Complementarity: The Psycho-Physical Problem in Spinoza and in the Succession of Niels Bohr". In: R. Kennington (ed.): *The Philosophy of Baruch Spinoza*. Washington: The Catholic University of America Press. pp. 121–130.

Jonas, Hans (1987): *Macht oder Ohnmacht der Subjektivität?* Frankfurt (Main): Suhrkamp.

Jones, Martin R. and Clifton, Robert K. (1993): "Against Experimental Metaphysics". *Midwest Studies in Philosophy* **18**, pp. 295–316.

Jordan, Pascual (1951): *Verdrängung und Komplementarität*. Hamburg-Bergedorf: Stromverlag. Second edition. First edition 1947.

Kafatos, Menas and Nadeau, Robert (1990): *The Conscious Universe. Part and Whole in Modern Physical Theory*. New York: Springer.

Kant, Immanuel (1998): *Critique of Pure Reason. Translated by Paul Guyer and Allen W. Wood*. Cambridge: Cambridge University Press.

Kanitscheider, Bernulf (1976): *Vom absoluten Raum zur dynamischen Geometrie*. Zürich: Bibliographisches Institut.

Kanitscheider, Bernulf (1988): *Das Weltbild Albert Einsteins*. München: Beck.

Kanitscheider, Bernulf (1993): *Von der mechanistischen Welt zum kreativen Universum*. Darmstadt: Wissenschaftliche Buchgesellschaft.

Kaplan, David (1989a); "Demonstratives. An Essay on the Semantics, Logic, Metaphysics, and Epistemology of Demonstratives and Other Indexicals". In: J. Almog, J. Perry and H. Wettstein (eds.): *Themes from Kaplan*. Oxford: Oxford University Press. pp. 481–563.

Kaplan, David (1989b): "Afterthoughts". In: J. Almog, J. Perry and H. Wettstein (eds.): *Themes from Kaplan*. Oxford: Oxford University Press. pp. 565–614.

Katz, Jerrold J. (1990): *The Metaphysics of Meaning*. Cambridge (Massachusetts): MIT Press.

Keeling, S. V. (1934): "Cartesian Mechanism". *Philosophy* **9**, pp. 51–66. Reprinted in Georges J. D. Moyal (ed.): *René Descartes. Critical Assessments. 4 volumes*, London: Routledge 1991. Vol. 4. pp. 67–80.

Kenny, Anthony (1968): *Descartes. A Study of his Philosophy*. New York: Random House.

Kim, Jaegwon (1993): *Supervenience and Mind: Selected Philosophical Essays*. Cambridge: Cambridge University Press.

Kirsch, I. and Hyland, M. E. (1987): "How Thoughts Affect the Body: A Metatheoretical Framework". *Journal of Mind and Behavior* **8**, pp. 417–434.

Kistler, Max (1998): "Reducing Causality to Transmission". *Erkenntnis* **48**, pp. 1–24.

Kistler, Max (1999): *La causalité et les lois de la nature*. Paris: Vrin.

Kobayashi, Michio (1993): *La philosophie naturelle de Descartes*. Paris: Vrin.

Kober, Michael (1998a): "Kripkensteins Bedeutung". *Philosophisches Jahrbuch* **105**, pp. 45–59.

Kober, Michael (1998b): "Kripkenstein Meets the Chinese Room: Looking for the Place of Meaning from a Natural Point of View". *Inquiry* **41**, pp. 317–332.

Kochen, Simon (1985): "A New Interpretation of Quantum Mechanics". In: P. J. Lahti and P. Mittelstaedt (eds.): *Symposium on the Foundations of Modern Physics. 50 Years of the Einstein-Podolsky-Rosen Gedankenexperiment*. Singapore: World Scientific. pp. 151–169.

Kochen, Simon and Specker, Ernst (1967): "The Problem of Hidden Variables in Quantum Mechanics". *Journal of Mathematics and Mechanics* **17**, pp. 59–87. Reprinted in Clifford A. Hooker (ed.): *The Logico-Algrebraic Approach to Quantum Mechanics. Volume 1*. Dordrecht: Reidel 1975, pp. 293–328.

Kowalski, Tomasz and Placek, Tomasz (1999): "Outcomes in Branching Space-Time and GHZ-Bell Theorems". *British Journal for the Philosophy of Science* **50**, pp. 349–375.

Krause, Décio and French, Steven (1995): "A Formal Framework for Quantum Non-Individuality". *Synthese* **102**, pp. 195–214.

Kripke, Saul A. (1980): *Naming and Necessity*. Oxford: Blackwell.

Kripke, Saul A. (1982): *Wittgenstein on Rules and Private Language*. Oxford: Blackwell.

Krips, Henry (1982): "Epistemological Holism: Duhem or Quine?". *Studies in History and Philosophy of Science* **13**, pp. 251–264.

Kronz, Frederick M. (1990): "Hidden Locality, Conspiracy and Superluminal Signals". *Philosophy of Science* **57**, pp. 420–444.

Kuhlmann, Meinard (1999): "Quanta and Tropes. Trope Ontology as Descriptive Metaphysics of Quantum Field Theory". In: U. Meixner and P. Simons (eds.): *Metaphysics in the Post-Metaphysical Age. Papers of the 22nd International Wittgenstein Symposium. Contributions of the Austrian Ludwig Wittgenstein Society. Vol. VII (1).* Kirchberg: Austrian Ludwig Wittgenstein Society. pp. 338–343.

Kuhn, Thomas S. (1970): *The Structure of Scientific Revolutions.* Chicago: University of Chicago Press. Second edition. First edition 1962.

Lance, Mark and O'Leary-Hawthrone, John (1997): *The Grammar of Meaning.* Cambridge: Cambridge University Press.

Landsman, Nicholas P. (1995): "Observation and Superselection in Quantum Mechanics". *Studies in History and Philosophy of Modern Physics* **26B**, pp. 45–73.

Langton, Rae and Lewis, David (1998): "Defining 'Intrinsic'". *Philosophy and Phenomenological Research* **58**, pp. 333–345.

Laudisa, Federico (1995): "Einstein, Bell, and Nonseparable Realism". *British Journal for the Philosophy of Science* **46**, pp. 309–329.

Laudisa, Federico (1999): "A Note on Nonlocality, Causation, and Lorentz Invariance". *Philosophy of Science* **66**, pp. S72–81. Supplementary Volume Proceedings of the Philosophy of Science Association 1998. Part I: Contributed Papers. Edited by Don Howard.

Laurikainen, K. V. (1988): *Beyond the Atom. The Philosophical Thought of Wolfgang Pauli.* Berlin: Springer.

Lecrivain, A. (1986): "Spinoza and Cartesian Mechanics". In: M. Grene and D. Nails (eds.): *Spinoza and the Sciences.* Dordrecht: Reidel. pp. 15-60.

Lehner, Christoph (1997): "What it feels like to be in a Superposition and Why". *Synthese* **110**, pp. 191–216.

Lehrer, Keith (1990): *Theory of Knowledge.* Boulder (Colorado): Westview Press.

Lemmon, E. J. (1967): "Comments on Donald Davidson's 'The Logical Form of Action Sentences'". In: N. Rescher (ed.): *The Logic of Decision and Action.* Pittsburgh: University of Pittsburgh Press. pp. 96–103.

Lepore, Ernest, Fodor, Jerry, Devitt, Michael, Rey, Georges, McLaughlin, Brian P., Brandom, Robert B. and Churchland, Paul M. (1993): "Book Symposium on Jerry Fodor / Ernest Lepore 'Holism. A Shopper's Guide'". *Philosophy and Phenomenological Research* **53**, pp. 637–682.

Levi, Isaac (1996): "Review of Robert B. Brandom, Making It Explicit, Cambridge (Massachusetts): Harvard University Press 1994". *Journal of Philosophy* **93**, pp. 145–158.

Lewis, David (1983a): *Philosophical Papers.* Vol. 1. Oxford: Oxford University Press.

Lewis, David (1983b): "New Work for a Theory of Universals". *Australasian Journal of Philosophy* **61**, pp. 343–377.

Lewis, David (1986a): *Philosophical Papers*. Vol. 2. Oxford: Oxford University Press.

Lewis, David (1986b): *On the Plurality of Worlds*. Oxford: Blackwell.

Lewis, David (1997): "Finkish Dispositions". *Philosophical Quarterly* **47**, pp. 145–158.

Linsky, Leonard (ed.) (1971): *Reference and Modality*. London: Oxford University Press.

Liu, Chuang (1996): "Holism vs. Particularism: A Lesson from Classical and Quantum Physics". *Journal for General Philosophy of Science / Zeitschrift für allgemeine Wissenschaftstheorie* **27**, pp. 267–279.

Lockwood, Michael (1989): *Mind, Brain and the Quantum. The Compound 'I'*. Oxford: Blackwell.

Lockwood, Michael (1996): "'Many-Minds' Interpretations of Quantum Mechanics". *British Journal for the Philosophy of Science* **47**, pp. 159–188.

Loewer, Barry (1996): "Humean Supervenience". *Philosophical Topics* **24**, pp. 101–127.

Lormand, Eric (1996): "How to be a Meaning Holist". *Journal of Philosophy* **93**, pp. 51–73.

Lowe, E. Jonathan (1994): "Ontological Dependency". *Philosophical Papers* **23**, pp. 31–48.

Lowe, E. Jonathan (1998): "Entity, Identity and Unity". *Erkenntnis* **48**, pp. 191–208.

Mainzer, Klaus (1988): *Symmetrien der Natur. Ein Handbuch zur Natur- und Wissenschaftsphilosophie*. Berlin: de Gruyter.

Malpas, Jeff E. (1992): *Donald Davidson and the Mirror of Meaning. Holism, Truth, Interpretation*. Cambridge: Cambridge University Press.

Manning, R. N. (1995): "Interpreting Davidson's Omniscient Interpreter". *Canadian Journal of Philosophy* **25**, pp. 335–374.

Margenau, Henry (1944): "The Exclusion Principle and its Philosophical Importance". *Philosophy of Science* **11**, pp. 187–208.

Margenau, Henry (1950): *The Nature of Physical Reality. A Philosophy of Modern Physics*. New York: McGraw-Hill.

Margenau, Henry (1984): *The Miracle of Existence*. Woodbridge (Connecticut): Ox Bow Press.

Marshall, T. W., Santos, E. and Selleri, F. (1985): "On the Compatibility of Local Realism with Atomic Cascade Experiments". In: G. Tarozzi and A. v. d. Merwe (eds.): *Open Questions in Quantum Physics. Invited Papers on the Foundations of Microphysics*. Dordrecht: Reidel. pp. 87–101.

Massey, Gerald J. (1992): "The Indeterminacy of Translation: A Study in Philosophical Exegesis". *Philosophical Topics* **20**, pp. 317–345.

Mathews, Freya (1989): "Some Reflections on Spinoza's Theory of Substance". *Philosophia* **19**, pp. 3–21.

Maudlin, Tim (1994): *Quantum Non-Locality and Relativity*. Oxford: Blackwell.

Maudlin, Tim (1996): "Space-Time in the Quantum World". In: J. T. Cushing, A. Fine and S. Goldstein (eds.): *Bohmian Mechanics and Quantum Theory: an Appraisal*. Dordrecht: Kluwer. pp. 285–307.

Maudlin, Tim (1997): "Describing the World in the Wave Function". *The Monist* **80**, pp. 3–23.

Mayer, Verena (1997): *Semantischer Holismus. Eine Einführung*. Berlin: Akademie-Verlag.

McCulloch, Gregory (1989): *The Game of the Name*. Oxford: Oxford University Press.

McCulloch, Gregory (1995): *The Mind and its World*. London: Routledge.

McDowell, John (1984): "Wittgenstein on Following a Rule". *Synthese* **58**, pp. 325–363.

McDowell, John (1986): "Singular Thought and the Extent of Inner Space". In: P. Pettit and J. McDowell (eds.): *Subject, Thought, and Context*. Oxford: Oxford University Press. pp. 137–168.

McDowell, John (1992): "Meaning and Intentionality in Wittgenstein's later Philosophy". In: P. A. French, T. E. J. Uehling and H. K. Wettstein (eds.): *The Wittgenstein Legacy. Midwest Studies in Philosophy. Volume 17*. Notre Dame: University of Notre Dame Press. pp. 40–52.

McDowell, John (1996): *Mind and World. With a New Introduction*. Cambridge (Massachusetts): Harvard University Press. First edition 1994.

McDowell, John (1997): "Brandom on Representation and Inference". *Philosophy and Phenomenological Research* **57**, pp. 157–162.

McDowell, John (1998): "Having the World in View: Sellars, Kant, and Intentionality". *Journal of Philosophy* **95**, pp. 431–490.

McDowell, John, Brandom, Robert B., Collins, Arthur W., Peacocke, Christopher, Rorty, Richard and Wright, Crispin (1998): "Book Symposium on John McDowell 'Mind and World'". *Philosophy and Phenomenological Research* **58**, pp. 365–431.

McGinn, Colin (1984): *Wittgenstein on Meaning. An Interpretation and Evaluation*. Oxford: Blackwell. Aristotelian Society Series Volume 1.

McGinn, Colin (1989): *Mental Content*. Oxford: Blackwell.

Mellor, D. Hugh (1971): *The Matter of Chance*. Cambridge: Cambridge University Press.

Mellor, D. Hugh (1981): *Real Time*. Cambridge: Cambridge University Press.

Mellor, D. Hugh (1996): "Natural Kinds". In: A. Pessin and S. Goldberg (eds.): *The Twin Earth Chronicles. Twenty Years of Reflection on Hilary Putnam's "The Meaning of 'Meaning'"*. Armonk, New York: Sharpe. pp. 69–80. First published in the *British Journal for the Philosophy of Science* **28** (1977), pp. 299–312.

Mermin, N. David (1986): "The EPR Experiment—Thoughts about the "Loophole"". In: D. M. Greenberger (ed.): *New Techniques and Ideas in Quantum Measurement Theory*. New York: New York Academy of Sciences. pp. 422–427. (Annals of the New York Academy of Sciences 480).

Mermin, N. David (1990): "Simple Unified Form for the Major No-Hidden-Variables Theorems". *Physical Review Letters* **65**, pp. 3373–3376.

Mermin, N. David (1998): "What is Quantum Mechanics Trying to Tell Us?". *American Journal of Physics* **66**, pp. 753–767.

Meyer-Abich, Klaus Michael (1965): *Korrespondenz, Individualität und Komplementarität. Eine Studie zur Geistesgeschichte der Quantentheorie in den Beiträgen Niels Bohrs.* Wiesbaden: Steiner.

Millikan, Ruth Garrett (1990): "Truth Rules, Hoverflies, and the Kripke–Wittgenstein Paradox". *Philosophical Review* **99**, pp. 323–353.

Miscevic, Nenad (1996): "Computationalism and Rule Following". *Proceedings of the Aristotelian Society* **96**, pp. 215–229.

Miscevic, Nenad (1996/97): "The Skeptic and the Hoverfly (A teleo-computational view of rule-following)". *Acta Analytica* **16**, pp. 171–187.

Misner, Charles W. (1974): "Some Topics for Philosophical Inquiry Concerning the Theories of Mathematical Geometrodynamics and of Physical Geometrodynamics". In: K. F. Schaffner and R. S. Cohen (eds.): *PSA 1972. Proceedings of the 1972 Biennial Meeting of the Philosophy of Science Association.* Dordrecht: Reidel. pp. 7–29. Boston Studies in the Philosophy of Science. Volume 20.

Misner, Charles W., Thorne, Kip S. and Wheeler, John A. (1973): *Gravitation.* San Francisco: Freeman.

Mittelstaedt, Peter (1998a): *The Interpretation of Quantum Mechanics and the Measurement Process.* Cambridge: Cambridge University Press.

Mittelstaedt, Peter (1998b): "Can EPR-Correlations be used for the Transmission of Superluminal Signals?". *Annalen der Physik* **7**, pp. 710–715.

Monk, Nicholas A. M. (1997): "Conceptions of Space–Time: Problems and Possible Solutions". *Studies in History and Philosophy of Modern Physics* **28**, pp. 1–34.

Moulines, Carlos Ulises (1986): "The Ways of Holism". *Noûs* **20**, pp. 313–330.

Murdoch, Dugald (1987): *Niels Bohr's Philosophy of Physics.* Cambridge: Cambridge University Press.

Nadler, Steven M. (1989): *Arnauld and the Cartesian Philosophy of Ideas.* Manchester: Manchester University Press.

Nagel, Ernest (1961): *The Structure of Science. Problems in the Logic of Scientific Explanation.* London: Routledge.

Nagel, Thomas (1986): *The View from Nowhere.* Oxford: Oxford University Press.

Nason, J. W. (1946): "Leibniz's Attack on the Cartesian Doctrine of Extension". *Journal of the History of Ideas* **7**, pp. 447–483.

Needham, Paul (2000): "Duhem and Quine". *Dialectica* **54**, pp. 109–132.

Nerlich, Graham (1994): *What Spacetime Explains. Metaphysical Essays on Space and Time.* Cambridge: Cambridge University Press.

Newton, Isaac (1952): *Opticks. Ed. I. B. Cohen.* New York: Dover.

Newton, Issac (1961): *Correspondence. Volume II. Ed. H. W. Turnbull.* Cambridge: Cambridge University Press.

Okasha, Samir (2000): "Holism about Meaning and about Evidence: in Defence of W. V. Quine". *Erkenntnis* **52**, pp. 39–61.

Omnès, Roland (1994a): *The Interpretation of Quantum Mechanics*. Princeton: Princeton University Press.

Omnès, Roland (1994b): *Philosophie de la science contemporaine*. Paris: Gallimard.

Pagin, Peter (1997): "Is Compositionality Compatible with Holism?". *Mind and Language* **12**, pp. 11–33.

Papineau, David (1979): *Theory and Meaning*. Oxford: Oxford University Press.

Papineau, David (1993): *Philosophical Naturalism*. Oxford: Blackwell.

Pascazio, S. (1988): "Variable Detection Probability Models for Einstein–Podolsky–Rosen-Type Experiments". In: F. Selleri (ed.): *Quantum Mechanics versus Local Realism*. New York: Plenum. pp. 391–411.

Paty, Michel (1986): "Einstein and Spinoza". In: M. Grene and D. Nails (eds.): *Spinoza and the Sciences*. Dordrecht: Reidel. pp. 267–302.

Paty, Michel (1988): "Einstein et Spinoza". In: R. Bouveresse (ed.): *Spinoza. Science et religion*. Paris: Vrin. pp. 183–207.

Pauli, Wolfgang (1948): "Editorial". *Dialectica* **2**, pp. 307–311.

Pauli, Wolfgang (1952): "Der Einfluss archetypischer Vorstellungen auf die Bildung naturwissenschaftlicher Theorien bei Kepler". In: C. G. Jung and W. Pauli (eds.): *Naturerklärung und Psyche*. Zürich: Rascher. pp. 109–194.

Pauli, Wolfgang (1961): *Aufsätze und Vorträge über Physik und Erkenntnistheorie*. Braunschweig: Vieweg.

Peacocke, Christopher (1981): "Reply: Rule-Following: the Nature of Wittgenstein's Arguments". In: S. H. Holtzman and C. M. Leich (eds.): *Wittgenstein: to Follow a Rule*. London: Routledge. pp. 72–95.

Peacocke, Christopher (1992): *A Study of Concepts*. Cambridge (Massachusetts): MIT Press.

Pearle, Philip (1996): "Wavefunction Collapse Models with Nonwhite Noise". In: R. Clifton (ed.): *Perspectives on Quantum Reality*. Dordrecht: Kluwer. pp. 93–109.

Penrose, Roger (1989): *The Emperor's New Mind. Concerning Computers, Minds, and the Laws of Physics*. Oxford: Oxford University Press.

Penrose, Roger (1994): *Shadows of the Mind*. Oxford: Oxford University Press.

Penrose, Roger (1997): *The Large, the Small and the Human Mind. Roger Penrose with Abner Shimony, Nancy Cartwright and Stephen Hawking. Edited by Malcolm Longair*. Cambridge: Cambridge University Press.

Perler, Dominik (1996): *Repräsentation bei Descartes*. Frankfurt (Main): Klostermann.

Pessin, Andrew and Goldberg, Sanford (eds.) (1996): *The Twin Earth Chronicles. Twenty Years of Reflection on Hilary Putnam's "The Meaning of 'Meaning'"*. Armonk, New York: Sharpe.

Pettit, Philip (1991): "Realism and Response-Dependence". *Mind* **100**, pp. 587–626.

Pettit, Philip (1996): *The Common Mind. An Essay on Psychology, Society, and Politics*. Oxford: Oxford University Press. Paperback edition with new postscript. First edition 1993.

Pettit, Philip (1998a): "Practical Belief and Philosophical Theory". *Australasian Journal of Philosophy* **76**, pp. 15–33.

Pettit, Philip (1998b): "Noumenalism and Response-Dependence". *Monist* **81**, pp. 112–132.

Pettit, Philip (1998c): "Terms, Things and Response-Dependence". *European Review of Philosophy* **3**, pp. 55–66.

Pettit, Philip (1998d): "Defining and Defending Social Holism". *Philosophical Explorations* **1**, pp. 169–184.

Pettit, Philip (1999): "A Theory of Normal and Ideal Conditions". *Philosophical Studies* **96**, pp. 21–44.

Placek, Tomasz (forthcoming): "Stochastic Outcomes in Branching Space–Time: Analysis of Bell's Theorem". *British Journal for the Philosophy of Science.*

Popescu, Sandu and Rohrlich, Daniel (1992): "Generic Quantum Nonlocality". *Physics Letters* **A166**, pp. 293–297.

Popper, Karl R. (1967): "Quantum Mechanics without "The Observer"". In: M. Bunge (ed.): *Quantum Theory and Reality.* Berlin: Springer. pp. 7–44.

Popper, Karl R. (1982): *Quantum Theory and the Schism in Physics.* Totowa (New Jersey): Rowman and Littlefield.

Popper, Karl R. (1989): *Objective Knowledge. An Evolutionary Approach.* Oxford: Oxford University Press. First edition 1972.

Popper, Karl R. (1990): *A World of Propensities.* Bristol: Thoemmes.

Price, Huw (1996): *Time's Arrow and Archimedes' Point. New Directions for the Physics of Time.* Oxford: Oxford University Press.

Price, Huw (1998): "Two Paths to Pragmatism II". *European Review of Philosophy* **3**, pp. 109–147.

Primas, Hans (1983): *Chemistry, Quantum Mechanics and Reductionism.* Berlin: Springer. Second edition. First edition 1981.

Primas, Hans (1993): "The Cartesian Cut, the Heisenberg Cut, and Disentangled Observers". In: K. V. Laurikainen and C. Montonen (eds.): *Symposium on the Foundations of Modern Physics 1992. The Copenhagen Interpretation and Wolfgang Pauli*. Singapore: World Scientific. pp. 245–269.

Putnam, Hilary (1975): *Mind, Language and Reality. Philosophical Papers Volume 2.* Cambridge: Cambridge University Press.

Putnam, Hilary (1981): *Reason, Truth and History.* Cambridge: Cambridge University Press.

Putnam, Hilary (1983): *Realism and Reason. Philosophical Papers Volume 3.* Cambridge: Cambridge University Press.

Putnam, Hilary (1990): *Realism with a Human Face. Ed. James Conant.* Cambridge (Massachusetts): Harvard University Press.

Quine, Willard Van Orman (1951): "Two Dogmas of Empiricism". *Philosophical Review* **60**, pp. 20–43.

Quine, Willard Van Orman (1960): *Word and Object*. Cambridge (Massachusetts): MIT Press.

Quine, Willard Van Orman (1969): *Ontological Relativity and Other Essays*. New York: Coulmbia University Press.

Quine, Willard Van Orman (1975): "On Empirically Equivalent Systems of the World". *Erkenntnis* **9**, pp. 313–328.

Quine, Willard Van Orman (1980): *From a Logical Point of View*. Cambridge (Massachusetts): Harvard University Press. Second edition, revised. First edition 1953.

Quine, Willard Van Orman (1981): *Theories and Things*. Cambridge (Massachusetts): Harvard University Press.

Quine, Willard Van Orman (1986a): "Reply to Hilary Putnam". In: L. E. Hahn and P. A. Schilpp (eds.): *The Philosophy of W. O. Quine*. La Salle: Open Court. pp. 427–431.

Quine, Willard Van Orman (1986b): "Reply to Jules Vuillemin". In: L. E. Hahn and P. A. Schilpp (eds.): *The Philosophy of W. O. Quine*. La Salle: Open Court. pp. 619–622.

Quine, Willard Van Orman (1991): "Two Dogmas in Retrospect". *Canadian Journal of Philosophy* **21**, pp. 265–274.

Quine, Willard Van Orman (1993): "In Praise of Observation Sentences". *Journal of Philosophy* **90**, pp. 107–116.

Recanati, François (1993): *Direct Reference. From Language to Thought*. Oxford: Blackwell.

Redhead, Michael L. G. (1983): "Quantum Field Theory for Philosophers". In: P. D. Asquith and T. Nickles (eds.): *Proceedings of the 1982 Biennial Meeting of the Philosophy of Science Association. Volume 2*. East Lansing: Philosophy of Science Association. pp. 57–99.

Redhead, Michael L. G. (1987): *Incompleteness, Nonlocality, and Realism. A Prolegomenon to the Philosophy of Quantum Mechanics*. Oxford: Oxford University Press.

Redhead, Michael L. G. (1988): "A Philosopher Looks at Quantum Field Theory". In: H. R. Brown and R. Harré (eds.): *Philosophical Foundations of Quantum Field Theory*. Oxford: Oxford University Press. pp. 9–23.

Redhead, Michael L. G. (1989a): "The Nature of Reality". *British Journal for the Philosophy of Science* **40**, pp. 429–441.

Redhead, Michael L. G. (1989b): "Nonfactorizability, Stochastic Causality and Passion-at-a-Distance". In: J. T. Cushing and E. McMullin (eds.): *Philosophical Consequences of Quantum Theory. Reflections on Bell's Theorem*. Notre Dame: University of Notre Dame Press. pp. 145–153.

Redhead, Michael L. G. (1992): "Propensities, Correlations, and Metaphysics". *Foundations of Physics* **22**, pp. 381–394.

Redhead, Michael L. G. (1995a): *From Physics to Metaphysics. The Tarner Lectures*. Cambridge: Cambridge University Press.

Redhead, Michael L. G. (1995b): "More Ado about Nothing". *Foundations of Physics* **25**, pp. 123–137.

Redhead, Michael L. G. (1995c): "The Vacuum in Relativistic Quantum Field Theory". In: D. Hull, M. Forbes and R. M. Burian (eds.): *Proceedings of the 1994 Biennial Meeting of the Philosophy of Science Association. Volume 2*. East Lansing: Philosophy of Science Association. pp. 77–87.

Redhead, Michael L. G. and La Rivière, Patrick (1997): "The Relativistic EPR Argument". In: R. S. Cohen, M. A. Horne and J. Stachel (eds.): *Potentiality, Entanglement and Passion-at-a-Distance. Quantum Mechanical Studies for Abner Shimony. Volume 2*. Dordrecht: Kluwer. pp. 207–215.

Redhead, Michael L. G. and Teller, Paul (1992): "Particle Labels and the Theory of Indistinguishable Particles in Quantum Mechanics". *British Journal for the Philosophy of Science* **43**, pp. 201–218.

Rhees, Rush (1966): "Can There Be a Private Language?". In: G. Pitcher (ed.): *Wittgenstein. The 'Philosophical Investigations'*. New York: Anchor. pp. 267–285. First published in *Proceedings of the Aristotelian Society*, Supplementary Volume **28** (1954), pp. 77–94.

Rice, Lee C. (1975): "Spinoza on Individuation". In: M. Mandelbaum and E. Freeman (eds.): *Spinoza: Essays in Interpretation*. La Salle: Open Court. pp. 195-214.

Robinson, Don (1992): "On Healey's Holistic Interpretation of Quantum Mechanics". *International Studies in the Philosophy of Science* **6**, pp. 227–240.

Robinson, Don (1994): "Can Superselection Rules Solve the Measurement Problem?". *British Journal for the Philosophy of Science* **45**, pp. 79–92.

Rohrlich, Fritz (1986): "Reality and Quantum Mechanics". In: D. M. Greenberger (ed.): *New Techniques and Ideas in Quantum Measurement Theory*. New York: The New York Academy of Sciences. pp. 373–381.

Rohrlich, Fritz (1987): *From Paradox to Reality. Our new Concepts of the Physical World*. Cambridge: Cambridge University Press.

Rohs, Peter (1996): *Feld – Zeit – Ich. Entwurf einer feldtheoretischen Transzendentalphilosophie*. Frankfurt (Main): Klostermann.

Rohrlich, Fritz (1987): *From Paradox to Reality. Our new Concepts of the Physical World*. Cambridge: Cambridge University Press.

Rombach, Heinrich (1966): *Substanz, System, Struktur*. Vol. 2. Freiburg: Alber.

Rorty, Richard (1967): "Relations, Internal and External". In: P. Edwards (ed.): *The Encyclopedia of Philosophy, Volume 7*. New York: Macmillan. pp. 125–133.

Rorty, Richard (1980): *Philosophy and the Mirror of Nature*. Oxford: Blackwell.

Rorty, Richard (1982): *Consequences of Pragmatism*. Minneapolis: University of Minnesota Press.

Rorty, Richard (1998): *Truth and Progress. Philosophical Papers. Volume 3*. Cambridge: Cambridge University Press.

Rovelli, Carlo (1996): "Relational Quantum Mechanics". *International Journal of Theoretical Physics* **35**, pp. 1637–1678.

Rudder Baker, Lynne (1995): *Explaining Attitudes. A Practical Approach to the Mind*. Cambridge: Cambridge University Press.

Russell, Bertrand (1954): *The Analysis of Matter*. London: Allen and Unwin. First edtion 1927.

Russell, Bertrand (1956): *Logic and Knowledge. Essays 1901–1950. Ed. R. Ch. Marsh*. London: Allen and Unwin.

Russell, Bertrand (1969): *The ABC of Relativity. Edited by Felix Pirani*. London: Allen and Unwin. First edition 1925.

Rust, Alois (1996): *Wittgensteins Philosophie der Psychologie*. Frankfurt (Main): Klostermann.

Sachs, Mendel (1976): "Maimonides, Spinoza, and the Field Concept in Physics". *Journal of the History of Ideas* **37**, pp. 125–131.

Salmon, Wesley (1998): *Causality and Explanation*. Oxford: Oxford University Press.

Savellos, Elias E. and Yalçin, Ümit D. (eds.) (1995): *Supervenience. New Essays*. Cambridge: Cambridge University Press.

Schantz, Richard (1996): *Wahrheit, Referenz und Realismus. Eine Studie zur Sprachphilosophie und Metaphysik*. Berlin: de Gruyter.

Scheibe, Erhard (1964): *Die kontingenten Aussagen in der Physik*. Frankfurt (Main): Athenäum.

Scheibe, Erhard (1973): *The Logical Analysis of Quantum Mechanics*. Oxford: Pergamon Press.

Scheibe, Erhard (1991): "Substances, Physical Systems, and Quantum Mechanics". In: G. Schurz and G. J. W. Dorn (eds.): *Advances in Scientific Philosophy. Essays in Honour of Paul Weingartner*. Amsterdam: Rodopi. pp. 215–229.

Scheibe, Erhard (1997/99): *Die Reduktion physikalischer Theorien. Ein Beitrag zur Einheit der Physik. 2 Bände*. Berlin: Springer.

Scholz, Oliver Robert (1999): *Verstehen und Rationalität. Untersuchungen zu den Grundlagen von Hermeneutik und Sprachphilosophie*. Frankfurt (Main): Klostermann.

Schrödinger, Erwin (1935a): "Discussion of Probability Relations between Separated Systems". *Proceedings of the Cambridge Philosophical Society* **31**, pp. 555–563.

Schrödinger, Erwin (1935b): "Die gegenwärtige Situation in der Quantenmechanik". *Naturwissenschaften* **23**, pp. 807–812, 823–828, 844–849.

Searle, John R. (1983): *Intentionality*. Cambridge: Cambridge University Press.

Sedgwick, Sally (1997): "McDowell's Hegelianism". *European Journal of Philosophy* **5**, pp. 21–38.

Sellars, Wilfrid (1953): "Inference and Meaning". *Mind* **62**, pp. 313–338.

Sellars, Wilfrid (1963): *Science, Perception and Reality*. London: Routledge.

Sellars, Wilfrid (1979): *Naturalism and Ontology*. Reseda: Ridgeview.

Shanks, Niall (1993): "Quantum Mechanics and Determinism". *Philosophical Quarterly* **43**, pp. 20–37.

Shimony, Abner (1989): "Search for a World View which can Accomodate our Knowledge of Microphysics". In: J. T. Cushing and E. McMullin (eds.): *Philosophical Consequences of Quantum Theory. Reflections on Bell's Theorem.* Notre Dame: University of Notre Dame Press. pp. 25–37.

Shimony, Abner (1993): *Search for a Naturalistic World View. Volume 2: Natural Science and Metaphysics.* Cambridge: Cambridge University Press.

Shimony, Abner (1999): "Holism". In: H. Atmanspacher, A. Amann and U. Müller-Herold (eds.): *On Quanta, Mind and Matter. Hans Primas in Context.* Dordrecht: Kluwer. pp. 233–246.

Simons, Peter M. (1987): *Parts. A Study in Ontology.* Oxford: Oxford University Press.

Simons, Peter M. (1994): "Particulars in Particular Clothing: Three Trope Theories of Substance". *Philosophy and Phenomenological Research* **54**, pp. 553–575.

Skryms, Brian (1984): "EPR: Lessons for Metaphysics". In: P. A. French et al. (eds.): *Causation and Causal Theories. Midwest Studies in Philosophy IX.* Minneapolis: University of Minnesota Press. pp. 245–255.

Smart, J. J. C. (1972): "Space–Time and Individuals". In: R. Rudner and I. Scheffler (eds.): *Logic and Art: Essays in Honor of Nelson Goodman.* Indianapolis: Bobbs-Merrill. pp. 3–20.

Smart, J. J. C. (1992): "Wittgenstein, Following a Rule, and Scientific Psychology". In: E. Ullmann-Margalit (ed.): *The Scientific Enterprise.* Dordrecht: Kluwer. pp. 123–137.

Smith, Barry and Mulligan, Kevin (1982): "Pieces of a Theory". In: B. Smith (ed.): *Parts and Moments. Studies in Logic and Formal Ontology.* München: Philosophia Verlag. pp. 15–109.

Smith, C. U. M. (1976): *The Problem of Life.* London: Macmillan.

Smith, Gerrit J. and Weingard, Robert (1987): "A Relativistic Formulation of the Einstein–Podolsky–Rosen Paradox". *Foundations of Physics* **17**, pp. 149–171.

Smith, Michael (1994): *The Moral Problem.* Oxford: Blackwell.

Smith, Michael and Stoljar, Daniel (1998): "Global Response-Dependence and Noumenal Realism". *Monist* **81**, pp. 85–111.

Smuts, Jan C. (1926): *Holism and Evolution.* London: Macmillan.

Soames, Scott (1998a): "Skepticism about Meaning: Indeterminacy, Normativity, and the Rule-Following Paradox". In: A. A. Kazmi (ed.): *Meaning and Reference.* Calgary: University of Calgary Press. pp. 211–249. *Canadian Journal of Philosophy* Supplementary Volume **23** (1997).

Soames, Scott (1998b): "Facts, Truth Conditions, and the Skeptical Solution to the Rule-Following Paradox". In: J. E. Tomberlin (ed.): *Philosophical Perspectives 12: Language, Mind, and Ontology.* Oxford: Blackwell. pp. 313–348.

Specker, Ernst (1960): "Die Logik nicht gleichzeitig entscheidbarer Aussagen". *Dialectica* **14**, pp. 239–246. English translation in Clifford A. Hooker (ed.): *The Logico-Algrebraic Approach to Quantum Mechanics.* Volume 1. Dordrecht: Reidel 1975, pp. 135–140.

Spinoza, Baruch de (1925): *Opera. Ed. C. Gebhardt.* Vol. 1–4. Heidelberg: Winter.

Spohn, Wolfgang (1997): "The Character of Color Terms: A Materialist View". In: W. Künne, A. Newen & M. Anduschus (eds.): *Direct Reference, Indexicality, and Propositional Attitudes*. Stanford: CSLI Publications. pp. 351–379.

Stachel, John (1974): "The Rise and Fall of Geometrodynamics". In: K. F. Schaffner and R. S. Cohen (eds.): *PSA 1972. Proceedings of the 1972 Biennial Meeting of the Philosophy of Science Association*. Dordrecht: Reidel. pp. 31–54. Boston Studies in the Philosophy of Science. Volume 20.

Stamatescu, Ion-Olimpiu (1996): "Stochastic Collapse Models". In: D. Giulini, E. Joos, C. Kiefer, J. Kupsch, I.-O. Stamatescu and H. D. Zeh (eds.): *Decoherence and the Appearance of a Classical World in Quantum Theory*. Berlin: Springer. pp. 249–267.

Stapp, Henry P. (1993): *Mind, Matter, and Quantum Mechanics*. Berlin: Springer.

Stegmüller, Wolfgang (1986): *Kripkes Deutung der Spätphilosophie Wittgensteins. Kommentarversuch über einen versuchten Kommentar*. Stuttgart: Kröner.

Stent, Gunter S. (1993): "Die Autonomie des Menschen". In: E. P. Fischer (Hg.): *Neue Horizonte 92/93*. München: Piper. S. 59–79.

Stöckler, Manfred (1984): *Philosophische Probleme der relativistischen Quantenmechanik*. Berlin: Duncker and Humblot.

Stöckler, Manfred (1988a): "Individualität, Identität, Ununterscheidbarkeit. Überlegungen zum Gegenstandsbegriff in der Quantenfeldtheorie". *Conceptus* **22**, pp. 5–29.

Stöckler, Manfred (1988b): *Philosophische Probleme der Elementarteilchenphysik*. Giessen: Habilitationsschrift.

Stöckler, Manfred (1992): "Plädoyer für einen pragmatisch eingeschränkten Reduktionismus". In: W. Deppert, H. Kliemt, B. Lohff and J. Schaefer (eds.): *Wissenschaftstheorien in der Medizin. Ein Symposium*. Berlin: de Gruyter. pp. 157–182.

Stöltzner, Michael (1999): "Über zwei Formen von Realismus in der Quantentheorie". *Journal for General Philosophy of Science / Zeitschrift für allgemeine Wissenschaftstheorie* **30**, pp. 289–316.

Strawson, Peter F. (1959): *Individuals*. London: Routledge. First edition 1959.

Stroud, Barry (1968): "Transcendental Arguments". *Journal of Philosophy* **65**, pp. 241–256.

Suppes, Patrick (1985): "Davidson's Views on Psychology as a Science". In: B. Vermazen and M. B. Hintikka (eds.): *Essays on Davidson: Actions and Events*. Oxford: Oxford University Press. pp. 183–194.

Teller, Paul (1979): "Quantum Mechanics and the Nature of Continuous Physical Qualities". *Journal of Philosophy* **76**, pp. 345–361.

Teller, Paul (1984): "The Projection Postulate: a new Perspective". *Philosophy of Science* **51**, pp. 369–395.

Teller, Paul (1986): "Relational Holism and Quantum Mechanics". *British Journal for the Philosophy of Science* **37**, pp. 71–81.

Teller, Paul (1987): "Space–Time as a Physical Quantity". In: R. Kargon and P. Achinstein (eds.): *Kelvin's Baltimore Lectures and Modern Theoretical Physics*. Cambridge (Massachusetts): MIT Press. pp. 425–447.

Teller, Paul (1989): "Relativity, Relational Holism, and the Bell Inequalities". In: J. T. Cushing and E. McMullin (eds.): *Philosophical Consequences of Quantum Theory. Reflections on Bell's Theorem.* Notre Dame: University of Notre Dame Press. pp. 208–223.

Teller, Paul (1991): "Substance, Relations, and Arguments about the Nature of Space-Time". *Philosophical Review* **100**, pp. 363–397.

Teller, Paul (1995): *An Interpretative Introduction to Quantum Field Theory.* Princeton: Princeton University Press.

Teller, Paul (1998): "Quantum Mechanics and Haecceities". In: E. Castellani (ed.): *Interpreting Bodies. Classical and Quantum Objects in Modern Physics.* Princeton: Princeton University Press. pp. 114–141.

Thalos, Mariam (1998): "A Trouble with Superselection Accounts of Measurement". *Philosophy of Science* **65**, pp. 518–544.

Thompson, Evan (1991): "Is Internal Realism a Philosophy of Scheme and Content?". *Metaphilosophy* **22**, pp. 212–230.

Tittel, W., Brendel, J., Gisin, B., Herzog, T., Zbinden, H. and Gisin, N. (1998): "Experimental Demonstration of Quantum-Correlations over more than 10 Kilometers". *Physical Review* **A57**, pp. 3229–3232.

Treder, H.-J. (1982): "Descartes' Physik der Hypothesen, Newtons Physik der Prinzipen und Leibnizens Physik der Prinzipe". *Studia Leibnitiana* **14**, pp. 278–286.

Tuomela, Raimo (1988): "The Myth of the Given and Realism". *Erkenntnis* **29**, pp. 181–200.

Uchiyama, S. (1995): "Local Reality: Can it Exist in the EPR–Bohm Gedankenexperiment?". *Foundations of Physics* **25**, pp. 1561–1575.

Vallentyne, Peter (1997): "Intrinsic Properties Defined". *Philosophical Studies* **88**, pp. 209–219.

van Cleve, James (1992): "Semantic Supervenience and Referential Indeterminacy". *Journal of Philosophy* **89**, pp. 344–361.

van Erkelens, Herbert (1991): "Wolfgang Pauli and the Spirit of Matter". In: P. J. Lahti and P. Mittelstaedt (eds.): *Symposium on the Foundations of Modern Physics 1990. Quantum Measurement Theory and its Philosophical Implications.* Singapore: World Scientific. pp. 425–439.

van Fraassen, Bas C. (1991): *Quantum Mechanics: An Empiricist View.* Oxford: Oxford University Press.

van Zandt, J. D. (1986): "Res extensa and the Space–Time Continuum". In: M. Grene and D. Nails (eds.): *Spinoza and the Sciences.* Dordrecht: Reidel. pp. 249–266.

von Neumann, Johann (1932): *Mathematische Grundlagen der Quantenmechanik.* Berlin: Springer.

von Neumann, John (1955): *Mathematical Foundations of Quantum Mechanics. Translated by R. T. Beyer.* Princeton: Princeton University Press.

von Savigny, Eike (1994/95): *Wittgensteins „Philosophische Untersuchungen". Ein Kommentar für Leser. Zwei Bände*. Frankfurt (Main): Klostermann. Zweite Auflage. Erste Auflage 1988/89.

von Savigny, Eike (1996): *Der Mensch als Mitmensch. Wittgensteins „Philosophische Untersuchungen"*. München: Deutscher Taschenbuch Verlag.

von Savigny, Eike (ed.) (1998): *Ludwig Wittgenstein. Philosophische Untersuchungen*. Berlin: Akademie Verlag.

Vuillemin, Jules (1986): "On Duhem's and Quine's Theses". In: L. E. Hahn and P. A. Schilpp (eds.): *The Philosophy of W. O. Quine*. La Salle: Open Court. pp. 595–618.

Wagner, Steven J. (1986): "Quine's Holism". *Analysis* **46**, pp. 1–6.

Wald, Robert M. (1986): "Correlations and Causality in Quantum Field Theory". In: R. Penrose and C. J. Isham (eds.): *Quantum Concepts in Space and Time*. Oxford: Oxford University Press. pp. 293–301.

Weihs, Gregor, Weinfurter, Harald and Zeilinger, Anton (1997): "Towards a Long-Distance Bell Experiment with Independent Observers.". In: R. S. Cohen, M. A. Horne and J. Stachel (eds.): *Experimental Metaphysics. Quantum Mechanical Studies for Abner Shimony. Volume 1*. Dordrecht: Kluwer. pp. 239–246.

Wheeler, John A. (1962a): *Geometrodynamics*. New York: Academic Press.

Wheeler, John A. (1962b): "Curved Empty Space as the Building Material of the Physical World: an Assessment". In: E. Nagel, P. Suppes and A. Tarski (eds.): *Logic, Methodology and Philosophy of Science. Proceedings of the 1960 International Congress*. Stanford: Stanford University Press. pp. 361–374.

Wieland, Wolfgang (1982): *Platon und die Formen des Wissens*. Göttingen: Vandenhoeck and Ruprecht.

Willaschek, Marcus (1999): *Über den mentalen Zugang zur Welt. Realismus, Skeptizismus und Intentionalität*. Habilitationsschrift Münster. Forthcoming Frankfurt (Main): Klostermann.

Williams, Meredith (1999): *Wittgenstein, Mind and Meaning. Toward a Social Conception of Mind*. London: Routledge.

Wilson, George M. (1998): "Semantic Realism and Kripke's Wittgenstein". *Philosophy and Phenomenological Research* **58**, pp. 99–122.

Wittgenstein, Ludwig (1952): *Philosophische Untersuchungen. Eds. G.E.M. Anscombe, G.H. von Wright, Rush Rhees. In: Ludwig Wittgenstein. Werkausgabe in 8 Bänden. Band 1*. Frankfurt (Main): Suhrkamp 1984.

Wittgenstein, Ludwig (1958): *Philosophical Investigations. Translated by G. E. M. Anscombe*. Oxford: Blackwell. Second edition. First edition 1953.

Wittgenstein, Ludwig (1978): *Remarks on the Foundations of Mathematics. Edited by G. H. von Wright, R. Rhees and G. E. M. Anscombe. Translated by G. E. M. Anscombe*. Cambridge (Massachusetts): MIT Press.

Woolhouse, Roger S. (1993): *Descartes, Spinoza, Leibniz. The Concept of Substance in Seventeenth Century Metaphysics*. London: Routledge.

Woolhouse, Roger S. (1994): "Descartes and the Nature of Body". *British Journal for the History of Philosophy* **2.1**, pp. 19–33.

Wright, Crispin (1980): *Wittgenstein on the Foundations of Mathematics*. London: Duckworth.

Wright, Crispin (1993): *Realism, Meaning and Truth*. Oxford: Blackwell. Second edition. First edition 1987.

Wright, Crispin (1996): "Human Nature? (Review of John McDowell: Mind and World, Cambridge (Massachusetts): Harvard University Press 1994)". *European Journal of Philosophy* **4**, pp. 235–254.

Young, James O. (1992): "Holism and Meaning". *Erkenntnis* **37**, pp. 309–325.

Zeh, H. Dieter (1996): "The Program of Decoherence: Ideas and Concepts". In: D. Giulini, E. Joos, C. Kiefer, J. Kupsch, I.-O. Stamatescu and H. D. Zeh (eds.): *Decoherence and the Appearance of a Classical World in Quantum Theory*. Berlin: Springer. pp. 5–34.

INDEX OF NAMES

INDEX OF SUBJECTS

218. M. Kusch, *Foucault's Strata and Fields*. An Investigation into Archaeological and Genealogical Science Studies. 1991 ISBN 0-7923-1462-X
219. C.J. Posy, *Kant's Philosophy of Mathematics*. Modern Essays. 1992 ISBN 0-7923-1495-6
220. G. Van de Vijver, *New Perspectives on Cybernetics*. Self-Organization, Autonomy and Connectionism. 1992 ISBN 0-7923-1519-7
221. J.C. Nyíri, *Tradition and Individuality*. Essays. 1992 ISBN 0-7923-1566-9
222. R. Howell, *Kant's Transcendental Deduction*. An Analysis of Main Themes in His Critical Philosophy. 1992 ISBN 0-7923-1571-5
223. A. García de la Sienra, *The Logical Foundations of the Marxian Theory of Value*. 1992 ISBN 0-7923-1778-5
224. D.S. Shwayder, *Statement and Referent*. An Inquiry into the Foundations of Our Conceptual Order. 1992 ISBN 0-7923-1803-X
225. M. Rosen, *Problems of the Hegelian Dialectic*. Dialectic Reconstructed as a Logic of Human Reality. 1993 ISBN 0-7923-2047-6
226. P. Suppes, *Models and Methods in the Philosophy of Science: Selected Essays*. 1993 ISBN 0-7923-2211-8
227. R. M. Dancy (ed.), *Kant and Critique: New Essays in Honor of W. H. Werkmeister*. 1993 ISBN 0-7923-2244-4
228. J. Woleński (ed.), *Philosophical Logic in Poland*. 1993 ISBN 0-7923-2293-2
229. M. De Rijke (ed.), *Diamonds and Defaults*. Studies in Pure and Applied Intensional Logic. 1993 ISBN 0-7923-2342-4
230. B.K. Matilal and A. Chakrabarti (eds.), *Knowing from Words*. Western and Indian Philosophical Analysis of Understanding and Testimony. 1994 ISBN 0-7923-2345-9
231. S.A. Kleiner, *The Logic of Discovery*. A Theory of the Rationality of Scientific Research. 1993 ISBN 0-7923-2371-8
232. R. Festa, *Optimum Inductive Methods*. A Study in Inductive Probability, Bayesian Statistics, and Verisimilitude. 1993 ISBN 0-7923-2460-9
233. P. Humphreys (ed.), *Patrick Suppes: Scientific Philosopher*. Vol. 1: Probability and Probabilistic Causality. 1994 ISBN 0-7923-2552-4
234. P. Humphreys (ed.), *Patrick Suppes: Scientific Philosopher*. Vol. 2: Philosophy of Physics, Theory Structure, and Measurement Theory. 1994 ISBN 0-7923-2553-2
235. P. Humphreys (ed.), *Patrick Suppes: Scientific Philosopher*. Vol. 3: Language, Logic, and Psychology. 1994 ISBN 0-7923-2862-0
    Set ISBN (Vols 233–235) 0-7923-2554-0
236. D. Prawitz and D. Westerståhl (eds.), *Logic and Philosophy of Science in Uppsala*. Papers from the 9th International Congress of Logic, Methodology, and Philosophy of Science. 1994 ISBN 0-7923-2702-0
237. L. Haaparanta (ed.), *Mind, Meaning and Mathematics*. Essays on the Philosophical Views of Husserl and Frege. 1994 ISBN 0-7923-2703-9
238. J. Hintikka (ed.), *Aspects of Metaphor*. 1994 ISBN 0-7923-2786-1
239. B. McGuinness and G. Oliveri (eds.), *The Philosophy of Michael Dummett*. With Replies from Michael Dummett. 1994 ISBN 0-7923-2804-3
240. D. Jamieson (ed.), *Language, Mind, and Art*. Essays in Appreciation and Analysis, In Honor of Paul Ziff. 1994 ISBN 0-7923-2810-8
241. G. Preyer, F. Siebelt and A. Ulfig (eds.), *Language, Mind and Epistemology*. On Donald Davidson's Philosophy. 1994 ISBN 0-7923-2811-6
242. P. Ehrlich (ed.), *Real Numbers, Generalizations of the Reals, and Theories of Continua*. 1994 ISBN 0-7923-2689-X

243. G. Debrock and M. Hulswit (eds.), *Living Doubt.* Essays concerning the epistemology of Charles Sanders Peirce. 1994 ISBN 0-7923-2898-1
244. J. Srzednicki, *To Know or Not to Know.* Beyond Realism and Anti-Realism. 1994 ISBN 0-7923-2909-0
245. R. Egidi (ed.), *Wittgenstein: Mind and Language.* 1995 ISBN 0-7923-3171-0
246. A. Hyslop, *Other Minds.* 1995 ISBN 0-7923-3245-8
247. L. Pólos and M. Masuch (eds.), *Applied Logic: How, What and Why.* Logical Approaches to Natural Language. 1995 ISBN 0-7923-3432-9
248. M. Krynicki, M. Mostowski and L.M. Szczerba (eds.), *Quantifiers: Logics, Models and Computation.* Volume One: Surveys. 1995 ISBN 0-7923-3448-5
249. M. Krynicki, M. Mostowski and L.M. Szczerba (eds.), *Quantifiers: Logics, Models and Computation.* Volume Two: Contributions. 1995 ISBN 0-7923-3449-3
Set ISBN (Vols 248 + 249) 0-7923-3450-7
250. R.A. Watson, *Representational Ideas from Plato to Patricia Churchland.* 1995 ISBN 0-7923-3453-1
251. J. Hintikka (ed.), *From Dedekind to Gödel.* Essays on the Development of the Foundations of Mathematics. 1995 ISBN 0-7923-3484-1
252. A. Wiśniewski, *The Posing of Questions.* Logical Foundations of Erotetic Inferences. 1995 ISBN 0-7923-3637-2
253. J. Peregrin, *Doing Worlds with Words.* Formal Semantics without Formal Metaphysics. 1995 ISBN 0-7923-3742-5
254. I.A. Kieseppä, *Truthlikeness for Multidimensional, Quantitative Cognitive Problems.* 1996 ISBN 0-7923-4005-1
255. P. Hugly and C. Sayward: *Intensionality and Truth.* An Essay on the Philosophy of A.N. Prior. 1996 ISBN 0-7923-4119-8
256. L. Hankinson Nelson and J. Nelson (eds.): *Feminism, Science, and the Philosophy of Science.* 1997 ISBN 0-7923-4162-7
257. P.I. Bystrov and V.N. Sadovsky (eds.): *Philosophical Logic and Logical Philosophy.* Essays in Honour of Vladimir A. Smirnov. 1996 ISBN 0-7923-4270-4
258. Å.E. Andersson and N-E. Sahlin (eds.): *The Complexity of Creativity.* 1996 ISBN 0-7923-4346-8
259. M.L. Dalla Chiara, K. Doets, D. Mundici and J. van Benthem (eds.): *Logic and Scientific Methods.* Volume One of the Tenth International Congress of Logic, Methodology and Philosophy of Science, Florence, August 1995. 1997 ISBN 0-7923-4383-2
260. M.L. Dalla Chiara, K. Doets, D. Mundici and J. van Benthem (eds.): *Structures and Norms in Science.* Volume Two of the Tenth International Congress of Logic, Methodology and Philosophy of Science, Florence, August 1995. 1997 ISBN 0-7923-4384-0
Set ISBN (Vols 259 + 260) 0-7923-4385-9
261. A. Chakrabarti: *Denying Existence.* The Logic, Epistemology and Pragmatics of Negative Existentials and Fictional Discourse. 1997 ISBN 0-7923-4388-3
262. A. Biletzki: *Talking Wolves.* Thomas Hobbes on the Language of Politics and the Politics of Language. 1997 ISBN 0-7923-4425-1
263. D. Nute (ed.): *Defeasible Deontic Logic.* 1997 ISBN 0-7923-4630-0
264. U. Meixner: *Axiomatic Formal Ontology.* 1997 ISBN 0-7923-4747-X
265. I. Brinck: *The Indexical 'I'.* The First Person in Thought and Language. 1997 ISBN 0-7923-4741-2
266. G. Hölmström-Hintikka and R. Tuomela (eds.): *Contemporary Action Theory.* Volume 1: Individual Action. 1997 ISBN 0-7923-4753-6; Set: 0-7923-4754-4

267. G. Hölmström-Hintikka and R. Tuomela (eds.): *Contemporary Action Theory.* Volume 2: Social Action. 1997 ISBN 0-7923-4752-8; Set: 0-7923-4754-4
268. B.-C. Park: *Phenomenological Aspects of Wittgenstein's Philosophy.* 1998 ISBN 0-7923-4813-3
269. J. Paśniczek: *The Logic of Intentional Objects.* A Meinongian Version of Classical Logic. 1998 Hb ISBN 0-7923-4880-X; Pb ISBN 0-7923-5578-4
270. P.W. Humphreys and J.H. Fetzer (eds.): *The New Theory of Reference.* Kripke, Marcus, and Its Origins. 1998 ISBN 0-7923-4898-2
271. K. Szaniawski, A. Chmielewski and J. Woleński (eds.): *On Science, Inference, Information and Decision Making.* Selected Essays in the Philosophy of Science. 1998 ISBN 0-7923-4922-9
272. G.H. von Wright: *In the Shadow of Descartes.* Essays in the Philosophy of Mind. 1998 ISBN 0-7923-4992-X
273. K. Kijania-Placek and J. Woleński (eds.): *The Lvov–Warsaw School and Contemporary Philosophy.* 1998 ISBN 0-7923-5105-3
274. D. Dedrick: *Naming the Rainbow.* Colour Language, Colour Science, and Culture. 1998 ISBN 0-7923-5239-4
275. L. Albertazzi (ed.): *Shapes of Forms.* From Gestalt Psychology and Phenomenology to Ontology and Mathematics. 1999 ISBN 0-7923-5246-7
276. P. Fletcher: *Truth, Proof and Infinity.* A Theory of Constructions and Constructive Reasoning. 1998 ISBN 0-7923-5262-9
277. M. Fitting and R.L. Mendelsohn (eds.): *First-Order Modal Logic.* 1998 Hb ISBN 0-7923-5334-X; Pb ISBN 0-7923-5335-8
278. J.N. Mohanty: *Logic, Truth and the Modalities from a Phenomenological Perspective.* 1999 ISBN 0-7923-5550-4
279. T. Placek: *Mathematical Intiutionism and Intersubjectivity.* A Critical Exposition of Arguments for Intuitionism. 1999 ISBN 0-7923-5630-6
280. A. Cantini, E. Casari and P. Minari (eds.): *Logic and Foundations of Mathematics.* 1999 ISBN 0-7923-5659-4 set ISBN 0-7923-5867-8
281. M.L. Dalla Chiara, R. Giuntini and F. Laudisa (eds.): *Language, Quantum, Music.* 1999 ISBN 0-7923-5727-2; set ISBN 0-7923-5867-8
282. R. Egidi (ed.): *In Search of a New Humanism.* The Philosophy of Georg Hendrik von Wright. 1999 ISBN 0-7923-5810-4
283. F. Vollmer: *Agent Causality.* 1999 ISBN 0-7923-5848-1
284. J. Peregrin (ed.): *Truth and Its Nature (if Any).* 1999 ISBN 0-7923-5865-1
285. M. De Caro (ed.): *Interpretations and Causes.* New Perspectives on Donald Davidson's Philosophy. 1999 ISBN 0-7923-5869-4
286. R. Murawski: *Recursive Functions and Metamathematics.* Problems of Completeness and Decidability, Gödel's Theorems. 1999 ISBN 0-7923-5904-6
287. T.A.F. Kuipers: *From Instrumentalism to Constructive Realism.* On Some Relations between Confirmation, Empirical Progress, and Truth Approximation. 2000 ISBN 0-7923-6086-9
288. G. Holmström-Hintikka (ed.): *Medieval Philosophy and Modern Times.* 2000 ISBN 0-7923-6102-4
289. E. Grosholz and H. Breger (eds.): *The Growth of Mathematical Knowledge.* 2000 ISBN 0-7923-6151-2

SYNTHESE LIBRARY

290. G. Sommaruga: *History and Philosophy of Constructive Type Theory*. 2000
ISBN 0-7923-6180-6
291. J. Gasser (ed.): *A Boole Anthology*. Recent and Classical Studies in the Logic of George Boole. 2000 ISBN 0-7923-6380-9
292. V.F. Hendricks, S.A. Pedersen and K.F. Jørgensen (eds.): *Proof Theory*. History and Philosophical Significance. 2000 ISBN 0-7923-6544-5
293. W.L. Craig: *The Tensed Theory of Time*. A Critical Examination. 2000 ISBN 0-7923-6634-4
294. W.L. Craig: *The Tenseless Theory of Time*. A Critical Examination. 2000
ISBN 0-7923-6635-2
295. L. Albertazzi (ed.): *The Dawn of Cognitive Science*. Early European Contributors. 2001
ISBN 0-7923-6799-5
296. G. Forrai: *Reference, Truth and Conceptual Schemes*. A Defense of Internal Realism. 2001
ISBN 0-7923-6885-1
297. V.F. Hendricks, S.A. Pedersen and K.F. Jørgensen (eds.): *Probability Theory*. Philosophy, Recent History and Relations to Science. 2001 ISBN 0-7923-6952-1
298. M. Esfeld: *Holism in Philosophy of Mind and Philosophy of Physics*. 2001
ISBN 0-7923-7003-1
299. E.C. Steinhart: *The Logic of Metaphor*. Analogous Parts of Possible Worlds. 2001
ISBN 0-7923-7004-X

Previous volumes are still available.

KLUWER ACADEMIC PUBLISHERS – DORDRECHT / BOSTON / LONDON

GPSR Compliance
The European Union's (EU) General Product Safety Regulation (GPSR) is a set of rules that requires consumer products to be safe and our obligations to ensure this.

If you have any concerns about our products, you can contact us on

ProductSafety@springernature.com

In case Publisher is established outside the EU, the EU authorized representative is:

Springer Nature Customer Service Center GmbH
Europaplatz 3
69115 Heidelberg, Germany

www.ingramcontent.com/pod-product-compliance
Ingram Content Group UK Ltd.
Pitfield, Milton Keynes, MK11 3LW, UK
UKHW021103170926
15229UKWH00007B/29

9789401717885